AF388968

Selected Papers from the First International Symposium on Future ICT (Future-ICT 2019) in Conjunction with 4th International Symposium on Mobile Internet Security (MobiSec 2019)

Selected Papers from the First International Symposium on Future ICT (Future-ICT 2019) in Conjunction with 4th International Symposium on Mobile Internet Security (MobiSec 2019)

Editors

Giovanni Pau

Hsing-Chung Chen

Fang-Yie Leu

Ilsun You

MDPI • Basel • Beijing • Wuhan • Barcelona • Belgrade • Manchester • Tokyo • Cluj • Tianjin

Editors
Giovanni Pau
Kore University of Enna
Italy

Hsing-Chung Chen
Asia University
Taiwan

Fang-Yie Leu
Tunghai University
Taiwan

Ilsun You
Soonchunhyang University
Korea

Editorial Office
MDPI
St. Alban-Anlage 66
4052 Basel, Switzerland

This is a reprint of articles from the Special Issue published online in the open access journal *Sensors* (ISSN 1424-8220) (available at: https://www.mdpi.com/journal/sensors/special_issues/Future_ICT_2019).

For citation purposes, cite each article independently as indicated on the article page online and as indicated below:

LastName, A.A.; LastName, B.B.; LastName, C.C. Article Title. *Journal Name* **Year**, *Volume Number*, Page Range.

ISBN 978-3-0365-0728-6 (Hbk)
ISBN 978-3-0365-0729-3 (PDF)

© 2021 by the authors. Articles in this book are Open Access and distributed under the Creative Commons Attribution (CC BY) license, which allows users to download, copy and build upon published articles, as long as the author and publisher are properly credited, which ensures maximum dissemination and a wider impact of our publications.

The book as a whole is distributed by MDPI under the terms and conditions of the Creative Commons license CC BY-NC-ND.

Contents

About the Editors

Giovanni Pau (Professor) received a B.S. in telematic engineering from the University of Catania, Italy; a M.S. (cum laude) in telematic engineering; and a Ph.D. from the Kore University of Enna, Italy. He is currently an Associate Professor with the Faculty of Engineering and Architecture, Kore University of Enna. He is the author or coauthor of more than 65 refereed articles published in journals and conference proceedings. His research interests include wireless sensor networks, fuzzy logic controllers, intelligent transportation systems, green communications, and network security. He has been involved in the organization of several international conferences as the session co-chair and technical program committee member. He serves as a leading guest editor for Special Issues of several international journals. He is an Editorial Board Member and an Associate Editor of several journals such as IEEE ACCESS, Wireless Networks (Springer), EURASIP *Journal on Wireless Communications and Networking* (Springer), *Wireless Communications and Mobile Computing* (Hindawi), and *Future Internet* (MDPI).

Hsing-Chung Chen (Professor) received a Ph.D. in Electronic Engineering from National Chung Cheng University, Taiwan, in 2007. From February 2008 to July 2018, he was sn Assistant Professor and Associate Professor in the Department of Computer Science and Information Engineering at Asia University, Taiwan. Since August 2018 until July 2019, he was a Full Professor in the Department of Computer Science and Information Engineering at Asia University, Taiwan. From August 2019–present, he has been a Distinguished Full Professor with the Department of Computer Science and Information Engineering, Asia University, Taiwan. From August 2019–July 2020, he was Chairman with the Department of Computer Science and Information Engineering, Asia University, Taiwan. Since May 2014–present, he has also been the Research Consultant of Dept. of Medical Research, China Medical University Hospital, China Medical University Taichung, Taiwan. In addition, since Feb 2017, he has been the Permanent Council Member of Taiwan Domain Names Association (Taiwan DNA), Taiwan. He was also the Program Committee Chair of APNIC44 (September 2017) organized by Asia-Pacific Network Information Centre (APNIC). He has been awarded Best Paper Awards by BWCCA2018, MobiSec2017 and BWCCA2016, individually. He was awarded the Best Journal Paper Award by Association Algorithm & Computation Theory (AACT). Currently, his research interests include information and communication security, cyberspace security, blockchain network security, internet of things, application engineering and security, mobile and wireless networks protocols, medical and bio-information signal image processing, artificial intelligence and soft computing, and applied cryptography.

Fang-Yie Leu (Distinguished Professor) received his B.S., M.S., and Ph.D. degrees all from National Taiwan University of Science and Technology, Taiwan, in 1983, 1986, and 1991, respectively. His research interests include wireless communication, network security, grid applications, and sensor networks. He is currently a distinguished professor with the Computer Science Department, and the chairperson of the Big Data Program, Tunghai University, Taiwan. He also acts as an Editorial Board Member of at least 7 journals and serves as the TPC member of at least 10 international conferences. Prof. Leu now organizes MCNCS and CWECS international workshops. He is an IEEE member and was also a visiting scholar of Pittsburg University.

Ilsun You (Professor) received M.S. and Ph.D. degrees in computer science from Dankook University, Seoul, Korea, in 1997 and 2002, respectively. He received a second Ph.D. from Kyushu University, Japan, in 2012. From 1997 to 2004, he was at THINmultimedia Inc., Internet Security Co., Ltd. and Hanjo Engineering Co., Ltd. as a research engineer. Now, he is a full professor with the Department of Information Security Engineering, Soonchunhyang University. Prof. You is the EiC of the *Journal of Wireless Mobile Networks, Ubiquitous Computing, and Dependable Applications* (JoWUA) and *Journal of Internet Services and Information Security* (JISIS). He is on the Editorial Board for *Information Sciences* (INS), *Journal of Network and Computer Applications* (JNCA), IEEE *Access, Intelligent Automation & Soft Computing* (AutoSoft), *International Journal of Ad Hoc and Ubiquitous Computing* (IJAHUC), *Computing and Informatics* (CAI), and *Journal of High Speed Networks* (JHSN). He has especially focused on 4/5G security, security for wireless networks and mobile internet, IoT security, and so forth, publishing more than 180 papers in these areas. He is a Fellow of the IET and a Senior member of the IEEE.

Editorial

Selected Papers from the First International Symposium on Future ICT (Future-ICT 2019) in Conjunction with the 4th International Symposium on Mobile Internet Security (MobiSec 2019)

Giovanni Pau [1], Hsing-Chung Chen [2], Fang-Yie Leu [3] and Ilsun You [4,*]

[1] Faculty of Engineering and Architecture, Kore University of Enna, 94100 Enna, Italy; giovanni.pau@unikore.it
[2] Department of Computer Science and Information Engineering, Asia University, Taichung 41354, Taiwan; cdma2000@asia.edu.tw
[3] Department of Computer Science, Tunghai University, Taichung 40799, Taiwan; leufy@thu.edu.tw
[4] Department of Information Security Engineering, Soonchunhyang University, 22 Soonchunhyang-ro, Shinchang-myeon, Asan-si 31538, Choongchungnam-do, Korea
*   Correspondence: ilsunu@gmail.com

check for updates

Citation: Pau, G.; Chen, H.-C.; Leu, F.-Y.; You, I. Selected Papers from the First International Symposium on Future ICT (Future-ICT 2019) in Conjunction with 4th International Symposium on Mobile Internet Security (MobiSec 2019). *Sensors* **2021**, *21*, 265. https://doi.org/10.3390/s21010265

Received: 25 December 2020
Accepted: 29 December 2020
Published: 3 January 2021

Publisher's Note: MDPI stays neutral with regard to jurisdictional claims in published maps and institutional affiliations.

Copyright: © 2021 by the authors. Licensee MDPI, Basel, Switzerland. This article is an open access article distributed under the terms and conditions of the Creative Commons Attribution (CC BY) license (https://creativecommons.org/licenses/by/4.0/).

The International Symposium on the Future ICT (Future-ICT 2019) in conjunction with the 4th International Symposium on Mobile Internet Security (MobiSec 2019) has been held on 17–19 October 2019 in Taichung, Taiwan. The symposium provided academic and industry professionals an opportunity to discuss the latest issues and progress in advancing smart applications based on future ICT and its relative security. The symposium aimed to publish high-quality papers strictly related to the various theories and practical applications concerning advanced smart applications, future ICT, and related communications and networks. Furthermore, it was expected that the symposium and its publications could be a trigger for further related research and technology improvements in this subject matter.

The authors of conference papers falling in the Sensors' scope at this symposium have been invited to submit the extended versions to this Special Issue for publication. Moreover, new papers strictly related to the conference themes have also been welcome. Among the 18 submissions, the guest editors picked 12 high-level contributions for publication after several rounds of review carried out by invited experts.

The authors of [1] introduce a new algorithm and its systolic composition for digital normalized cross-correlation, based on the statistical characteristic of an inner-product. A relationship between the inner-product in cross-correlation and a first-order moment is acquainted. Subsequently, digital normalized cross-correlation is molded in a novel estimation method that essentially comprises the first-order moment. As the first-order moment's algorithm can be realized by systolic structure, the authors design a systolic array for normalized cross-correlation with a seldom multiplier for its fast hardware implementation. The comparison with other approaches highlights the encouraging performance of the proposed method.

A direct byte read (DRB) durable hybrid RAM disk (DHRD) scheme for hybrid storage systems, composed of RAM disk and SSD, is presented in [2]. The proposed solution implements a byte-range interface, agreeable with present interfaces, and can be practiced with buffered writes. Experimental evaluations are conducted utilizing various benchmarks that apply to various systems delivering dense I/O operations. For the hybrid storage device, the proposed scheme plays three to five times quicker throughputs than other approaches and can also diminish the accomplishment times of multimedia files read and write processing.

The authors of [3] advance a simplistic option for determining the fine aerosols with a diameter of less than 2.5 microns ($PM_{2.5}$) concentration, in which a set of image processing schemes and simple linear regression are applied. The suggested approach practices images with a high and low $PM_{2.5}$ concentration to differentiate these images. Examinations are carried out to validate the proposed approach employing an image data set and an open $PM_{2.5}$ concentration data set. The results show that the proposed approach delivers the best performance compared to other solutions.

A census transform with the Haar wavelet (CTHW) method, enhancing the efficiency with a wavelet transform, and an adaptive window census transform (AWCT), to allow the conversion window size adjusted for every point, are introduced in [4]. The suggested CTHW can produce a more favorable result with a small window size and be pleasantly employed to a low computational resource system. Besides, AWCT delivers better performance in lessening the running times with satisfactory quality.

The authors of [5] propose an effective multidimensional secret data-embedding scheme based on the mini SuDoKu matrix. The reference matrix is extended to multi-dimension to achieve even higher embedding capacity while still maintaining adequate security and efficiency. The proposed scheme is compared with other solutions, and the empirical results show that it achieves higher dB in terms of image quality and two bits per pixel in terms of the embedding capacity. Moreover, the proposed algorithm's time consumption is smaller than half of the conventional approach.

The authors of [6] propose the improvement for Error-correction codes (ECC) capability inside the page and RAID parity management outside the page, counting on lossless data compression. The adaptive ECC method can lessen the size of a source length in relationship to the compression ratio. The experimental results prove that adaptive ECC can improve error recovery's efficacy, thus achieving high reliability.

A secret image sharing solution based on a new maze matrix is presented in [7]. A pair of different cover images are utilized to carry secret data, and a pair of shadow images are formed following the supervision of the maze matrix. The secret data is obtained if both true shadows are manifested. Performance evaluations show that the detection ratio is 43% for cases in which a single shadow is tampered with, while it is 72% for cases in which both shadows tamper.

The authors of [8] propose a multi-labeled hierarchical classification (MLHC) learning model with hierarchical groups that attack employing a machine-learning algorithm to identify freak operations of the in-vehicle network. The suggested method can execute prompt decisions about an attack or favorable circumstances for in-vehicle networks by learning the CAN traffic, and it can record further accurate knowledge when an attack is recognized. The simulation outcomes reveal that the proposed approach delivers excellent efficiency.

A second-order Markov model for predicting the frame-level communication failures in an Industrial Wireless Sensor Network (IWSN), based on the preliminary tracks collected in a real-world factory, is introduced in [9]. The proposed model affords a more detailed explanation of the communication quality in IWSNs than traditional approaches. The simulation results prove that the suggested method increases the expected communication reliability compared to that obtained employing the original independent error model.

The authors of [10] suggest a new signature method employing certificateless public key cryptography (CL-PKC) to produce and validate a message's signature in an IoT environment. The advanced system is a certificateless aggregate arbitrated signature, and the gateway aggregates the signatures of messages created by the device group to lessen the volume of the whole signature. Validations show that the authors' solution can resolve the difficulties caused by public key replacement attacks and malicious key generation center (KGC), adding arbitrated signatures of the gateway to increase non-repudiation.

A global navigation satellite system (GNSS)-based crowd-sensing policy for distinct geographic regions, useful to determine how many targets are in precise topographical ranges or whether a target is in a definite territory, is introduced in [11]. The approach presented by the authors is based on the coordinates of latitude and longitude produced by GNSS to discover the positions of these coordinates. The data records, including latitude and longitude in a popular social networking service platform, are employed in simulations, and the obtained results are encouraging.

The authors of [12] present an architecture for WSNs based on Sub1G-Hz and a star topology to reduce sensing nodes' power consumption. The laboratory results determine that nodes' packet error rate can adequately be commanded near a target value, proving beneficial communication reliability and keeping energy consumption low.

Finally, the guest editors are grateful to the Editor-in-Chief and the editorial staff of Sensors for accepting their special issue proposal and the kind cooperation, patience, and active engagement.

Funding: This research received no external funding.

Conflicts of Interest: The authors declare no conflict of interest.

References

1. Pan, C.; Lv, Z.; Hua, X.; Li, H. The Algorithm and Structure for Digital Normalized Cross-Correlation by Using First-Order Moment. *Sensors* **2020**, *20*, 1353. [CrossRef] [PubMed]
2. Baek, S.H.; Park, K.W. A Durable Hybrid RAM Disk with a Rapid Resilience for Sustainable IoT Devices. *Sensors* **2020**, *20*, 2159. [CrossRef] [PubMed]
3. Liaw, J.J.; Huang, Y.F.; Hsieh, C.H.; Lin, D.C.; Luo, C.H. PM2.5 Concentration Estimation Based on Image Processing Schemes and Simple Linear Regression. *Sensors* **2020**, *20*, 2423. [CrossRef] [PubMed]
4. Liaw, J.J.; Lu, C.P.; Huang, Y.F.; Liao, Y.H.; Huang, S.C. Improving Census Transform by High-Pass with Haar Wavelet Transform and Edge Detection. *Sensors* **2020**, *20*, 2357. [CrossRef] [PubMed]
5. Horng, J.H.; Xu, S.; Chang, C.C.; Chang, C.C. An Efficient Data-Hiding Scheme Based on Multidimensional Mini-SuDoKu. *Sensors* **2020**, *20*, 2739. [CrossRef] [PubMed]
6. Lim, S.H.; Park, K.W. Compression-Assisted Adaptive ECC and RAID Scattering for NAND Flash Storage Devices. *Sensors* **2020**, *20*, 2952. [CrossRef] [PubMed]
7. Chang, C.C.; Horng, J.H.; Shih, C.S.; Chang, C.C. A Maze Matrix-Based Secret Image Sharing Scheme with Cheater Detection. *Sensors* **2020**, *20*, 3802. [CrossRef] [PubMed]
8. Park, S.; Choi, J.Y. Hierarchical Anomaly Detection Model for In-Vehicle Networks Using Machine Learning Algorithms. *Sensors* **2020**, *20*, 3934. [CrossRef] [PubMed]
9. Yu, Y.S.; Chen, Y.S. A Measurement-Based Frame-Level Error Model for Evaluation of Industrial Wireless Sensor Networks. *Sensors* **2020**, *20*, 3978. [CrossRef] [PubMed]
10. Lee, D.H.; Yim, K.; Lee, I.Y. A Certificateless Aggregate Arbitrated Signature Scheme for IoT Environments. *Sensors* **2020**, *20*, 3983. [CrossRef] [PubMed]
11. Lin, C.B.; Hung, R.W.; Hsu, C.Y.; Chen, J.S. A GNSS-Based Crowd-Sensing Strategy for Specific Geographical Areas. *Sensors* **2020**, *20*, 4171. [CrossRef] [PubMed]
12. Hung, C.W.; Zhang, H.J.; Hsu, W.T.; Zhuang, Y.D. A Low-Power WSN Protocol with ADR and TP Hybrid Control. *Sensors* **2020**, *20*, 5767. [CrossRef] [PubMed]

 *sensors*

Article

A GNSS-Based Crowd-Sensing Strategy for Specific Geographical Areas

Chuan-Bi Lin [1], Ruo-Wei Hung [2], Chi-Yueh Hsu [3] and Jong-Shin Chen [1,*]

[1] Department of Information and Communication Engineering, ChaoYang University of Technology, Taichung 413310, Taiwan; cblin@cyut.edu.tw
[2] Department of Computer Science and Information Engineering, Chaoyang University of Technology, Taichung 413310, Taiwan; rwhung@cyut.edu.tw
[3] Department of Leisure Services Management, Chaoyang University of Technology, Taichung 413310, Taiwan; cyhsu@cyut.edu.tw
* Correspondence: jschen26@cyut.edu.tw

Received: 2 June 2020; Accepted: 22 July 2020; Published: 27 July 2020

Abstract: Infectious diseases, such as COVID-19, SARS, MERS, etc., have seriously endangered human safety, economy, and education. During the spread of epidemics, restricting the range of activities of personnel is one of the options for the prevention and treatment of infectious diseases. A global navigation satellite system (GNSS), it can provide accurate coordinates of latitude and longitude to targets with GNSS receivers. However, it is not common to use GNSS coordinates to represent positions in social life. For epidemic management, it is important to know the locations (and addresses) of targets, especially in social life. When there are many targets, it is not easy to efficiently map these coordinates to locations. Therefore, we propose a GNSS-based crowd-sensing strategy for specific geographical areas that can be used to calculate how many targets are in specific geographical areas or whether a target is in a specific area. This strategy is based on the coordinates of latitude and longitude provided by GNSS to find the locations of these coordinates. As simulated data, the data records containing latitude and longitude in a well-known social networking service platform are used. The strategy is also available for mining hot spots or hot areas.

Keywords: epidemic management; GNSS/GPS; infectious disease; isolation; social networking service; hot spot

---

1. Introduction

A satellite navigation system with global coverage is called a global navigation satellite system (GNSS). It allows small electronic receivers to find its position in longitude and latitude coordinates. Current international GNSS standards for International Civil Aviation address only two core constellations: the U.S. Global Positioning System (GPS) and the Global Navigation Satellite System (GLONASS) [1,2]. These systems allow small electronic receivers to determine their location with the values of longitude and latitude to high precision using time signals transmitted along a line of sight by radio from satellites. Moreover, GPS receivers released in 2018 that use the L5 band [3] can have much higher accuracy. Many studies, that have used different techniques to improve observation precisions of GNSS positioning [4–6], can support this study. In this study, the corresponding locations of the latitude and longitude coordinates provided by a GNSS are evaluated. We also assume there is a large number of targets. Each target has a GNSS receiver that can acquire its current latitude and longitude coordinates and delivers its coordinates to a server. The coordinates of targets can be acquired by a computer manner. Our proposition is to assume that, when spaces of specific areas are large, such as New York City, Wuhan City China, and Taipei City, the number of targets to evaluate their locations, such as the population of a city, is also large.

During the spread of an epidemic, such as COVID-19 [7–9], it is an effective option to limit the movement range of people in order to control the epidemic. Determining the locations of targets can help determine whether people are in quarantine. There are two topics in this research. The first item is whether the target is within a certain range, which can be used to determine whether the quarantined person is in the isolation zone. The second item is the number of targets is in a specific area, thereby, it can be used to determine whether too many people are likely to cause infectious diseases. For example, in Taiwan, the four days from 2 April to 5 April 2020, are a consecutive holiday. During this period, there were a large number of tourists. The National Health Command Center (NHCC) in Taiwan [9] estimates that there were 1.5 million visits in 11 scenic spots through the telecommunications operator signal. Since the average stay time of these visits exceeded 15 min, the command center was worried about triggering a cluster infection of COVID-19 and urgently issued a national police call to the 11 scenic spots to call for evacuation, as well as the health management of tourists.

In general, a location with latitude and longitude is termed as a geographical point. There are very many geographical points generated by various social network services (SNSs) of the platforms, such as Facebook, Twitter, Google, and Foursquare. Check-in for some targets is a location-based service. It provides a mechanism to record users who have visited these geographical points. Among these platforms, Facebook is the largest one with regard to the number of users. In Facebook, these points are termed as "places". In other words, the check-in places are the locations that people actually visited. So, using these points as experimental samples to explore the crowd distribution is reasonable and credible. The Facebook penetration rate in Taiwan is the highest in the world. Here, the number of daily users reached 13 million of approximately 23 million. Accordingly, Taiwan is an appropriate selection as the experimental area, where the main island has an area of 35,808 square kilometers.

In light of above discussions, this study proposes solutions to the following:

1. For real-time coordinates of targets, this strategy can be used to determine the number of targets in the specified area in time and can also be used to determine whether the target is in the specified area.
2. For historical coordinates of targets, this strategy can be used to determine areas in which the targets can easily gather or which locations are the hot spots.

The rest of the paper is organized as follows. In Section 2, the research related to our study is introduced. In Section 3, the system architecture and the problems are described, in which the proposition is formalized. In Section 4, the crowd-sensing strategy is proposed. In Section 5, a demonstration is given by taking the area of Taiwan Island as the experimental area and the Facebook check-in places as the targets and examples. In Section 6, a discussion about the performance is described. Finally, a conclusion is given in Section 7.

2. Related Work

Accurate latitude and longitude positioning supports our research to evaluate the corresponding social locations. In [4–6], there are different techniques to improve observation precisions of GNSS positioning. In [10], the authors focused on the integrated methodology of GNSS and device-to-device measurements. The simulation and experimental results demonstrated that the integrated methodology outperforms the nonintegrated one. In [11], a two-step approach is studied, namely, computing first the Fisher Information Matrix (FIM) for the channel parameters, and then transforming it into the FIM of the position, rotation, and clock-bias. The analysis demonstrated the advantages of the hybrid positioning in terms of (1) localization accuracy, (2) coverage, (3) precise rotation estimation, and (4) clock-error estimation. In [12], this study presented a wideband/multiband quad-antenna system for 4G/5G/GPS metal-frame mobile phones. The merit of the proposed antenna system is that a quad-antenna system is achieved under the condition of a metal frame and, without using any decoupling structure, the desired bands for 4G/5G/GPS are covered.

The Internet environment has generated a large number of geographical points, such as Facebook check-in places [13–15], Google Maps places, Foursquare check-in places, etc. These places in Social Networking Services (SNS) are special kinds of geographical points. However, each SNS point contains not only a geographical coordinate and some contents to introduce this point. Accordingly, there are many studies focused on these points with their contents. In [16], the paper aimed to assess the role that interactive technology can play in enhancing urban governance to meet social needs. In [17], a real-time Google Maps-based arterial traffic information system for urban streets is presented. In [18], the authors proposed the reuse of up-to-date and low-cost place data from social media applications for land use mapping purposes by Foursquare place data. In [19], the study aimed to explore Foursquare mobility networks and investigate the phenomena of clustering venues across the cities. In [20], the study aimed to inform on how scientific researchers could utilize data generated in location-based social networks to attain a deeper understanding of human mobility. In [21], the authors proposed to find the geographical points related to a special folk belief. In [22], the author utilized the Markov Chain model with their proposed activity detection method to predict the activity category of the user's next check-in location. In [23], an urban tourism check algorithm is proposed. It can find those who are tourists and find out where the people come from and the route of their visit. In [24], the structure analysis of place networks is explored, in which vertices of geographic places, while the links between places are formed, are based on the user's check-in history. In [25], a new feature fusion-based prediction approach is proposed, based on carefully designed feature extraction methods. If these points can be mapped to social locations, not only coordinates, these applications could have further enhancements. To acquire the social locations of geographical points is also a promising topic, such as in epidemic management. In computational geometry, this is the point-in-polygon (PIP) problem [26–30]. These studies provide methods to evaluate that a point is inside an area or is not-inside an area without error. In [29], a computer-friendly method was proposed for the PIP problem.

In [30], the authors developed a PIP algorithm that can evaluate that a point is inside or is not-inside an area. Based on this PIP algorithm, the method to find all points within a specific geographical area was developed. It first plans a range that can cover the entire specific area. Then, it finds all the points in the planned range. For each point, the PIP algorithm is used to calculate whether this point is in the specified area. Because the PIP method is verified, this method can be applied to evaluate all the points in the specified area without error. However, this method is inefficient when the area is large with a large number of points. The planned range is often much larger than the specific area. This method selects a large number of points that are not in this specific area and confirms their locations by PIP. If the planned range can be smaller, the points that need to be confirmed by PIP will be reduced a lot. In addition, if certain points are known in the area or known outside the area, there is no need to confirm through PIP. In this way, the performance will be improved a lot.

3. System Architecture and Problem Description

The architecture of the system assumes that the target has a GNSS receiver. It can convert the received signal into coordinate data and send it to storage, such as databases, according to the transmission mechanism of the wired or wireless network. The target will be represented by a geographic point, referring to a point that has its latitude and longitude coordinate. In geometry, a point with coordinate is termed as a geographical point. Moreover, a geographical area (a specific area) is a polygon. A polygon is composed of edges with geographical points. These edges enclose a measurable interior [21]. Moreover, for a geographical point p, $x(p)$ and $y(p)$ are used to represent the coordinate of p. For a geographical area A with n points, $a_0, a_1, \ldots$, and a_n are used to represent these n points, where $a_0 = a_n$.

The architecture of this system can be simplified into a geographic area A and a set P of geographical points. Because the area of area A is large and the size of set P is also very large, it is impossible in the field of computers to load all the points of P into the variables in the programming language and then calculate whether each point is in the range of A. It must be planned to only take points in part of the area in A at a time and these points are affordable by computers. The type of this area is chosen as a circle. There are two reasons. The first reason is that some well-known platforms (servers), such as Facebook, also provide geographical points in a circle manner. The other reason is to follow the access method of storages. The storage in general is a database system. To retrieve some records from the storage, it must be obtained through the access mechanism of the database system, such as structured query language (SQL). The circle-area manner can be mapped to the corresponding SQL commands. However, the configuration of the circle manner is not easy to fully cover a geographical area. Here the cellular architecture using a regular hexagonal configuration, instead of the circular configuration, can be applied. The cell architecture is shown in Figure 1, in which Figure 1a provides the layout of a cell and Figure 1b provides the relative positions of cell c and its six neighbors. If the radius of the circle is r, then the length of the regular hexagonal side is r. For convenience, a cell is used to represent a regular hexagon. There are six cells around a cell, termed as the neighbors of this cell. For a cell c, cells c_{N0}, c_{N1}, ... , and c_{N5} are used to represent the six neighbors of cell c. The relative positions of cell c and its six neighbors can be expressed by an oblique coordinate architecture with 60 degrees (i.e., the angle between the x-axis and the y-axis is 60 degrees). Moreover, for cell c with radius r, Table 1 provides the coordinates of each vertex. Table 2 provides the coordinates of each neighbor.

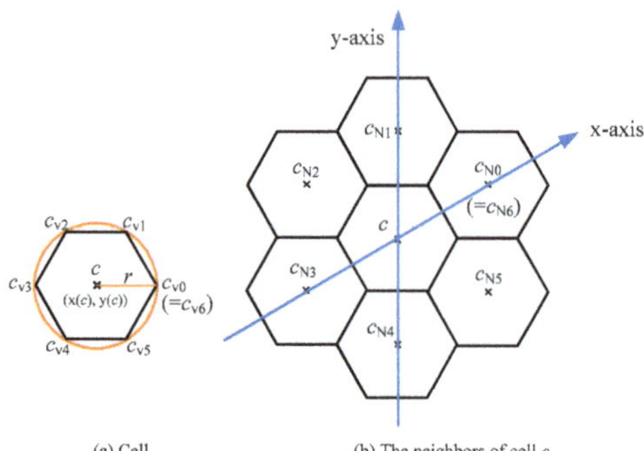

(a) Cell   (b) The neighbors of cell c

Figure 1. Cell architecture, a cell is a regular hexagon area. (**a**) For cell c with radius r, c_{V0}, c_{V1}, c_{V2}, c_{V3}, c_{V4}, c_{V5} are its six vertices. Moreover, c_{V6} represents vertex c_{V0}. (**b**) The neighbors of cell c represent as c_{N0}, c_{N1}, c_{N2}, c_{N3}, c_{N4}, c_{N5}. Moreover, cell c_{N6} represents cell c_{N0}.

Table 1. The vertices of cell c. For a cell c with radius r, the coordinates of each vertex are as follows.

Vertex	Coordinate
c_{V0} $(=c_{V6})$	$(x(c) + r, y(c))$
c_{V1}	$(x(c) + 0.5r, y(c) + 0.5\sqrt{3}r)$
c_{V2}	$(x(c) - 0.5r, y(c) + 0.5\sqrt{3}r)$
c_{V3}	$(x(c) - r, y(c))$
c_{V4}	$(x(c) - 0.5r, y(c) - 0.5\sqrt{3}r)$
c_{V5}	$(x(c) + 0.5r, y(c) - 0.5\sqrt{3}r)$

Table 2. The neighbors of cell c. For a cell c with radius r, the coordinates of each neighbor are as follows.

Neighbor	Coordinate
c_{N0} $(=c_{N6})$	$(x(c) + 1.5r, y(c) + 0.5\sqrt{3}r)$
c_{N1}	$(x(c), y(c) + r)$
c_{N2}	$(x(c) - 1.5r, y(c) - \sqrt{3}r)$
c_{N3}	$(x(c) - 1.5r, y(c) - \sqrt{3}r)$
c_{N4}	$(x(c), y(c) + - \sqrt{3}r)$
c_{N5}	$(x(c) + 1.5r, y(c) - 0.5\sqrt{3}r)$

Moreover, if expressed by the geographic coordinate system (i.e., use latitude and longitude values to describe a coordinate with cell c), the offset coordinates of the six neighbors are shown in Table 2. Notably, the distance r is transformed by the Cartesian coordinate. Every point that is expressed in ellipsoidal coordinates can be expressed as rectilinear x y z (Cartesian) coordinates. Cartesian coordinates simplify many mathematical calculations. The Cartesian systems of different data are not equivalent. The distance between the points of longitude 121 and latitude 21 to the point of longitude 122 and latitude 21 is about 103 km. The distance between the points of longitude 122 and latitude 21 to the point of longitude 122 and latitude 22 is about 111 km. If it is configured with a radius of 1 km, r can be set to about 0.009 (=1/111) near this area.

A cell c is a regular hexagon area, defined as (c, r), where $(x(c), y(c))$ is the geographical coordinate, and r is the length of the regular hexagonal side. Moreover, $V(c) = \{c_{V0}, c_{V1}, c_{V2}, c_{V3}, c_{V4}, c_{V5}\}$ is defined as the set of six vertices and $NB(c) = \{c_{N0}, c_{N1}, c_{N2}, c_{N3}, c_{N4}, c_{N5}\}$ is defined as the set of six neighboring cells. Table 1 provides the coordinates of each vertex in $V(c)$ and Table 2 provides the coordinates of each neighbor in $NB(c)$.

Hereafter, the word "geographical point" is simply termed as a "point" and the word "geographical area" is simply termed as an "area". Evaluating whether a point is inside or is not-inside is the PIP problem. It can count the intersections of the polygon with the ray of this point. A ray of a point is starting from this point to any fixed direction. That the number of intersections between the polygon and the ray is odd indicates this point is inside this polygon. Otherwise, it indicates this point is not-inside this polygon. This method can evaluate the locations of points with respect to an area without error. To implement this method, it must first be able to determine whether two lines are intersected. In [29], a computer-friendly method was proposed to do it. Suppose there are two lines. The first line crosses over both point o_1 and point o_2. The other line crosses over both point o_3 and point o_4. First, ρ, as shown in (1), can be evaluated. If the value of ρ equal 0, these two lines are parallel. Next, κ_1 and κ_2, as, respectively, shown in (2) and (3), can be evaluated. If the value of κ_1 is between 0 and 1, the intersection is between o_1 and o_2. If the value of κ_2 is between 0 and 1, the intersection is between o_3 and o_4. In other words, if edge (o_1, o_2) and edge (o_3, o_4) are intersected, the value of κ_1 is between 0 and 1 and the value of κ_2 is also between 0 and 1.

$$\rho = (x(o_1) - x(o_2)) \cdot (y(o_3) - y(o_4)) - (y(o_1) - y(o_2)) \cdot (x(o_3) - x(o_4)) \tag{1}$$

$$\kappa_1 = ((x(o_1) - x(o_3)) \cdot (y(o_3) - y(o_4)) - (y(o_1) - y(o_3)) \cdot (x(o_3) - x(o_4)))/\rho \tag{2}$$

$$\kappa_2 = -(((x(o_1) - x(o_2)) \cdot (y(o_1) - y(o_3)) - (y(o_1) - y(o_2)) \cdot (x(o_1) - x(o_3)))/\rho \tag{3}$$

The overlap of an area A with a cell c, where the space of A is far greater than the space of c, is also considered. There are two overlapping cases between A and c. The first case is the space of A is overlapped with the space of c. Non-first relationship is the second relationship. The first case can be evaluated when all six points are inside A and there are no intersection points among the edges of c and the edges of A.

As shown in Figure 2, an area is composed of a convex polygon with six points, $a_0, a_1, \ldots,$ and a_5. The rays of p_1, p_2, and p_3 within this area, respectively, have 2, 1, and 0 intersection(s). Since the ray of a point has odd intersections within the area, it means that the point is inside the area; otherwise, the point is not-inside the area. Then, both point p_1 and point p_3 are not-inside the area and point p_2 is inside the area. Continually, there are three cells, c_1, c_2, and c_3. Since all vertices of c_1 are in this area, two vertices of c_2 are in this area, and no vertices of c_3 are in this area, both c_1 and c_2 overlap with this area and c_3 does not overlap with this area.

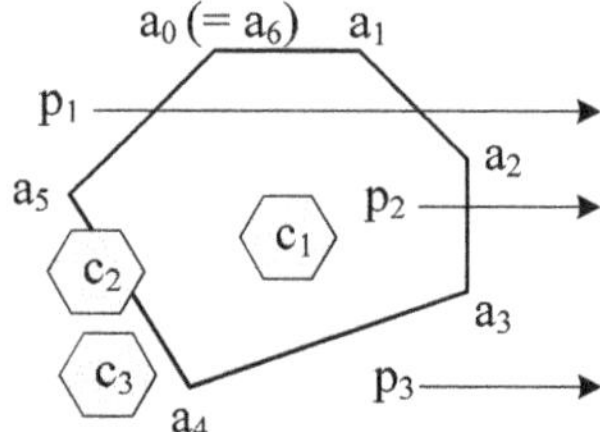

Figure 2. Example of an area, points, and cells. This area is a polygon composed by points a_0, a_1, ... , and a_5. To this area, point p_1 and point p_3 are not-inside and point p_2 is inside, both cells c_1 and c_2 are overlapped and c_3 is not overlapped. Moreover, cell c_1 is also termed as inside this area.

4. Crowd-Sensing Strategy

Given an area $A = \{a_0, a_1, \dots , a_n\}$ and a large number of points P, where P are stored in a database and can be accessed by using SQL commands, this strategy includes two step cell allocation and point acquisition. It can achieve the evaluation all of points in P inside area A by applying the two steps.

4.1. Cell Allocation

The first step applies algorithms EEI, PIA, and CAO to achieve cell allocation. Algorithm 1 provides algorithm EEI. It is an Edge–Edge Intersection algorithm, where inputs are two edges (o_1, o_2) and (o_3, o_4) and the output is a value of 0 or 1. If edge (o_1, o_2) and edge (o_3, o_4) are intersected, it results by returning 1. Otherwise, it results by returning 0. In this algorithm, it first calculates the value of ρ according to (1). If the value of ρ is 0, the two edges are parallel. If the value of ρ is not 0, it calculates the values of κ_1 and κ_2. If both κ_2 and κ_1 are between 0 and 1, it indicates that the two edges have an intersection.

Algorithm 1 EEI (o_1, o_2, o_3, o_4)

---

1.   {

2. $\rho := (x(o_1) - x(o_2)) \cdot (y(o_3) - y(o_4)) - (y(o_1) - y(o_2)) \cdot (x(o_3) - x(o_4));$

3.   if $\rho = 0$ then return 0;

4.   else

5.   {

6. $\kappa_1 := ((x(o_1) - x(o_3)) \cdot (y(o_3) - y(o_4)) - (y(o_1) - y(o_3)) \cdot (x(o_3) - x(o_4)))/\rho;$

7. $\kappa_2 := - (((x(o_1) - x(o_2)) \cdot (y(o_1) - y(o_3)) - (y(o_1) - y(o_2)) \cdot (x(o_1) - x(o_3)))/\rho;$

8.   if $(\kappa_1 \geq 0$ and $\kappa_1 \leq 1)$ and $(\kappa_2 \geq 0$ and $\kappa_2 \leq 1)$ then return 1;

9.   else return 0;

10.   }

11.   }

Algorithm 2 provides the Point Inside Area (PIA) algorithm, whose inputs are a point p and an area A and its output is a value of 0 or 1. If point p is inside area A, it results by returning value 1. Otherwise, it results by returning value 0. In this algorithm, variables o_1 and o_2 are used to represent an edge of area A and variables o_3 and o_4 are used to represent the incremental extension from p toward the x-axis until it is greater than the coordinate values of $x(o_1)$ and $x(o_1)$, as in line 6 (i.e., $x(o_4)$ is set as $(\max(x(o_1), x(o_2)) + 1)$. Variable count is used to record the number of intersections between the extension line of p and the edge of A. If the number of intersections is odd, it means that p is inside A, otherwise p is not-inside A.

Algorithm 2 PIA (p, A)

---

1. {
2.   *count*: = 0;
3.   for i: = 0 to $n - 1$ do
4.   {
5. o_1: = a_i; o_2: = a_{i+1};
6. o_3: = p; o_4: =(max(x(o_1), x(o_2)) + 1, y(p));
7.   if (EEI(o_1, o_2, o_3, o_4) == 1) then *count*: = *count* + 1;
8.   }
9.   if (*count* % 2 == 1) then return 1;
10.   else return 0;
11. }

Algorithm 3 provides the Cell-Area Overlap (CAO) algorithm, whose inputs are a point p and an area A and its output is a value of 0 or 1. In this algorithm, each vertex of the cell is taken out independently, and is then calculated for whether it is inside A. The variable count is used to record how many vertices of cell c are inside A. If any of the six vertices of cell c are inside A, c and A are overlapping. Otherwise, c and A are non-overlapping.

Algorithm 3 CAO (c, A)

---

1. {
2.   *count*: = 0;
3.   for i: =0 to 5 do
4.   {
5. p:= c_{vi};
6.   if (PIA(p, A) == 1) then *count*: = *count* + 1;
7.   }
8.   if (*count* > 0) then return 1;
9.   else return 0;
10. }

Algorithm 4 provides the Cell Allocation (CA) algorithm, whose inputs are an area A and a cell c_s. The allocation starts with cell c_s, called the seed cell, and then expands to its neighboring cells NB (c_s). Then, the neighbors of c_s continue to extend the allocation of cells. The allocation is done until area A is completely covered by cells. In this algorithm, variable C records the allocated cells and variable P records the reference cells. In each iteration (in lines 5 to 14), each cell c_p will be selected in sequence from P. Then, the neighbors of the c_p are calculated using NB(c_p) (in line 7). For each cell, c_{NB} of NB(c_p), if cell c_{NB} is overlapped with A and c_{NB} is not included in C, c_{NB} will be added into C (i.e., C: = $C \cup \{c_{NB}\}$). Additionally, c_{NB} is added to N (i.e., C: = $\cup\{c_{NB}\}$). At end of the iteration, the value of N is assigned to p to start a new iteration. Therefore, this algorithm is stopped when P is empty (in line 3) and then it returns the allocated cells C.

Algorithm 4 CA (c_s, A)

---

1.   {
2. $P := \{c_s\}$, $N := \varnothing$, $C := \{c_s\}$;
3.   while ($P \neq \varnothing$)
4.   {
5.   for each $c_p \in P$ do
6.   {
7. $NB := NB(c_p)$;
8.   for each $c_{NB} \in NB$ do
9.   {
10.   if (CAO(A, c_{NB}) == 1 && $c_{NB} \notin C$) then
11.   {
12. $N := N \cup \{c_{NB}\}$, $C := C \cup \{c_{NB}\}$;
13.   }
14.   }
15. $P := N$, $N := \varnothing$;
16.   }
17.   return C;
18.   }

As shown in Figure 3, the irregular area is composed of 2055 points, and the numbers on cells index the order of cell allocation. In this example, c_i is used to represent the cell, numbered as i. Initially, cell c_0 is allocated. In the first iteration, the six neighbors $c_1, c_2, \ldots$, and c_6 are included in C. Herein, C contains seven cells, $c_0, c_1, c_2, \ldots$, and c_6, and P contains six cells, $c_1, c_2, \ldots$, and c_6. In the second iteration, cells $c_1, c_2, \ldots$, and c_6 will be the reference cells. For instance, the neighbors of c_1 contains cells c_7, c_8, c_2, c_0, c_6, and c_9. However, c_2, c_0, and c_6 are already included in C. Only c_7, c_8, c_9 are included in C. Continually, in the third iteration, cells c_{19} to c_{32} are included to C and in fourth iteration, cells c_{33} to c_{37} are included in C. In the fifth iteration, cells c_{33} to c_{37} are the reference cells P. Since all of the non-overlapped neighbors of $c_{33}, c_{34}, \ldots$, or c_{37} are included in C, no cell are included in C and P is empty (in line 3). Finally, the allocation is finished by returning the allocation C.

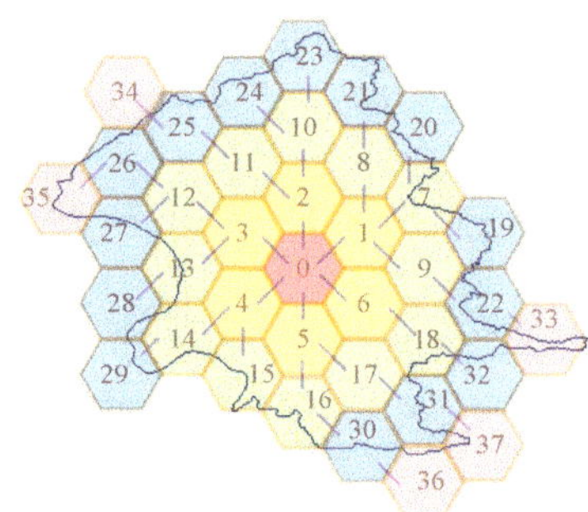

Figure 3. Example of Cell Allocation to an area. The allocation is started from c_0. In each iteration, some cells will be allocated. After five iterations, the allocation is finished.

4.2. Point Acquisition

Points in the local database can be obtained with the similar SQL command in (4), where 'P' is the table that stores all of geographic points, 'x' and 'y' are the fields for storing latitude and longitude values, and r is the radius of the circular area. In addition, the SQRT (v) function is used to calculate the square root of the value v, and the POWER (v, n) function is used to calculate the nth power of the value v.

$$\text{SELECT * FROM 'P' WHERE SQRT(POWER('x' − x(c),2) + POWER('y' − y(c), 2))} \leq r \qquad (4)$$

Algorithm 5 provides the Cell Inside Area (CIA) algorithm, whose inputs are a cell c and an area A and its output is a value of 1 or 0 that, respectively, indicates c inside A or c not-inside A. In this algorithm, each edge of c and each edge of A will be calculated to the number of intersections. If the number of intersections is 0, c is inside A, otherwise c is not inside A.

In summary of the above algorithm, the Crowd-Sensing (CS) algorithm is purposed. Algorithm 6 provides the CS algorithm. It is a Crowd-Sensing algorithm, where the input includes a seed of cell c_s, an area A, and all of points P and its output is a set P_A of all points inside A. In this algorithm, it first calculates the cell allocation C by applying CA (c_s, A), where C is the set of cells that can fully cover the space of area A. Then, for each cell c in C, it first extracts the geographic points Pc included in cell c from the database. There are two cases to deal with Pc. The first case based on cell c is inside area A, and Pc is directly included in P_A. The other case is based on cell c not-inside A. In this case, every point p in Pc will be evaluated with area A. If point p is inside area A, then p is included to P_A. For example, in Figure 3, because cells $c_0, c_1, \ldots, c_6, c_8, c_9, \ldots, c_{12}$ are inside this area, the points in this range are included in P_A. Other cells are not-inside this area, so the points in this range must be evaluated. In particular, our strategy only needs to evaluate the points in the outermost cells. In fact, the range of points to evaluate is very small. In general, if an area is allocated thousands of cells, only hundreds of cells are evaluated.

Algorithm 5 CIA (c, A)

```
1.     {
2.         count: = 0;
3.         for i: =0 to 5 do
4.         {
5.             for j: =0 to n-1 do
6.             {
7.                 o1: = cVi, o2: = cVi+1, o3: = aj, o4: = aj+1;
8.                 if (EEI(o1, o2, o3, o4) == 1 ) count: = count +1;
9.             }
10.        }
11.        if (count == 0) return 1;
12.        else return 0;
13.    }
```

Algorithm 6 CS (c_s, A)

```
1.     {
2.         P_A: = Ø;
3.         C: = CA (c_s, A);
4.         for each c ∈ C do
5.         {
6.             P_C: = (SELECT * FROM 'P' WHERE SQRT(POWER('x'-x(c),2) + POWER('y'-y(c), 2)) <= r)
7.             if (CIA(c, A) == 1) P_A: = P_A ∪ P_C;
8.             else
9.             {
10.                for each p∈ P_C do
11.                {
12.                    If (PIA(p, A) == 1) then P_A: = ∪{p};
13.                }
14.            }
15.        }
16.        return P_A;
17.    }
```

---

5. Demonstration

The geographical area of experiment is mainly Taiwan Island that is located between 120 degrees to 122 degrees east longitude and 22 degrees to 25 degrees north latitude. It has an area of 35,808 square kilometers with 23.7 million inhabitants. The points are based on the check-in places of Facebook. The points were acquired from Facebook platform in January 2017, for a total of 1,112,188. We used these points as targets and try to find the targets in specific areas. This area currently contains 6 special municipalities, 10 counties and 3 provincial cities. We demonstrate the results according to the 19 subareas. Figure 4 provides the distribution of 19 subareas, the special municipalities are numbered as 1–6, counties are numbered as 7–16, and provincial cities are numbered as 17–19. Each subarea is composed of hundreds to thousands of vertices. For instance, subarea 1 is composed of 2055 vertices. For each subarea, the proposed strategy was applied to evaluate the number of points inside it.

In Section 5.1, we first display the points in each area. In addition, the population density and the space density with points were also considered to find hot areas. In Section 5.2, we then display the points in each spot. The numbers of points in spots are also used to find hot spots. In addition, the distribution of spots are used to display the crowd distribution. In Section 5.3, we present the contributions of this study. In Section 5.3, we discuss the differences between this study and the previous study.

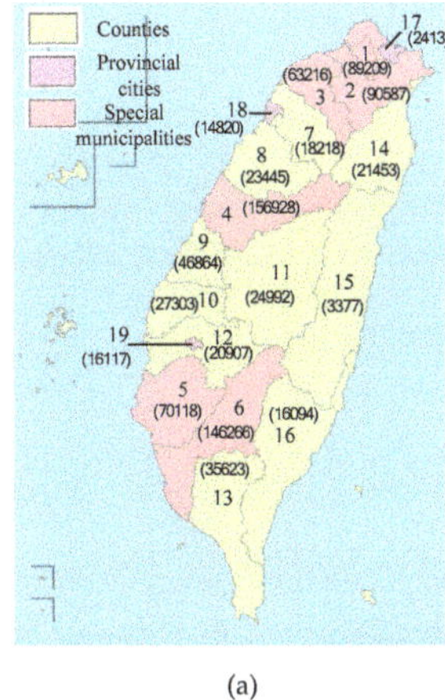

(a)

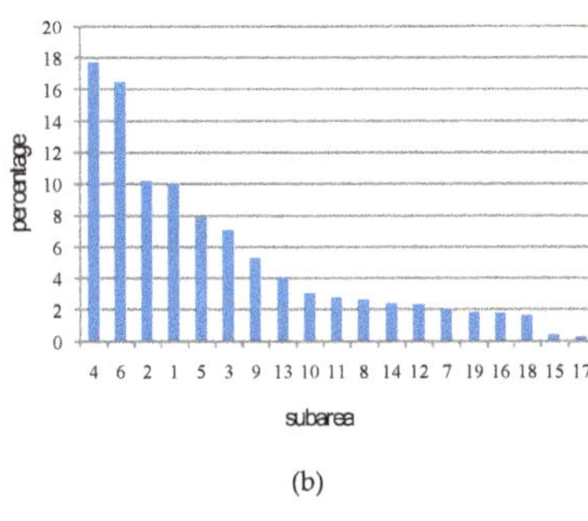

(b)

Figure 4. The experimental area with 887,950 points. This area currently contains 6 special municipalities, 10 counties and 3 provincial cities. These subareas are numbered as 1–19: (**a**) The number of points in each subarea. (**b**) The point ratio of each subarea to 16 subareas.

5.1. Points in Areas

In Figure 4a, the value in parentheses is the number of points of the subarea. For example, there are 156,928 points in the area numbered as 4. Moreover, there are a total of 887,950 points in these 19 areas, of which subareas 4, 6, 2, 1, 5, and 3 are the hot areas with the most points. Figure 4b provides the point ratio of each subarea to 16 subareas. There are 616,324 (69.41%) points in these six hot areas. Especially, all of the six hot areas are special municipalities. These results show that, through the coordinates, we can accurately calculate the targets in a specific administrative area, rather than only the information of the latitude and longitude coordinates. Then, the population and the space of hot area are taken into consideration.

Table 3 provides the population, space, points of each hot area, population density, and space density. For example, in subarea 1, the number of the population is 2,687,629, the space is 271.8 square kilometers, and the number of points is 89,209. Therefore, its population density is 3.32% (i.e., the value of 89,209/2,687,629 × 100%) and its space density is 328.21 (i.e., the value of 89,209/271.8). Among these six hot areas, subarea 1 especially has a much higher space density than the other five hot areas. For a specific area with positioned targets, the population and the space can be taken into consideration. It can get the proportion of the targets to the number of the population and the spatial proportion of the target object. This result can be used to determine whether the targets are too crowded in this area based on population or space. It can be used as a control for crowds, such as prohibiting more targets from entering this area or evacuating some targets from leaving this area. It can help to keep the number of targets in this area under control for social applications, such as traffic or epidemic control.

Table 3. Population density and space density based on number of points.

Subarea	Population	Space (km^2)	No. of Points	Population Density (#/Population × 100%)	Space Density (#/Space)
1	2,687,629	271.8	89,209	3.32	328.22
2	3,984,051	2052.57	90,587	2.27	44.13
3	2,171,127	1220.95	63,216	2.91	51.78
4	2,778,182	2214.90	156,928	5.65	70.85
5	1,886,267	2191.65	70,118	3.72	31.99
6	2,777,873	2951.85	146,266	5.27	49.55

#: number of points.

Figure 5 provides the distribution of points for subarea 1 and its neighborhood. The range has 93,549 points, where there are 89,029 points inside subarea 1 and 4520 points not-inside subarea 1. Most of the points are gathered in subarea 1. Moreover, the points are concentrated between longitude 121.45 to 121.6 and latitude 24.95 to 25.01. However, there are almost no geographic points between latitude 25.1 and 25.22. Obviously, the distribution of points is very uneven. Then, we continue to calculate the locations of the hot spots to reveal the concentrations.

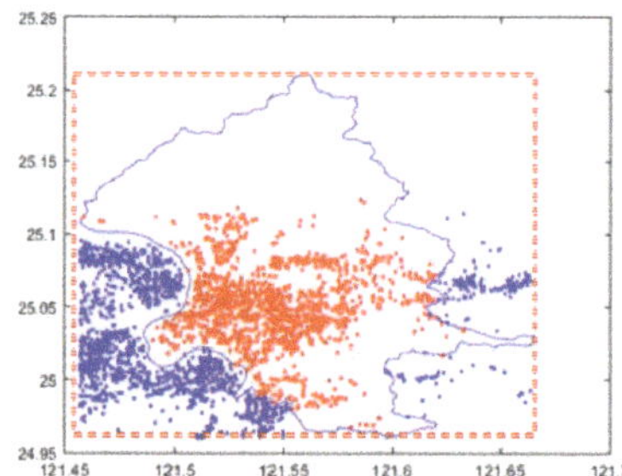

Figure 5. Point distribution of subarea 1 and its neighborhood. The range is longitude 121.45 to 121.67 and latitude 24.95 to 25.22. The number of points in this range is 93,549. Randomly, 1600 points inside subarea 1 and 1600 points not-inside subarea 1 are marked on this range.

5.2. Points in Spots

The area is divided into small areas, termed as spots, with a range of 0.01 degrees of latitude and longitude, and the area is slightly larger than 1 square kilometer. As shown in Table 4, there are a total of 15,573 spots. There are 15,347 spots whose numbers of points are less 1000. However, there are 12 spots whose numbers of points are more than 4000. Table 5 provides the details of the 12 hot spots, including the locations, the coordinates, and the numbers of points. The 12 hot spots are in subarea 4, subarea 6, subarea 9, and subarea 19. Moreover, subarea 4 contains six hot spots, subarea 6 contains four hot spots, subarea 9 contains one hot spot, and subarea 19 contains one hot spot. Figure 6 shows the distribution of points in hot spot one. There are 6024 points in this range of about one square kilometer. It means when we are in this hot spot, we are easily exposed to these targets (points). The analysis of hot spots helps to understand whether the target is concentrated in a certain small range, and it is also easy to monitor these spots.

Table 4. Spot statistics with points.

No. of Points	Spots	Percentage
less than 1000	15,347	98.55
1000–2000	152	0.98
2000–3000	49	0.31
3000–4000	13	0.08
4000–5000	4	0.03
5000–6000	7	0.04
6000–7000	1	0.01
Total	15,573	100

Table 5. The first 12 hot spots.

Hot Spot	Location	Coordinate	No. of Points
1	subarea 6	120.3, 22.67	6024
2	subarea 6	120.3, 22.62	5731
3	subarea 4	120.64, 24.18	5527
4	subarea 6	120.3, 22.63	5506
5	subarea 4	120.66, 24.16	5428
6	subarea 4	120.68, 24.16	5154
7	subarea 4	120.68, 24.15	5152
8	subarea 6	120.3, 22.64	5043
9	subarea 4	120.65, 24.16	4855
10	subarea 19	120.44, 23.48	4544
11	subarea 4	120.68, 24.14	4424
12	subarea 9	120.54, 24.08	4056

Figure 6. Distribution of points in hot spot one with 6024 points.

The first 10,000 spots, according to the number of geographic points, are taken into consideration. Then, these spots are classified into 10 groups, where each group is represented using brown color. The darker color represents more geographic points in this spot. The layout of the 10,000 spots is shown in Figure 7a. Comparing the relative positions of the 19 subareas in Figure 4a, it shows that hot spots are almost concentrated in the range of subarea 1 to subarea 6. These six areas are six municipalities in which the population densities are more than other administrative areas. Figure 7b shows the distribution of Highway in this area. The distribution of spots is consistent with the distribution of highway. The geographic points of this experiment are the check-in places of Facebook in Taiwan Island. These places are the historical records of Facebook users who have been to these places for a long time. Because Facebook users are very numerous in this experimental area, these points represent almost every point covered by social activities in the area. Therefore, areas with convenient transportation and high urbanization will have more geographic points. We indicated the spot, according to the number of geographic points, as a spot height map. This map, as shown in Figure 7a, coincides with the highway distribution and administrative area distribution of this area. These facts reveal that our research is a crowd-sensing strategy.

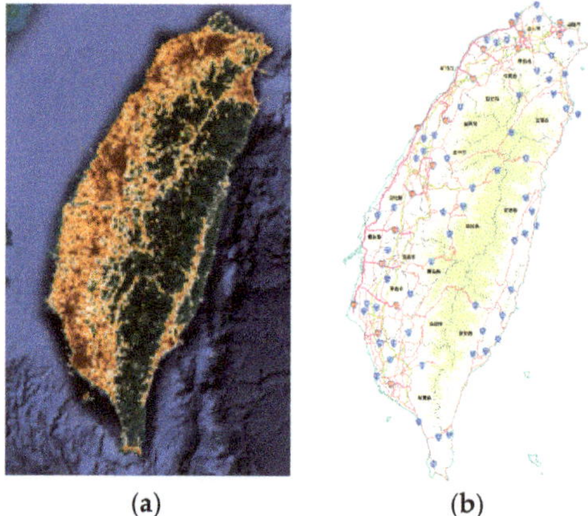

(a)   (b)

Figure 7. (**a**) Distribution of spots and (**b**) distribution of highways. The layout of (**a**), respectively, coincides with the layout of (**b**) and the layout of Figure 4a.

5.3. Summary

In [29], an error-free method is presented that can calculate whether a point is inside a polygon. For this, we planned the PIA algorithm, which can also evaluate whether a point is within an area or not without error. Therefore, the efficiency of this type of method is determined by how many points are needed to apply the PIA algorithm and not the accuracy. We take an example, shown in Figure 5, to illustrate the difference between our strategy and [30]. In [30], it first plans a rectangle range that can cover this area. The range is marked as a dashed line. It then finds all points in this range. In our experimental environment, there are 95,349 points. So, the number of executing PIA algorithms is 95,349. In our strategy, we planned a set of cells that could cover this area. The range of the cells is similar to the example shown in Figure 3 that is slightly larger than this area. In fact, we planned with a cell of about 1 square kilometer. This area is about 271.8 square kilometers, as shown in Subarea 1 of Table 3. In this example, the number of cells is about 350 and the number of points within these cells is about 90,000. These cells are classified into inside cells and not-inside cells through the CIA algorithm. Most points are inside cells. The points are surely also in this area without confirming locations by performing PIA. Only a few points located in the outermost cells (not-inside cells) need to confirm their locations. In the examples of Sections 5.1 and 5.2, the points that need to execute PIA rarely exceeds 10%.

The previous study [30] is very time-consuming to find out the points in large-scale areas, especially when the number of candidate points is very large. The main contribution of our study is that the points that PIA needs to confirm are reduced to very few. Therefore, our strategy has the ability to efficiently handle large-scale areas, such as countries and cities. In the computer field, when the amount of data is large, it is not feasible to process all the data at once. It is necessary to transfer to the batch manner. The design of cells is adaptive to this. It is also convenient to be implemented for epidemic prevention management.

6. Discussion

Let A be a specific area and P be the set of points. Our proposed strategy provided a solution to acquire all points in P inside A. The PIA algorithm can be used to evaluate whether point p is inside or not-inside area A. The time complexity is $O(n_A)$, where n_A is the number of vertices (or edges) of area A. The reason is that it needs to count the number of intersections among the ray of p and the n_A edges. The simple method to acquire all of points inside area A is to evaluate each point of P by PIP. The candidate points, which are points needing the evaluation of PIA, are all of the n_P points. The time complexity is $O(n_A \times n_P)$, where n_P is the number of points in set P. When n_P is a very large number, it is greatly time-consuming to acquire these points. In [30], an enhanced strategy was proposed. It first evaluates the boundary (i.e., a space of rectangle). This rectangle is fully covered by the space of area A. The points within the rectangle are evaluated. In our strategy, we allocate a set of cells to fully cover area A. The total space of these cells is slightly larger than the space of area A. Then, the cells are divided into inside cells and not-inside cells. Only points within not-inside cells must be evaluated. The not-inside cells are the cells that have intersections with area A. In fact, only points near the edges of area A are evaluated. The number of candidate points is reduced to a very small number. Therefore, our strategy is efficient to acquire the targets in an area.

7. Conclusions

In this study, Taiwan Island and the administrative areas are used as geographic areas, and check-in places on Facebook are used as geographic points to verify the proposed strategy. These are the actual data. Due to the fact that the spot height map, respectively, matches the distribution of administrative areas and matches the distribution of highways, it verifies the practicability of our strategy. This strategy mainly provides two techniques. The first technique is used to calculate whether the geographic points are in a specific area. The second technique is used to calculate the number of points in a specific area. In our demonstration, we first, respectively, analyzed the numbers of points of the 16 administrative areas. Then the populations of the administrative areas and the spaces of the administrative areas were included in the discussion. This provided these results: the number of geographic points per unit area and the ratio of geographic points to the population. This is the category of hot areas.

Because the numbers of geographic points in geographic areas are not enough to represent the distribution of geographic points, we planned a space of about 1 square kilometer as a unit, termed as a spot, and calculated the number of geographic points in each spot. This is a hotspot category. Of course, the size of a spot depends on the actual situation. Based on the above discussion, our strategy is a GNSS-based crowd-sensing strategy for specific areas. This research is very useful in many fields. We used non-real-time and real-time GNSS coordinates for the applications. Non-real-time coordinates (i.e., historical records), can be used to know where hot areas hot spots can develop. It can be deployed in advance for this area to avoid or mitigate future events. For real-time coordinates, it can be used to know where it is becoming a hot area or where it is becoming a hot spot. It can be deployed ahead of schedule to avoid or mitigate ongoing events. The acquisition of coordinates may have privacy constraints. To acquire current location information with user consent may be available. Observing privacy constraints to do more epidemic prevention management is the goal of our future work.

Author Contributions: Conceptualization, C.-B.L. and J.-S.C.; methodology, R.-W.H.; software, C.-Y.H.; validation, C.-B.L., R.-W.H., C.-Y.H., and J.-S.C.; formal analysis, R.-W.H.; investigation, C.-B.L. and J.-S.C.; writing, C.-B.L. and J.-S.C. All authors have read and agreed to the published version of the manuscript.

Funding: This research received no external funding.

Acknowledgments: This research was partially supported by the Ministry of Science and Technology, Taiwan (ROC), under contract no.: MOST 108-2410-H-324-007 and MOST 108-2637-E-324 -004.

Conflicts of Interest: The authors declare no conflict of interest.

References

1. *International Civil Aviation Organization Annex 10 to the Convention of International Civil Aviation*; International Civil Aviation Organization: Montreal, QC, Canada, 2007; Volume I.
2. Hegarty, C.J.; Chatre, E. Evolution of the Global Navigation Satellite System (GNSS). *Proc. IEEE* **2008**, *96*, 1902–1917. [CrossRef]
3. Leclère, J.; Landry, R., Jr.; Botteron, C. Comparison of L1 and L5 Bands GNSS Signals Acquisition. *Sensors* **2018**, *18*, 2779. [CrossRef] [PubMed]

4.   Zhang, Z.; Li, B.; Shen, Y.; Gao, Y.; Wang, M. Site-Specific Unmodeled Error Mitigation for GNSS Positioning in Urban Environments Using a Real-Time Adaptive Weighting Model. *Remote Sens.* **2018**, *10*, 1157. [CrossRef]

5.   Li, T.; Zhang, H.; Gao, Z.; Chen, Q.; Niu, X. High-accuracy positioning in urban environments using single-frequency multi-GNSS RTK/MEMS-IMU integration. *Remote Sens.* **2018**, *10*, 205. [CrossRef]

6.   Cai, C.; Pan, L.; Gao, Y. A precise weighting approach with application to combined L1/B1 GPS/BeiDou positioning. *J. Navig.* **2014**, *67*, 911–925. [CrossRef]

7.   Park, S.W.; Cornforth, D.M.; Dushoff, J.; Weitz, J.S. The time scale of asymptomatic transmission affects estimates of epidemic potential in the COVID-19 outbreak. *Epidemics* **2020**, *31*, 100392. [CrossRef] [PubMed]

8.   Wigginton, N.S.; Cunningham, R.M.; Katz, R.H.; Lidstrom, M.E.; Moler, K.A.; Wirtz, D.; Zuber, M.T. Moving academic research forward during COVID-19. *Science* **2020**, *368*, 1190–1192. [CrossRef] [PubMed]

9.   Wang, C.J.; Ng, C.Y.; Brook, R.H. Response to COVID-19 in Taiwan: Big Data Analytics, New Technology, and Proactive Testing. *JAMA* **2020**, *323*, 1341–1342. [CrossRef] [PubMed]

10.   Yin, L.; Ni, Q.; Deng, Z. A GNSS/5G integrated positioning methodology in D2D communication networks. *IEEE J. Sel. Areas Commun.* **2018**, *36*, 351–362. [CrossRef]

11.   Destino, G.; Saloranta, J.; Seco-Granados, G.; Wymeersch, H. Performance Analysis of Hybrid 5G-GNSS Localization. In Proceedings of the 2018 52nd Asilomar Conference on Signals, Systems, and Computers, Pacific Grove, CA, USA, 28–31 October 2018; pp. 8–12.

12.   Huang, D.; Du, Z.; Wang, Y. A Quad-Antenna System for 4G/5G/GPS Metal Frame Mobile Phones. *IEEE Antennas Wirel. Propag. Lett.* **2019**, *18*, 1586–1590. [CrossRef]

13.   Feitelson, D.G.; Frachtenberg, E.; Beck, K.L. Development and Deployment at Facebook. *IEEE Internet Comput.* **2013**, *17*, 8–17. [CrossRef]

14.   Lin, H.-T. Applying location based services and social network services onto tour recording. In Proceedings of the 2012 Ninth International Conference on Computer Science and Software Engineering (JCSSE), Bangkok, Thailand, 30 May–1 June 2012; pp. 197–200.

15.   Chen, J.S.; Hsu, C.Y.; Yang, C.Y.; Wei, C.C.; Ciang, H.G. A data mining method for Facebook social network: Take "New Row Mian (Beef Noodle)" in Taiwan for example. In Proceedings of the 2017 IEEE 8th International Conference on Awareness Science and Technology (iCAST), Taichung, Taiwan, 8–10 November 2017; pp. 165–169.

16.   Liu, H.K.; Hung, M.J.; Tse, L.H.; Saggau, D. Strengthening urban community governance through geographical information systems and participation: An evaluation of my Google Map and service coordination. *Aust. J. Soc. Issues* **2020**, *55*, 182–200. [CrossRef]

17.   Wu, Y.-J.; Wang, Y.; Qian, D. A google-map-based arterial traffic information system. In Proceedings of the 2007 IEEE Intelligent Transportation Systems Conference, Seattle, WA, USA, 30 September–3 October 2007; pp. 968–973.

18.   Spyratos, S.; Stathakis, D.; Lutz, M.; Tsinaraki, C. Using Foursquare place data for estimating building block use. *Environ. Plan. B Urban Anal. City Sci.* **2017**, *44*, 693–717. [CrossRef]

19.   Novović, O.; Grujić, N.; Brdar, S.; Govedarica, M.; Crnojević, V. Clustering Foursquare Mobility Networks to Explore Urban Spaces. In *World Conference on Information Systems and Technologies*; Springer: Cham, Switzerland, 2020; pp. 544–553.

20.   Noulas, A.; Scellato, S.; Mascolo, C.; Pontil, M. An empirical study of geographic user activity patterns in foursquare. In Proceedings of the Fifth International AAAI Conference on Weblogs and Social Media, Catalonia, Spain, 17–21 July 2011.

21.   Huang, Y.F.; Chen, J.S.; Lin, C.B. A Specific Targeted-Place Mining Method for a Famous Social Network: Take Wang-Ye Worship in Taiwan for Example. In Proceedings of the 2018 15th International Symposium on Pervasive Systems, Algorithms and Networks (I-SPAN), Yichang, China, 16–18 October 2018; pp. 263–266.

22.   Xia, T.; Shen, J.; Yu, X. Predicting human mobility using sina weibo check-in data. In Proceedings of the 2018 International Conference on Audio, Language and Image Processing (ICALIP), Shanghai, China, 16 July 2018; pp. 380–384.

23.   Yang, K.; Wan, W.; Xia, T.; He, X. Urban tourism research based on the social media check-in data. In Proceedings of the 4th International Conference on Smart and Sustainable City (ICSSC 2017), Shanghai, China, 5 June 2017; pp. 1–3.

24. Ding, X.; Xu, J.; Chen, G. Exploring structural analysis of place networks using check-in signals. In Proceedings of the 2013 IEEE Global Communications Conference (GLOBECOM), Atlanta, GA, USA, 9–13 December 2013; pp. 3194–3199.

25. Han, Y.; Yao, J.; Lin, X.; Wang, L. GALLOP: GlobAL feature fused LOcation Prediction for Different Check-in Scenarios. *IEEE Trans. Knowl. Data Eng.* **2017**, *29*, 1874–1887. [CrossRef]

26. Ding, J.; Wu, K.; Guan, H.; Wang, D.; Rui, T. Point-in-polygon algorithm based on monolithic calculation for included angle of half plane continuous chains. In Proceedings of the 2010 18th International Conference on Geoinformatics, Beijing, China, 18–20 June 2010; pp. 1–4.

27. Kularathne, D.; Jayarathne, L. Point in Polygon Determination Algorithm for 2-D Vector Graphics Applications. In Proceedings of the 2018 National Information Technology Conference (NITC), Colombo, Sri Lanka, 2–4 October 2018; pp. 1–5.

28. Ochilbek, R. A new approach (extra vertex) and generalization of Shoelace Algorithm usage in convex polygon (Point-in-Polygon). In Proceedings of the 2018 14th International Conference on Electronics Computer and Computation (ICECCO), Kaskelen, Kazakhstan, 29 November–1 December 2018; pp. 206–212.

29. Antonio, F. Faster line segment intersection. In *Graphics Gems III (IBM Version)*; Morgan Kaufmann: Burlington, MA, USA, 1992; pp. 199–202.

30. Chang, S.C.; Huang, H.Y.; Huang, Y.F.; Yang, C.Y.; Hsu, C.Y.; Chen, J.S. An efficient geographical place mining strategy for social networking services. In Proceedings of the 2019 IEEE International Conference on Consumer Electronics-Taiwan (ICCE-TW), Ilan, Taiwan, 20–22 May 2019; pp. 1–2.

© 2020 by the authors. Licensee MDPI, Basel, Switzerland. This article is an open access article distributed under the terms and conditions of the Creative Commons Attribution (CC BY) license (http://creativecommons.org/licenses/by/4.0/).

Article

A Certificateless Aggregate Arbitrated Signature Scheme for IoT Environments

Dae-Hwi Lee [1], Kangbin Yim [2] and Im-Yeong Lee [1,*]

[1] Department of Computer Science and Engineering, Soonchunhyang University, Asan 31538, Korea; leedh527@sch.ac.kr

[2] Department of Information Security Engineering, Soonchunhyang University, Asan 31538, Korea; yim@sch.ac.kr

* Correspondence: imylee@sch.ac.kr; Tel.: +82-41-530-1323

Received: 16 June 2020; Accepted: 15 July 2020; Published: 17 July 2020

Abstract: The Internet of Things (IoT) environment consists of numerous devices. In general, IoT devices communicate with each other to exchange data, or connect to the Internet through a gateway to provide IoT services. Most IoT devices participating in the IoT service are lightweight devices, in which the existing cryptographic algorithm cannot be applied to provide security, so a more lightweight security algorithm must be applied. Cryptographic technologies to lighten and provide efficiency for IoT environments are currently being studied a lot. In particular, it is necessary to provide efficiency for computation at a gateway, a point where many devices are connected. Additionally, as many devices are connected, data authentication and integrity should be fully considered at the same time, and thus digital signature schemes have been proposed. Among the recently studied signature algorithms, the certificateless signature (CLS) based on certificateless public key cryptography (CL-PKC) provides efficiency compared to existing public key-based signatures. However, in CLS, security threats, such as public key replacement attacks and signature forgery by the malicious key generation center (KGC), may occur. In this paper, we propose a new signature scheme using CL-PKC in generating and verifying the signature of a message in an IoT environment. The proposed scheme is a certificateless aggregate arbitrated signature, and the gateway aggregates the signatures of messages generated by the device group to reduce the size of the entire signature. In addition, it is designed to be safe from security threats by solving the problems caused by public key replacement attacks and malicious KGC, and adding arbitrated signatures of the gateway to strengthen non-repudiation.

Keywords: IoT; certificateless signature; aggregate signature; arbitrated signature; public key replace attack

1. Introduction

The Internet of things (IoT) means an environment or technology in which heterogeneous devices are connected to the Internet. The devices participating in the IoT environment can be connected to the Internet to provide various services to users. Thus, the services provided to users through such environments and technologies can also be called IoT. Servers process data collected by "things" (end devices), such as sensors and actuators, and users are provided with services through their smartphones. The most common IoT service structures are things and gateways, storage or servers, and consumers. It consists of things that collect data or perform commands, a gateway that is an intermediate that transmits data collected from things, a storage or server that stores and analyzes data in the form of data desired by users, and consumers who use it. Recently, with the advent of an IoT-based hyper-connected society, technological innovation is ongoing in various fields. In particular, as the fifth-genereation (5G) telecommunications standard has recently attracted attention, the number of IoT devices connected to the Internet will increase rapidly, and various services can be provided [1–4].

With the development of communication technology, weight reduction and mass production of devices became possible. Since then, following smart homes, more devices are evolving into large-scale IoT services, such as smart buildings, factories, and cities that connect to the Internet. Objects participating in the IoT service environment are each connected to the Internet to communicate with other objects. The feature of the IoT environment is that all devices need to be connected to the Internet, so small devices, such as electrical outlets and gas valves in smart homes, need to include communication capabilities. Therefore, it is composed of ultra-light and low-power technologies, unlike the existing environment. It is also difficult to apply existing public key infrastructure (PKI) security technologies to IoT devices, so it is necessary to use lightweight cryptographic algorithms that work well in this new environment of ultra-light and low-power devices [5,6].

In particular, a technology for providing integrity for messages in an IoT environment is essential [7,8]. Figure 1 shows a scenario for the data signing and verification process in an IoT environment. Sensors create a message, create a signature, and place them in cloud-like storage through a gateway. Thereafter, the consumer (i.e., the verifier) who needs the message can secure the integrity of that message through the message and the signature. Since the IoT environment is composed of a wireless communication network, it is possible to provide a secure service only by providing integrity for commands sent by a user to a device or for data collected from a device. Of course, since the IoT is a lightweight environment, the signature must also be lightweight to be applied to that IoT environment. Therefore, in this paper, we analyze the certificateless (CL) signature (CLS) and CL aggregate signature (CL-AS), a signature scheme using the lightweight CL public key cryptography (PKC, hence CL-PKC), which is suitable for IoT environments. In addition, we propose a CL aggregate arbitrated signature (AAS, hence CL-AAS) that applies the gateway arbitrated signature technique to enhance its non-repudiation and the aggregate signature technique to enhance its efficiency. In particular, it is designed to be safe against the forgery of signatures, including against public key replacement attacks that occur in CL-PKC-based cryptographic technologies, and to be suitable for IoT environments by avoiding the use of a large number of computational pairing operations.

Figure 1. Data signing and verification scenario in an Internet of things (IoT) environment.

The contributions to the proposed scheme in this paper are as follows.

- Analyze existing CL-AS schemes and design scenarios for secure public key replacement and malicious KGC attacks.
- In addition to the existing security requirements, the concept of the arbitrated signature for the non-repudiation function is applied considering the security of the aggregator that aggregates signatures.
- Aggregate signature is performed on messages and signatures of IoT devices, and the arbitrated signature of the gateway is also aggregated in the aggregate signature of IoT devices. Through this, we propose a secure and efficient CL-AAS scheme compared to the existing schemes.

More details on CL-PKC and security threats are covered in Section 2 on related work. Section 3 introduces the security requirements for each item in the cryptosystem, and Section 4 introduces the proposed scheme that satisfies those security requirements. Section 5 gives a comparative analysis of the proposed scheme, with Section 6 giving the paper's conclusions.

2. Background and Related Work

In this section, we consider background and related work, before suggesting the CL-PKC-based CL-AAS scheme to satisfy security requirements and provide efficiency in an IoT environment. Even before wireless sensor networks had become common, research had been conducted on digital signatures to provide message integrity. Some cryptography schemes have been used recently in IoT environments, such as the CLS and CL-AS used in the signature process proposed in this paper; we now examine these, including their security threats.

2.1. Elliptic Curve Cryptography and ECDLP

The elliptic curve cryptography (ECC) is a public key cryptography based on the elliptic curve theory, and provides a similar level of security while using a shorter key than the existing public key cryptography. Therefore, it is applied to various cryptographic algorithms used in IoT environments. The definition of the elliptic curve cipher is as follows.

Let F_q denote a finite field with a large prime order q. Let E_q denote an elliptic curve on F_q, which is specified by the equation: $y^2 = x^3 + ax + b \pmod{p}$, where $a, b \in F_q$ and $\left(4a^3 + 27b^2\right) \bmod p \neq 0$. Let the notation O denote a point of infinity, form the additive cyclic group G of the elliptic curve under the computation of point addition $T = U + V$ for $U, V \in G$ defined on the basis of a chord-and-tangent rule. Suppose P is a generator of the cyclic group G, and the order of G is q. Let $x \in Z_q^*$, and scalar multiplication is defined by the equation: $x \times P = (P + P + \ldots + P)$ (x times).

The ECC was designed based on the elliptic curve discrete logarithm problem (ECDLP). Given two random points $P, Q \in G$, the ECDLP is to find the integer $x \in Z_q^*$, where $Q = x \times P$.

2.2. Digital Signature

Digital signature is a security technology in which a signature has been changed into a digital form; it serves as proof of the identity of a digital electronic document's author. Digital signature is a cryptographic technology created by PKI-based public key cryptography (PKC) technology, starting with the authentication of messages [9–11]. Authentication can be divided into that of the user and that of the message. The former confirms the identity of a valid user and is a basic element for ensuring the responsibility of users. The latter may, for instance, provide assurance that the received message has not been tampered with. Digital signature consists of a signature using a private key, a verification using a public key, and provides a non-repudiation function, to ensure that it was sent from the sender. For digital signatures, the following five items must be satisfied:

- Signer authentication: The signer of the electronic document must be verifiable;
- Unforgeable: The electronic document cannot be forged;
- Non-reusable: The electronic signature cannot be used as a signature for another document;
- Unmodifiable: The content of the electronic document cannot be changed; and
- Non-repudiation: The signature of the electronic document cannot be denied.

Digital signatures began with PKC and various other types have been developed, including blind, arbitrated, group, multi-, and one-time signatures, and a variety of AS schemes that aggregate multiple signatures into one [9,11–18].

2.3. Certificateless PKC (CL-PKC)

There is a problem with PKC in that it is difficult to apply in an environment requiring ultra-light and low-power devices, such as the IoT. In particular, PKI, upon which PKC is based, uses a certificate, to verify the public key for authentication of the user, and the user's public key. For this reason, the computational overhead is exceptionally large for managing keys, signatures, and certificates and their distribution, verification, and revocation.

To solve these problems of PKC, identity-based cryptography (IBC) was developed [19]. Since IBC uses a public key based on a known user's identifier, it can solve the problems of key distribution, certificate verification, and memory overhead by eliminating the public key verification process. However, identity-based encryption has a problem with key escrow [20,21]. In IBC, there is a key generation center (KGC) that receives a user's identifier to generate a private key based on it and returns this to the user. The direct generation of the user's private key in this way can expose the user's key to the KGC afterwards: The key escrow problem.

One proposed scheme to solve the key escrow problem in IBC is CL-PKC [22], in which the KGC does not generate all public and private keys but only partially generates them and returns them to the user. This "partial" key is called a partial secret key (PSK), and the user creates his or her full key pair using this. Since the full key pair is not generated by the KGC but by the user, CL-PKC can solve the key escrow problem, uses a smaller key than the existing PKI, and does not incur the overhead for public key verification; it is thus suitable for ultra-light and low-power environments. Research has been conducted on various aspects of CL-PKC, such as authentication and key agreement [23,24], signatures [25,26], and encryption [27,28]. The main difference between the structures of CL-PKC and its predecessor, PKC, is that there is no certificate for public key verification in the former and, in this sense, the term "certificateless" is used.

2.4. Certificateless Aggregate Signature (CL-AS)

Before explaining CL-AS, we describe CLS, which is the basis for it. CLS is a signature technique using CL-PKC: The signature for a message is generated using the private key of the signer, generated via CL-PKC, and the verifier verifies it using the public key and identifier of the signer and the public key of the KGC. CLS and CL-PKC are currently active research topics [22,25,26].

CLS is a scheme for signing a single message, whereas CL-AS is a scheme for creating an aggregated signature for multiple messages. If there are multiple senders, the signature on the generated messages will generate the signature of each sender. That is, N signatures are generated for N messages generated by N senders. The verifier must perform verification of N messages individually, using N public keys of N senders. Thus, the number of individual verifications will, likewise, be N. CL-AS aggregates these N signatures into one signature. N public keys are used for verification, but the advantage of being able to verify all of the signatures for N messages in one step is that only one verification process is required. This can reduce the computational overhead on the verifier side and, on the storage side (messages and signatures must be stored), only one signature, not N of them, needs to be stored, thereby reducing memory overhead. The CL-AS scheme was proposed based on a pairing operation, but, recently, pairing-free schemes to reduce the number of operations have been proposed [29–34]. Figure 2 shows the structure of CL-AS.

The basic CL-AS schemes include a CL-PKC-based signature that receives a PSK after registering a user with a KGC. In general, the CL-based AS technique consists of the following eight steps. Among them, the setup and partial-private-key-extract steps are performed by the KGC, and the set-secret-value and set-public-key steps are performed by the user who generates the key. Thereafter, the signer who wants to generate the signature does so for the message with his key through the CL-sign step and can verify the message and signature through the CL-verify step. Signatures generated from multiple signers are sent to the aggregator through the CL-aggregate step, which aggregates them into one signature, and then on to the verifier. The verifier can acquire and verify messages through the CL-aggregate-verify step.

- Setup: The KGC generates public parameters and a master secret key using a security parameter as input.

- Partial-private-key-extract: The KGC generates the user's partial private key and partial public key using the public parameters, master secret key, and the user's personal identification information, and delivers these keys to the user.

- Set-secret-value: The user creates his own secret information and secret key by inputting public parameters and user identification information.

- Set-public-key: The user sets the public key by entering the public parameters, his partial public key, and secret information.

- CL-sign: Among the users who generated the key, the user who wants to sign the message becomes a signer, and signs the message using his private key. The message and its signature are sent to the verifier.

- CL-verify: The verifier verifies the integrity of individual messages and signatures using the signer's public key.

- CL-aggregate: The aggregator, which receives the messages and signatures from multiple signers, aggregates these signatures into a single one for multiple messages, to reduce their overall size, and outputs this.

- CL-aggregate-verify: Upon receiving a message and an aggregated signature, the verifier can verify the signature using the signer's public key, verify the user who created the signature, and verify the integrity of the message.

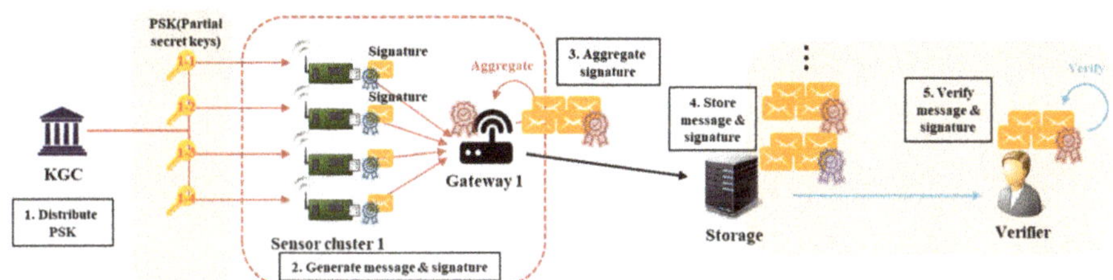

Figure 2. Structure of the general certificateless aggregate signature process. KGC: key generation center.

2.5. Security Threat of CL-AS

CL-AS has several advantages, but it also has problems with the forgery of messages and signatures. The public key used in CL-PKC does not use a certificate, so the user's identifier and public key cannot be authenticated. Because of this, CLS has a weaker non-repudiation function than PKI, and a malicious attacker can conduct an attack in which another user's public key is replaced. This CLS public key replacement attack is a method of forging the signature transmitted by device A to device B and replacing the public key of user A with a public key generated by the attacker to verify that forged signature. This is an attack that occurs because it is not possible to authenticate whether the public key, which can bypass the verification of the signature generated using A's private key, is that of A or that of the attacker. It can be verified that the signature using the public key of device A, but the function of non-repudiation to verify the signature is actually signed by A is weak. The act of being able to verify by replacing the public key itself is an infringement of non-repudiation, and in order to prevent this, CL-PKC must be considered for a public key replacement attack. Additionally, unlike IBC, there is no problem of key escrow for the device's secret key; however another problem may also occur: The KGC can generate the partial key of A and forge user A's signature using a partial key generated by itself.

Therefore, the security model for CL-AS can be roughly divided into two types of attack model [29–34]. Each model is of a game by an attacker (A_I or A_{II}) communicating with a challenger to successfully forge a signature. A_I has the ability to arbitrarily replace the public key of a legitimate user without the system's master key. A_{II} cannot replace the public key of users but knows the master secret key of the KGC. Thus, each of them can perform different types of attacks.

2.6. Analysis of Existing CL-AS Schemes

CLS was first proposed by Al-Riyami et al. in 2003 [21]. Based on this, the recently proposed CL-AS has had many implementations published that are more efficient because they do not use pairing operations. Table 1 summarizes some of these variants and their security threats.

Qu et al. [29] proposed an efficient CL-AS scheme that does not use a pairing operation. Most commonly, CL-AS structures are based on the elliptic curve discrete logarithm problem (ECDLP). This scheme is one such and adds the user-generated key and PSK to the signature. However, since the identifier is not bound to the public key, there is a risk that the public key can be replaced.

Deng el al. [30] proposed a CL-AS scheme that prevents forgery of signatures by adding two kinds of signatures in one signature statement, by adding an RSA signature along with an ECDLP-based Schnorr signature. However, the size of the signature statement for the message is exceptionally large, and the signature verification overhead is disadvantageous because two types of signatures must be verified. In particular, the former signature is based on an exponential operation, unlike the latter signature, which uses an elliptic curve, so it has a large overhead compared to other schemes.

Cui et al. [31] proposed a scheme to prevent the transmission of a forged signature due to the replacement of the public key by adding a timestamp when sending the signature and message. However, since the identifier is not actually bound to the public key, it does not provide direct defense against a public key replacement attack.

Du et al. [32] proposed a scheme that was safe against public key replacement attacks by binding an identifier and a verification key to a public key, but there is a risk of key leakage and subsequent signature forgery.

Gayathri et al. [33] proposed a scheme to aggregate public parameters for signature verification. Previous schemes required N public parameters to verify N signatures, but Gayathri et al. could reduce the memory overhead by using a scheme that reduces the verification parameters. However, there is a risk of a public key replacement attack because the identifier is not bound to the public key.

Zhao et al. [34] proposed a scheme to prevent signature forgery by adding a value for verifying the signature directly to the overall transmitted value of the message and the aggregated signature. However, the message that is transmitted is exceptionally large, and there is still a risk of a public key replacement attack.

The above schemes suggest CL-AS for various environments. The security analysis and efficiency comparison of existing schemes are summarized in Sections 5 and 6. In general, the risk of public key replacement attack occurs when the user, identifier, and verifiable value are not bound to the PSK received via the KGC. In other words, the signature can be verified with the public key of the device A, but it occurs when the public key used to verify the signature cannot verify whether the public key generated by the object with the actual identifier A is correct. This means that the existing problem is related to non-repudiation, and this problem can be solved if the public key can verify the identity of the identifier A. To solve this, in recent CL-PKC schemes, the partial key is received from the KGC first, and the user does not generate the full key but, instead, first generates the verification key pair and sends the identifier and public key for verification to the KGC. Using this, the KGC binds the user's identifier, the public key, for verification, and the verification tag to the PSK, and then enables verification of the user's public key [35,36]. Additionally, in the existing CL-AS schemes, the aggregator that aggregates the signatures of the signer's messages serves solely to aggregate these signatures. The aggregator can be one of the signers or a third entity, depending on the environment. If it is a third entity, the problem of trusting it may occur, which can make the AS unreliable. Therefore, it is necessary that the aggregator in CL-AS has non-repudiation.

3. Security Requirements

- Integrity: The most important requirement for digital signatures, including CLSs, is integrity. In particular, in the IoT environment, since data are transmitted and received using a wireless communication network, it is particularly important to ensure integrity by signing important messages. In the existing CL-AS schemes, since the aggregator only aggregates the signature, the entity that verifies the signer's signature first is that aggregator. The integrity of the aggregate signature itself must be ensured, as it can also be an attack point.

- Prevention of key leakage: The reason for performing the signature is to ensure the integrity of the transmitted message, and the signer's signature key must not be leaked to the outside or be possible to derive via public parameters. If an attacker can derive or steal the signature key, they can forge the signature on messages generated by themselves, reducing the reliability of the IoT service, and create and transmit a malicious message that the attacker can have verified legitimately.

- Unforgeability: An attack on CL-PKC-based signatures is an attack with counterfeit signatures. As described in Section 2.4, forgery of signatures can occur through the public key replacement attack of adversary A_I or the generation of the signer's partial key using the KGC master key of adversary A_{II}. For adversary A_I, even if public key replacement is performed, it should not be possible to generate a valid signature. If the verifier could remove the private key portion of the signature using the replaced public key, the attack would succeed. In particular, since a public key certificate is not used in CL-PKC-based cryptographic protocols, it is essential to verify that the public key used for signature verification is the actual signer's public key, and the user's identifier and public key cannot be authenticated. So, the non-repudiation function must be strengthened. For adversary A_{II}, it should not be possible to generate a signature using only the signer's partial key. This means that both the PSK and the signer-generated key must be used when generating the signature. In addition, even if the signature is generated using both, the signature can be forged, so the verifier should not be able to verify the forged signature normally.

4. Proposed Scheme

In this paper, we propose CL-AAS, a scheme with an aggregated and arbitrated signature, for IoT environments. Figure 3 shows the proposed scenario. In the IoT environment, sensor devices act as signers to generate messages and directly generate signatures. Sensor devices gather to form a sensor cluster, and each cluster has a gateway. A message is generated from the sensor device, and each device signs the message through the private key generated by CL-PKC and sends it to the gateway, which simultaneously acts as an arbitrator and an aggregator.

As a feature of the proposed scheme, it is possible to strengthen the non-repudiation of the signature of the sensor device through the arbitrated signature of the gateway, and reduce both the size of the signature stored in storage and the verification overhead of the verifier through the AS. In the existing CLS scheme [37,38], an entity, such as an external time server or "helper", synchronizes with the signer to strengthen non-repudiation for the signature. In the IoT scenario proposed in this paper, the messages generated by the sensor device are aggregated and transmitted through the gateway, so we do not use other external entities but try to strengthen the non-repudiation by using the gateway itself. In other words, the gateway does not merely act as an aggregator for combining multiple signatures into a single one but also makes it an arbitrated signature.

The system parameters of the proposed scheme are as follows.

- ID_*: Identifier of entity;
- E: Elliptic curve on group G of prime order q;
- P: Generator of cyclic group G;
- pu_*, sv_*: Verification the public key and private key pair of entity;
- PU_*, PR_*: Full public key and private key pair of entity;
- msk: Master key of KGC;
- P_{Pub}: Public key of KGC ($P_{Pub} = msk \times P$);
- $D_* = (R_*, z_*)$: Partial key of the entity;
- $H_1(\cdot)$: Cryptographic hash function $\left(\{0,1\}^* \times G \times G \to Z_q^*\right)$;
- $H_2(\cdot)$: Cryptographic hash function $\left(\{0,1\}^* \times \{0,1\}^* \times G \times G \to Z_q^*\right)$; and

- $H_3(\cdot)$: Cryptographic hash function $\left(\{0,1\}^* \times \{0,1\}^* \times Z_q^* \times G \times G \rightarrow Z_q^*\right)$.

The proposed CL-AAS scheme consists of four phases: Setup, individual signing and verifying, aggregated arbitrated signing, and aggregated verifying. In the setup phase, the KGC sets the public parameters and distributes the participants' partial keys. In the individual signing and verifying phase, the participants (such as devices) use their partial keys to generate individual signatures and the aggregator verifies them. In the aggregated arbitrated signing phase, the messages and signatures of all the signer devices are turned into one signature and the gateway adds its arbitrated signature.

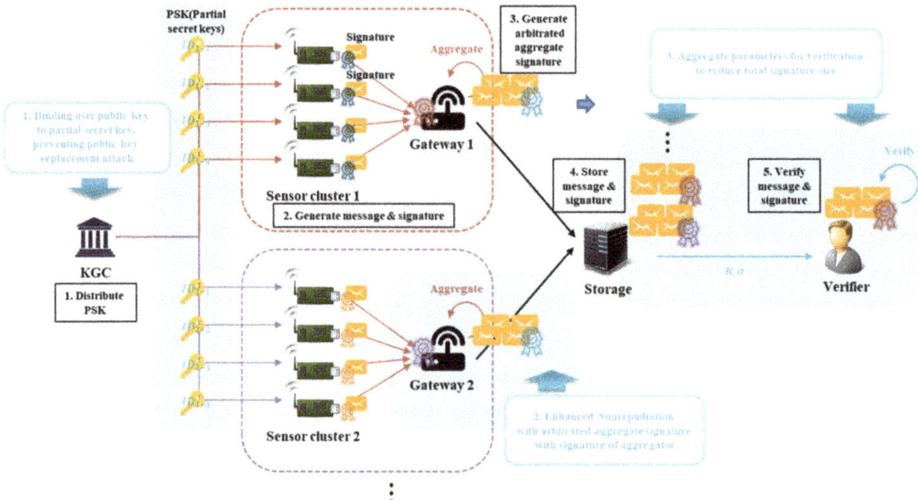

Figure 3. Scenario and advantages of the proposed scheme. KGC: key generation center.

Finally, in the aggregate verify phase, the signatures of the devices and the signature of the gateway's arbitrated signature are verified at once.

Each phase of the proposed scheme is modified from the eight algorithms described in Section 2.3. The set-secret-value and set-public-key algorithms are replaced with set-device-key and set-full-key ones, respectively. This is because the verification key pair of the user is generated first, not the partial key. Additionally, the AS (aggregated signature) and verification algorithms are replaced by CL-AA-sign to generate the AAS (aggregate arbitrated signature) and CL-AA-verify for aggregate verification. Therefore, this proposed scheme consists of the following eight algorithms for the KGC; A; gateway, G, acting as an aggregator and arbitrator; and verifier, V:

- Setup (k): The KGC creates public parameters and a master secret key with a security parameter, k, as input.
- Set-device-key $(params, ID_A)$: A generates a verification key pair from the public parameters, params, and A's public identifier, ID_A.
- Partial-private-key-extract $(params, msk, ID_S, pu_S)$: The KGC uses params, the master secret key, msk, ID_A, and the verification public key, pu_A, to generate the partial key, D_A, of A and transmits it to A.
- Set-full-key $(params, ID_A, D_A, pu_A, sv_A)$: A sets its full key pair, PU_A, PR_A, using params, D_A received from the KGC, and the verification key pair, sv_A, pu_A.
- CL-sign (m_A, ID_A, PR_A, PU_A): A becomes a signer, and signs a single message, m_A, using its private key, PR_A. Then, m_A and its signature are transmitted to G.

- CL-verify $(m_A, \sigma_A, ID_A, PU_A)$: Verification of m_A and its signature, σ_A, is performed using ID_A and the public key, PU_A. In the proposed scheme, the gateway performs verification, and the signatures of all received messages are verified through this process.
- CL-AA-sign $(m_1, \ldots, m_n, \sigma_1, \ldots, \sigma_n, ID_1, \ldots, ID_n, ID_G, PU_1, \ldots, PU_n)$: G, which has received messages and signatures from multiple devices, reduces the size of the signature. The signature is aggregated through the process, and an arbitrated signature is added: This algorithm outputs one signature that has been aggregated for multiple messages. To reiterate, G creates a single aggregated signature for all the signatures of the devices.
- CL-AA-verify $(m_1, \ldots, m_n, \sigma_{AS}, ID_1, \ldots, ID_n, ID_G, PU_1, \ldots, PU_n)$: When V receives the message and its aggregated signature from G, the signature and public keys can be used to verify the signature and, thus, the integrity of the message.

Figure 4 shows the four phases and eight algorithms of the proposed scheme.

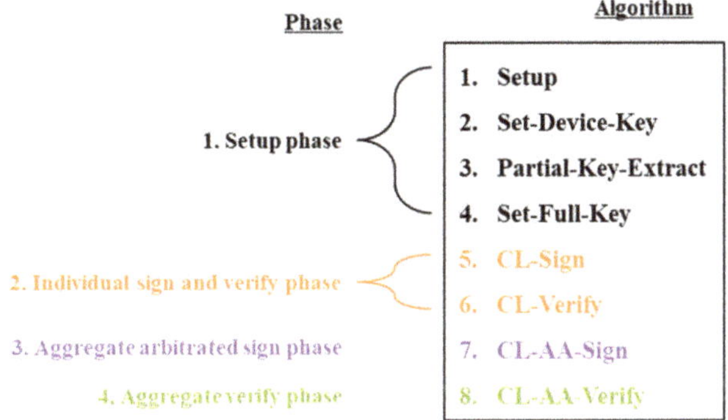

Figure 4. The relationship between the phases and algorithms of the proposed scheme. CL: certificateless, AA: aggregate arbitrated.

4.1. Setup Phase

In the setup phase, KGC first performs the setup algorithm using k (the security parameter) to generate the initial parameters, *msk* (the master secret key), and a master public key, P_{Pub}. The KGC is responsible for generating the partial keys of the devices after generating *params* (the public parameters). Subsequently, a verification key pair for each device is generated through the set-device-key algorithm; then, a partial key is generated for each by performing the partial-private-key-extract algorithm by sending information to the KGC. A device that receives a PSK generates its own full key pair through the set-full-key algorithm. Figure 5 shows the sequence diagram of the setup phase.

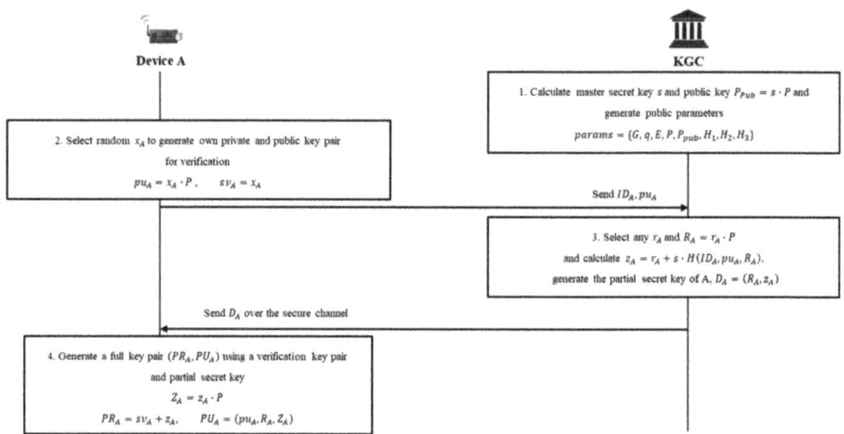

Figure 5. Sequence diagram of setup phase.

Step 1. The KGC selects k and generates *msk*. After that, $P_{pub} = msk \times P$ and params are generated as follows, through the setup(k) algorithm:

$$params = \left\{ G, q, E, P, P_{pub}, H_1, H_2, H_3 \right\}. \tag{1}$$

Step 2. A, which needs to receive a partial key from KGC, first creates its own verification public and private key pair, pu_A, sv_A, through the set-device-key (*params*, ID) algorithm. A selects $x_A \in_R Z_q^*$ and computes pu_A, sv_A as follows:

$$pu_A = x_A \times P, sv_A = x_A. \tag{2}$$

Step 3. A sends ID_A and pu_A (its verification public key) to the KGC, which performs the partial-private-key-extract (*params, msk, ID_A, pu_A*) algorithm to generate D_A (the partial key). The KGC selects $r_A \in_R Z_q^*$ and calculates the result of Equation (3). Then, the result of Equation (4) is calculated to generate a signature for the public key. The KGC transmits $D_A = (R_A, z_A)$ to A over a secure channel:

$$R_A = r_A \times P, \tag{3}$$

$$z_A = r_A + msk \times H_1(ID_A, pu_A, R_A). \tag{4}$$

Step 4. A, which has received D_A, creates PR_A (its full private key) and PU_A (its full public key) through set-full-key (*params, ID_A, D_A, pu_A, sv_A*) as follows:

$$Z_A = z_A \times P, \tag{5}$$

$$PR_A = sv_A + z_A, PU_A = (pu_A, R_A, Z_A). \tag{6}$$

4.2. Individual Signing and Verifying Phase

The individual signing and verifying phase use the CL-sign and CL-verify algorithms. A, which needs to generate a signature, becomes a signer and signs a message using its own key. Figure 6 shows the sequence diagram of the individual signing and verifying phase.

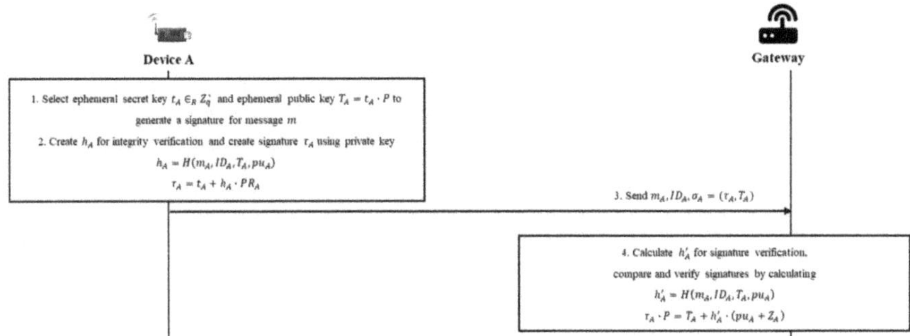

Figure 6. Sequence diagram of individual signing and verifying phase.

Step 1. A selects an ephemeral secret key, $t_A \in_R Z_q^*$, and calculates an ephemeral public key, $T_A = t_A \times P$.

Step 2. A needs to send a signed message to the arbitrator, G, to send the message and signature to V, to communicate. A calculates h_A and τ_A as follows, to generate the signature for m_A (the message):

$$h_A = H_2(m_A, ID_A, T_A, pu_A), \tag{7}$$

$$\tau_A = t_A + h_A \times PR_A. \tag{8}$$

Step 3. A sends m_A, $\sigma_A = (\tau_A, T_A)$ (the signature for m_A), and ID_A to G.

Step 4. G receives σ_A, ID_A, and m_A and performs the process of verifying σ_A. G calculates h'_A as in Equation (9), using the information from the message and that has been published by A, and verifies the validity of τ_A via Equation (10). If the validity of τ_A is verified, G completes verification of the individual message and its signature for A. G performs the signature not only for A but also for the messages and signatures of the other devices that will form the aggregate of signatures, as in Equations (9) and (10):

$$h'_A = H_2(m_A, ID_A, T_A, pu_A), \tag{9}$$

$$\tau_A \times P = T_A + h'_A \times (pu_A + Z_A). \tag{10}$$

The validity of the verified contents can be confirmed as follows:

$$\begin{aligned}
\tau_A \times P &= (t_A + h_A \times PR_A) \times P \\
&= t_A \times P + h_A \times PR_A \times P \\
&= T_A + h_A \times (pu_A + Z_A).
\end{aligned} \tag{11}$$

4.3. Aggregated Arbitrated Signing Phase

In the aggregated arbitrated signing phase, G aggregates signatures on messages and signatures received from N devices, and creates an arbitrated signature that has been verified, and adds it to the aggregated signature. An arbitrated signature is not simply a signature but is also the means of confirming that G has completed verification for the messages and signatures received from each included device; it is generated using G's own private key. Thus, an aggregated signature is generated, including the devices' signatures and the gateway-generated arbitrated signature. The aggregated arbitrated signing phase includes the CL-AA-sign algorithm. Figure 7 shows the sequence diagram of the aggregated arbitrated signing phase.

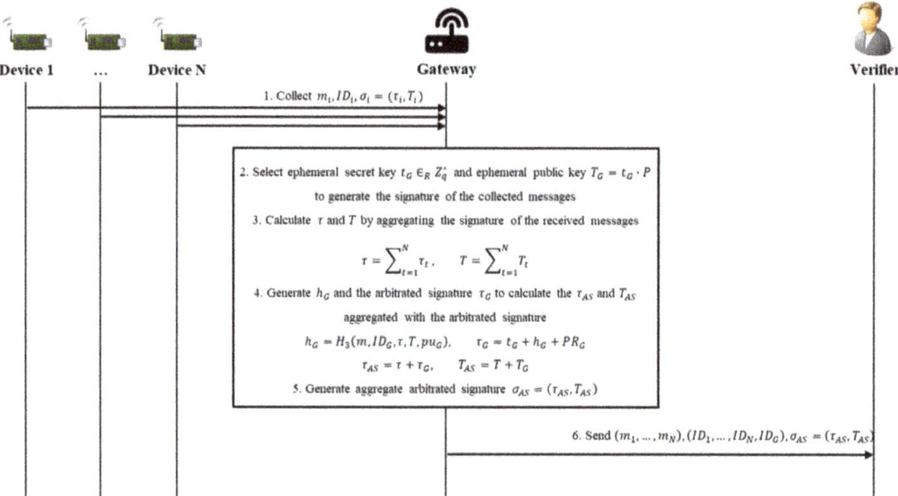

Figure 7. Sequence diagram of aggregated arbitrated signing phase.

Step 1. Each device sends the content of the message and signature created by itself (m_i, ID_i, and σ_i, where these relate to the ith device) to G, for which G collected and verified each signature during the signing and verifying phase.

Step 2. G selects an ephemeral secret key, $t_G \in_R Z_q^*$, and generates an ephemeral public key, $T_G = t_G \times P$, to generate the signature of the collected message, $m = (m_1,\ldots,m_N)$.

Step 3. G calculates the following to perform aggregation on the actual signature, τ_i, constituting the signature, σ_i, and the verification value, T_i:

$$\tau = \left(\sum_{t=1}^{N} \tau_t\right), T = \left(\sum_{t=1}^{N} T_t\right). \tag{12}$$

Step 4. G calculates the results of Equations (13) and (14) using its private key, PR_G, to generate the elements of the arbitrated signature to indicate that it has completed verification of each message. Then, the results of Equation (15) are calculated to generate τ_{AS} and T_{AS}:

$$h_G = H_3(m, ID_G, \tau, T, pu_G), \tag{13}$$

$$\tau_G = t_G + h_G \times PR_G, \tag{14}$$

$$\tau_{AS} = \tau + \tau_G, T_{AS} = T + T_G. \tag{15}$$

Step 5. Finally, G creates an AAS (aggregate arbitrated signature), $\sigma_{AS} = (\tau_{AS}, T_{AS})$, that aggregates the arbitrated signature of G and all the signatures of the devices.

Step 6. To verify σ_{AS}, $\{m, (ID_1,\ldots,ID_N,ID_G), \sigma_{AS}, (PU_1,\ldots,PU_N,PU_G)\}$ must be used. PU_i is the published public key of the ith device. Therefore, the gateway then sends $\{m, (ID_1,\ldots,ID_N,ID_G), \sigma_{AS}\}$ to V (the verifier requesting the message).

4.4. Aggregated Verifying Phase

In the aggregated verifying phase, V verifies the signature and message received from G, and verifies the signer's signature and G's arbitrated signature together. During this process, the initial signer and the arbitrator are verified simultaneously, strengthening the non-repudiation function.

This involves the final algorithm, CL-AA-verify, of the eight in this scheme. Figure 8 shows the sequence diagram of the aggregated verifying phase.

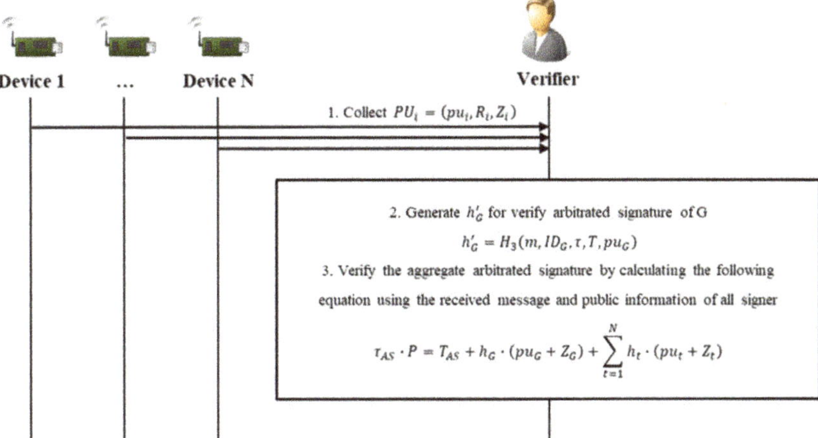

Figure 8. Sequence diagram of aggregated verifying phase.

Step 1. G can store the generated σ_{AS} and the message in a repository or send it directly to V. V performing the verification receives $\{m, (ID_1, \ldots, ID_N, ID_G), \sigma_{AS}\}$ from G and confirms the identifiers of the devices to obtain the public keys, $PU_i = (pu_i, R_i, Z_i)$.

Step 2. V calculates the value h'_G for the arbitrated signature verification of G as follows:

$$h'_G = H_3(m, ID_G, \tau, T, pu_G). \tag{16}$$

Step 3. V can verify the validity of σ_{AS} by calculating the result of Equation (17). If valid, V has completed verification of the N signer devices and G's arbitrated signature in one step:

$$\tau_{AS} \times P = T_{AS} + h_G \times (pu_G + Z_G) + \sum_{t=1}^{N} h_t \times (pu_t + Z_t). \tag{17}$$

The validity of the verified contents can be confirmed as follows:

$$
\begin{aligned}
\tau_{AS} \times P &= (\tau_1 \times P + \ldots \tau_N \times P + \tau_G \times P) \\
&= (t_1 + h_1 \times PR_1) \times P + \ldots + (t_N + h_N \times PR_N) \times P + (t_G + h_G \times PR_G) \times P \\
&= (T_1 + h_1 \times (pu_1 + Z_1)) + \ldots + (T_N + h_N \times (pu_N + Z_N)) + (T_G + h_G \times (pu_G + Z_G)) \\
&= (T_1 + \ldots + T_N + T_G) + (h_1 \times (pu_1 + Z_1)) + \ldots + (h_N \times (pu_N + Z_N)) + (h_G \times (pu_G + Z_G)) \\
&= T_{AS} + h_G \times (pu_G + Z_G) + \sum_{t=1}^{N} h_t \times (pu_t + Z_t).
\end{aligned}
\tag{18}
$$

5. Security Analysis

This section describes how the proposed scheme satisfies the security requirements presented in Section 3, including the requirements regarding integrity, key leakage, and forgery.

5.1. Integrity

In CLS or CL-AS, an existing CL-PKC-based signature scheme, a Schnorr signature, is used. In the proposed scheme, the individual signature value, τ_A in σ_A, of A is in the same form as the Schnorr signature, and the tag for verifying this is T_A. τ_A is created using A's private key, PR_A, and the tag's secret value (its ephemeral secret key), t_A, and can be verified through PU_A. Here, the value that is actually verified is the content of the hashed value, h_A. The values hashed from h_A are the message, m_A; the identifier, ID_A, of the device, A, that was the signer; the verification public key, pu_A; and the verification tag, T_A. Eventually, h_A is signed using PR_A to provide the integrity of the message and can be verified using pu_A and the partial public key, Z_A, which are elements of PU_A for verification. Therefore, it indicates that the signature, σ_A, was generated by A and was not forged in "the middle", between A and V.

In addition, the aggregated signature, σ_{AS}, generated by the gateway, G, using the individual signatures is added to the τ_i and T_i of the signers who generated the message, and to τ_G and T_G, the arbitrated signature values generated by G with the private key, PR_G. G's signature takes the form of a Schnorr signature, just like for the messages generated by the other devices, and can be verified in the same way. However, there is one difference: The content of the arbitrated signature being verified, h_G, is the message set, m, and the identifier, ID_G, of G and the aggregated devices' signature elements, τ and T. G can ultimately provide the integrity by signing the signature set itself of messages received from devices via σ_{AS}, and can be verified using the public key, PU_G, of G. If the message of the device is forged, the verification of the individual signature or AAS will fail, and only the normal message can be verified.

5.2. Prevention of Key Leakage

The signing key (full private key) and verification key (full public key) used in the proposed scheme are CL-PKC-based key pairs generated by the KGC and the device itself. It is assumed that when a device is first issued a partial key by the KGC, this is transmitted through a secure channel. In addition, all other messages and signatures that are transmitted are transmitted through a public channel. In the individual signing and verifying phase, the message sent to the public channel is the entire message, m_A, ID_A, and σ_A, and if they can derive the signing key, the attacker will succeed in leaking the key. In the aggregated arbitrated signing phase, the entire message transmitted to the public channel is m, $(ID_1, \ldots, ID_N, ID_G)$, and σ_{AS}, and if σ_{AS} can derive the signing key, the key can be successfully leaked.

First, σ_A is composed of $\tau_A = t_A + h_A \times PR_A$ and $T_A = t_A \times P$. The public key, PU_A, of A consists of pu_A, R_A, and Z_A, the values used in the verification of σ_A are pu_A and Z_A, and R_A is used to verify the validity of the PSK, z_A, generated by the KGC. Furthermore, it can be verified that the public key, Z_A, was made by A and the KGC.

The signature is verified as in Equation (10) and, even if the attacker knows the public key and other published information, obtaining the signature key from the signature is the same as the difficulty of solving the ECDLP in $(pu_A + Z_A) = (sv_A + z_A) \times P$. Therefore, it is difficult for an attacker to derive a public key using disclosed information. In particular, in the proposed method, since each value in the form of the signature key, $(sv_A + z_A)$, is used as a single value by adding them together, rather than independently using sv_A and z_A as in many existing schemes, this helps against leaking keys: There are fewer threats. Similarly, in σ_{AS}, calculating the private keys of each signature using τ_{AS} and T_{AS} is the same as solving the ECDLP problem, so it is difficult to leak or derive the signature key from the proposed scheme.

5.3. Unforgeability

As described in Section 3, the attack that can occur in a CL-PKC-based signature protocol is, in fact, an attack on unforgeability. A signature protocol is insecure if a tampered signature on any message can be verified (as if it were legitimate). Attacks on unforgeability can be divided into those by A_I and those by A_{II}.

5.3.1. Unforgeability from Adversary A_I

The adversary A_I has the ability to replace the public keys of other users with one generated by themselves. Due to the safety of the ECDLP, a private key corresponding to the public key of a user cannot be generated, but validation can be bypassed by replacing the public key alone. The public key replacement attack is mainly possible in existing schemes because the partial key, $D_A = (R_A, z_A)$, generated by the KGC is not related to the public key of the device. In short, it is a CL scheme, and thus lacks a certificate that can authenticate the public key of the signing device. By using this, it is possible to bypass the verification process of the signature, so that the forged signature will be verified by the verifier as if it were proper.

To solve the public key replacement attack, since it is a CL-PKC-based protocol (i.e., without a certificate), the binding between the public key and the identifier must be strengthened. When verifying a public key or verifying a signature signed with a private key, it is only necessary to confirm whether the user has used a public key or a signature made with a key created using a partial key received from the KGC. In summary, the partial public key, R_A, in the partial key, D_A, received from device A from the KGC is a tag for verifying the PSK, z_A, and if the public key, Z_A, of A created using z_A can be confirmed to belong to A, it can be said to be safe against key replacement attacks. In generating z_A, the value hashed from Equation (4) to H_1 later serves to verify the public key. If the public key is replaced, it is said that it is safe against public key replacement attacks if the verification of the signature cannot be bypassed using the public key replaced thus. On the other hand, when A's public key, PU_A, is replaced by the attacker's public key, PU'_A, the public key replacement attack is successful if the verifier successfully verifies the forged signature, σ'_A, for the message, m_A.

In the proposed scheme, the form of the individual signature is $\sigma_A = (\tau_A, T_A)$ and the message can be verified normally using the forged public key, $PU'_A = (pu'_A, R'_A, Z'_A)$, associated with the identifier of A_I. However, it should not be possible to generate the forged signature, $\sigma'_A = (\tau'_A, T'_A)$. The signature generation for m_A is according to Equation (8) and, since A_I cannot know t_A and PR_A, a valid τ_A cannot be generated. The signature verification is according to Equation (10) and can be generated using Equation (9) and PU_A. The published information is T_A, pu_A, and R_A and an attacker who wants to forge it can perform an attack by replacing the public key with pu'_A, R'_A, and Z'_A, and the attacker will try to bypass the verification of h_A by generating the same value as $pu'_A + Z'_A = h_A^{-1} \times P$. However, the verifier can confirm that the public key Z'_A has not been properly generated using $Z'_A = R'_A + H_1(ID_A, pu_A, R_A) \times P_{pub}$, which can be verified using the public key of the KGC. As for the $\sigma_{AS} = (\tau_{AS}, T_{AS})$, which is an AAS (aggregate arbitrated signature), it has the same form as the individual signature in Equation (14), so the verifier can correctly verify this even if A_I replaces G's public key. Therefore, A_I cannot forge the signature.

5.3.2. Unforgeability from Adversary A_{II}

The adversary A_{II} is a malicious KGC, and since they know *msk*, they have the ability to know all the partial keys of the participating devices. If A_{II} wants to forge A's signature, they will try to generate one from the partial key, since they lack the ability to replace the public key.

The partial key of A is $D_A = (R_A, z_A)$, where R_A is a partial public key and z_A is a PSK. The signature generation for the message, m_A, is via Equation (8) and, since the full private key, PR_A, generated by the signer, A, is used for this, the KGC cannot forge a signature using only z_A. It should be impossible to forge the signer's signature using only external public parameters, including in this A_{II} scenario.

Table 1. Security analysis of various certificateless aggregate signature schemes, including the proposed one.

	Qu et al. [29]	Deng et al. [30]	Cui et al. [31]	Du et al. [32]	Gayathri et al. [33]	Zhao et al. [34]	Proposed Scheme
Key leakage attack	O Cannot derive key	X Can derive key with public parameters	O Cannot derive key	X Can derive key with public parameters	O Cannot derive key	O Cannot derive key	O Cannot derive key
Forgery with public key replacement (A_I)	X No identifier binding to public key	X No identifier binding to signature	X No identifier binding to public key	O Binds identifier to public key	X No identifier binding to public key	X No identifier binding to public key	O Binds identifier to public key
Forgery with KGC master key (A_{II})	X Can forge due to public key replacement	O Uses two types of signature	X Can forge due to public key replacement	X Can forge due to key leakage	O Uses two types of signature	O Sends signature verification tag directly	O Uses gateway-arbitrated signature

O (X): scheme is strong (weak) in this category, KGC: key generation center.

In particular, in the proposed scheme, since the arbitrated signature is performed through a gateway called the arbitrator, it is possible to strengthen non-repudiation. The arbitrated signatures involve this arbitrator, between the signer and the verifier, to protect the validity of the signature and prevent repudiation of the signer; if the gateway performs its arbitrated signature properly, it can prevent forgery of the signature.

6. Efficiency Analysis

Another important requirement in the IoT environment is efficiency. In this environment, in which a large number of heterogeneous devices participate in communication, efficiency of the protocol is required so that it can operate even for devices with low computational performance. This includes reducing the amount of computation, and this section compares the existing schemes with the execution time of the proposed CL-AAS.

The simulation environment constructed in this paper is an Intel i5-4690 processor with 3.50 GHz, 16 GB memory, and Windows 10 operating system. Additionally, to provide security strength like 1024-bit RSA and ECC group, it uses the Koblitz elliptic curve $y^2 = x^3 + ax + b(mod\ p)$, where $a = 1$ and b is a 163-bit random prime defined on $F_{2^{163}}$. Table 2 is a comparison of the execution times with cryptographic operation. The proposed CL-AAS scheme provides computational efficiency compared to the existing [29–34] schemes, as shown in Figure 9, by a graph showing the total execution time according to the number of signatures being aggregated, and Table 3. As the number of messages and signatures being aggregated increases, the total times for the aggregated signature and for the verification process increase in direct proportion.

Table 2. Comparison of execution times with cryptographic operation.

Notations	Description	Run Time (ms)
T_{EM}	The execution time of scalar multiplication operation in ECC	0.4420
T_{EA}	The execution time of point addition operation in ECC	0.0018
T_h	The execution time of hash operation	0.0082
T_E	The execution time of scalar exponential operation	5.3100

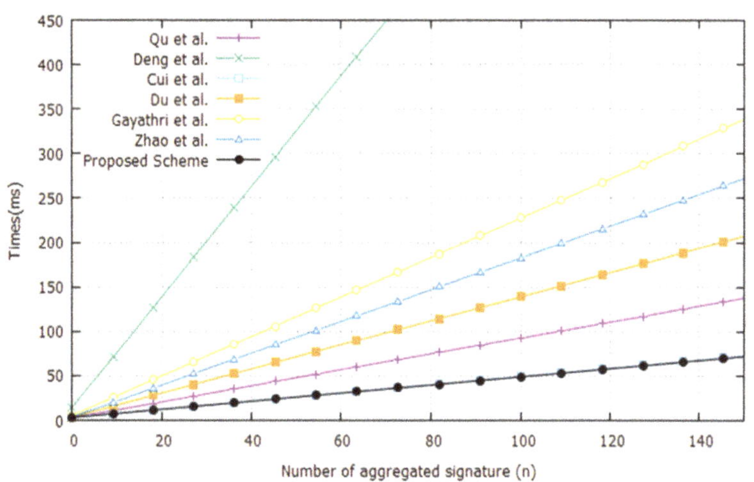

Figure 9. Comparison of execution time between proposed and existing schemes.

Table 3. Efficiency analysis of the proposed scheme.

	Qu et al. [29]	Deng et al. [30]	Cui et al. [31]	Du et al. [32]	Gayathri et al. [33]	Zhao et al. [34]	Proposed Scheme
Form of signature	$\sigma_i = (U_i, s_i)$	$\sigma_i = (T_i, B_i, r_i, R_i)$	$\sigma_i = (R_i, S_i)$	$\sigma_i = (S_i, v_i)$	$\sigma_i = (Y_{1i}, u_i, w_i)$	$\sigma_i = (R_i, \phi_i)$	$\sigma_i = (\tau_i, T_i)$
Signing operation	$1H + 2EA + 2EM$	$1H + 2E + 1EA + 3EM$	$H + EA + 2EM$	$2H + 2EA + 3EM$	$3H + 3EA + 5EM$	$2H + 2EA + 2EM$	$1H + 2EA + 2EM$
Verifying operation	$2EA + 3EM$	$E + 1EA + 4EM$	$2H + 2EA + 3EM$	$3H + 3EA + 3EM$	$2H + 3EA + 5EM$	$2H + 3EA + 4EM$	$1H + 2EA + 2EM$
Aggregating operation	nEA	$2nEA$	nEA	nEA	$3n(EA + EM)$	nEA	$1H + (2n+3)EA + 2EM$
Aggregated verifying operation	$n(1H + 4EA + 2EM) + 1EA + 1EM$	$n(1H + 2EA + 2EM + E) + 1EM$	$n(2H + 2EA + 1EM) + 2EM + 2EA$	$n(3H + 4EA + 3EM) + 2EA - 1EM$	$n(1H + 1EA + 2EM) + 2EA + 1EM$	$n(2H + 4EA + 4EM) + 3EA + 2EM$	$n(1H + 2EA + 1EM) + 1H + 3EA + 1EM$
Total operations	$(n+1)H + (5n+4)EA + (2n+6)EM$	$(n+1)H + (n+2)E + (4n+2)EA + (2n+8)EM$	$(2n+3)H + (3n+5)EA + (n+7)EM$	$(3n+5)H + (5n+7)EA + (3n+7)EM$	$(n+5)H + (4n+8)EA + (5n+11)EM$	$(2n+4)H + (5n+8)EA + (4n+8)EM$	$(n+3)H + (4n+10)EA + (n+7)EM$
Total operation time (ms, n = 100)	92.7874	635.1078	49.5076	139.1076	227.4574	182.9232	48.8766

H: One-way hash function, E: Modular exponential operation. EA: Elliptic curve addition operation, EM: Elliptic curve scalar multiple operation. See references for definitions of variables in the forms of the signatures.

In this proposed scheme, without using a pairing operation, compared with other pairing-free schemes, elliptic-curve cryptography-based addition and multiplication operations are efficiently applied to reduce the total operation time. In addition, since the tag, T, for verification is also aggregated for all the messages together, only the part of the public key that the verifier actually acquires and directly calculates is included. Because of this, storage, such as that of a gateway or server, can save space and the verification overhead for the verifier is reduced.

7. Conclusions

To maintain the integrity of messages transmitted in an increasingly large IoT service environment, digital signatures for messages are required. Digital signature protocols have been studied for a long time, and many studies are underway to make them suitable for such environments. They are being studied to satisfy various security requirements while respecting the "lightweight" nature of the IoT environment. Although research has been conducted to apply lightweight signature techniques, such as CL-AS, to IoT environments, solutions are needed for the problems of CL-PKC-based schemes, specifically, public key replacement attacks and malicious KGCs. In particular, it is necessary to study solutions that satisfy the requirements for both security and computational efficiency. Therefore, this paper proposes an efficient secure CL-AAS scheme.

The proposed scheme provides the integrity of messages transmitted in an IoT environment using the concepts of an arbitrated signature and an AS (aggregated signature). The role of the AS is to provide efficiency, and that of the arbitrated signature is to enhance non-repudiation by aggregating the arbitrated signatures of a gateway and its devices together. Through this, in this paper, we designed a secure scenario against existing security threats, and considered the security of the gateway, which is an intermediate to transmit data. The proposed scheme is designed to satisfy various security requirements (Section 3), such as such as public key replacement attack, malicious KGC attack, and key leakage. In the existing schemes, as shown in Table 1, there were problems with key leakage and forgery of the message and signature via attacks either by public key replacement or a malicious KGC. To solve this, non-repudiation was strengthened by applying the arbitrated signature of the gateway, and it is possible to provide efficiency by applying an AS to reduce the memory overhead and the verification overhead of the verifier.

In the future, not only as a simple signer and verifier but also in a more complex, grouped, and device-involved environments, the provision of a suitable security scheme is needed. The IoT service may transmit sensitive data, such as personal privacy, depending on the environment. In the future, research on practical security technologies to provide confidentiality and integrity for sensitive data should be conducted.

Author Contributions: Conceptualization, D.-H.L., K.Y. and I.-Y.L.; data investigation, D.-H.L.; analysis and validation, D.-H.L., K.Y. and I.-Y.L.; writing—original draft, D.-H.L.; writing—review and editing, D.-H.L. All authors have read and agreed to the published version of the manuscript.

Funding: This research was supported by the MSIT (Ministry of Science, ICT), Korea, under the ITRC (Information Technology Research Center) support program (IITP-2020-0-00403) supervised by the IITP (Institute for Information & Communications Technology Promotion). This research was also supported by the National Research Foundation of Korea (NRF) grant funded by the MSIT (NRF-2018R1A4A1025632).

Conflicts of Interest: The authors declare no conflict of interest.

References

1. Li, S.; Da Xu, L.; Zhao, S. 5G Internet of Things: A survey. *J. Ind. Inf. Integr.* **2018**, *10*, 1–9. [CrossRef]
2. Yassein, M.B.; Aljawarneh, S.; Al-Sadi, A. Challenges and features of IoT communications in 5G networks. In Proceedings of the 2017 International Conference on Electrical and Computing Technologies and Applications (ICECTA), Ras Al Khaimah, UAE, 21–23 November 2017.
3. Griffiths, F.; Ooi, M. The fourth industrial revolution-Industry 4.0 and IoT [Trends in Future I&M]. *IEEE Instrum. Meas. Mag.* **2018**, *21*, 29–43. [CrossRef]

4. Sadeghi, A.-R.; Wachsmann, C.; Waidner, M. Security and privacy challenges in industrial internet of things. In Proceedings of the 2015 52nd ACM/EDAC/IEEE Design Automation Conference (DAC), San Francisco, CA, USA, 8–12 June 2015.

5. Khajenasiri, I.; Estebsari, A.; Verhelst, M.; Gielen, G. A review on Internet of Things solutions for intelligent energy control in buildings for smart city applications. *Energy Procedia* **2017**, *111*, 770–779. [CrossRef]

6. Khatoun, R.; Zeadally, S. Cybersecurity and Privacy Solutions in Smart Cities. *IEEE Commun. Mag.* **2017**, *55*, 51–59. [CrossRef]

7. Mahmoud, R.; Yousuf, T.; Aloul, F.; Zualkernan, I. Internet of things (IoT) security: Current status, challenges and prospective measures. In Proceedings of the 2015 10th International Conference for Internet Technology and Secured Transactions (ICITST), London, UK, 14–16 December 2015.

8. Zhang, Z.K.; Cho, M.C.Y.; Wang, C.W.; Hsu, C.W.; Chen, C.K.; Shieh, S. IoT security: Ongoing challenges and research opportunities. In Proceedings of the 2014 IEEE 7th international conference on service-oriented computing and applications, Matsue, Japan, 17–19 November 2014.

9. Diffie, W.; Hellman, M. New directions in cryptography. *IEEE Trans. Inf. Theory* **1976**, *22*, 644–654. [CrossRef]

10. Goldwasser, S.; Micali, S.; Rivest, R.L. A Digital Signature Scheme Secure Against Adaptive Chosen-Message Attacks. *SIAM J. Comput.* **1988**, *17*, 281–308. [CrossRef]

11. Schnorr, C.P. Efficient identification and signatures for smart cards. In Proceedings of the Conference on the Theory and Application of Cryptology, Daejeon, Korea, 6–10 December 1989.

12. Chaum, D. Blind signature system. In Proceedings of the Advances in cryptology, Paris, France, 9–11 April 1984.

13. Chen, L.; Pedersen, T.P. New group signature schemes. In *Workshop on the Theory and Application of Cryptographic Techniques*; Springer: Berlin/Heidelberg, Germany, 1994.

14. Ateniese, G.; Camenisch, J.; Joye, M.; Tsudik, G. A Practical and Provably Secure Coalition-Resistant Group Signature Scheme. In *Annual International Cryptology Conference*; Springer: Berlin/Heidelberg, Germany, 2000.

15. Harn, L. Group-oriented (t, n) threshold digital signature scheme and digital multisignature. *IEE Proc. Comput. Digit. Tech.* **1994**, *141*, 307–313. [CrossRef]

16. Perrig, A. The BiBa One-Time Signature and Broadcast Authentication Protocol. Available online: https://dl.acm.org/doi/abs/10.1145/501983.501988 (accessed on 2 May 2020).

17. Boneh, D.; Gentry, C.; Lynn, B.; Shacham, H. Aggregate and Verifiably Encrypted Signatures from Bilinear Maps. Available online: https://link.springer.com/chapter/10.1007/3-540-39200-9_26 (accessed on 2 May 2020).

18. Zhang, L.; Zhang, F. A new certificateless aggregate signature scheme. *Comput. Commun.* **2009**, *32*, 1079–1085. [CrossRef]

19. Shamir, A. Identity-Based Cryptosystems and Signature Schemes. Available online: https://link.springer.com/chapter/10.1007/3-540-39568-7_5 (accessed on 2 May 2020).

20. Oh, J.; Lee, K.; Moon, S. How to Solve Key Escrow and Identity Revocation in Identity-Based Encryption Schemes. Available online: https://link.springer.com/chapter/10.1007/11593980_22 (accessed on 3 May 2020).

21. Yuen, T.H.; Susilo, W.; Mu, Y. How to construct identity-based signatures without the key escrow problem. *Int. J. Inf. Secur.* **2010**, *9*, 297–311. [CrossRef]

22. Al-Riyami, S.S.; Paterson, K.G. Certificateless Public Key Cryptography. Available online: https://link.springer.com/chapter/10.1007/978-3-540-40061-5_29 (accessed on 4 May 2020).

23. He, D.; Chen, J.; Hu, J. A pairing-free certificateless authenticated key agreement protocol. *Int. J. Commun. Syst.* **2012**, *25*, 221–230. [CrossRef]

24. Mandt, T.K.; Tan, C.H. Certificateless Authenticated Two-Party Key Agreement Protocols. Available online: https://link.springer.com/chapter/10.1007/978-3-540-77505-8_4 (accessed on 5 May 2020).

25. Yum, D.H.; Lee, P.J. Generic Construction of Certificateless Signature. Available online: https://link.springer.com/chapter/10.1007/978-3-540-27800-9_18 (accessed on 6 May 2020).

26. Huang, X.; Mu, Y.; Susilo, W.; Wong, D.S.; Wu, W. Certificateless Signature Revisited. Available online: https://link.springer.com/chapter/10.1007/978-3-540-73458-1_23 (accessed on 6 May 2020).

27. Dent, A.W. A survey of certificateless encryption schemes and security models. *Int. J. Inf. Secur.* **2008**, *7*, 349–377. [CrossRef]

28. Libert, B.; Quisquater, J.J. On Constructing Certificateless Cryptosystems from Identity Based Encryption. Available online: https://link.springer.com/chapter/10.1007/11745853_31 (accessed on 6 May 2020).

29. Qu, Y.; Mu, Q. An efficient certificateless aggregate signature without pairing. *Int. J. Electron. Secur. Digit. Forensics* **2018**, *10*, 188–203. [CrossRef]

30. Deng, L.; Yang, Y.; Chen, Y.; Wang, X. Aggregate signature without pairing from certificateless cryptography. *J. Internet Technol.* **2018**, *19*, 1479–1486.

31. Cui, J.; Zhang, J.; Zhong, H.; Shi, R.; Xu, Y. An efficient certificateless aggregate signature without pairings for vehicular ad hoc networks. *Inf. Sci.* **2018**, *451*, 1–15. [CrossRef]

32. Du, H.; Wen, Q.; Zhang, S. An Efficient Certificateless Aggregate Signature Scheme Without Pairings for Healthcare Wireless Sensor Network. *IEEE Access* **2019**, *7*, 42683–42693. [CrossRef]

33. Gayathri, N.B.; Thumbur, G.; Rajesh Kumar, P.; Rahman, M.Z.U.; Reddy, P.V.; Lay-Ekuakille, A. Efficient and Secure Pairing-Free Certificateless Aggregate Signature Scheme for Healthcare Wireless Medical Sensor Networks. *IEEE Internet Things J.* **2019**, *6*, 9064–9075. [CrossRef]

34. Zhao, Y.; Hou, Y.; Wang, L.; Kumari, S.; Khan, M.K.; Xiong, H. An efficient certificateless aggregate signature scheme for the Internet of Vehicles. *Trans. Emerg. Telecommun. Technol.* **2020**, *31*, e3708. [CrossRef]

35. Seo, S.-H.; Won, J.; Bertino, E. pCLSC-TKEM: A Pairing-free Certificateless Signcryption-tag Key Encapsulation Mechanism for a Privacy-Preserving IoT. *Trans. Data Priv.* **2016**, *9*, 101–130.

36. Yang, Q.; Zhou, Y.; Yu, Y. Leakage-Resilient Certificateless Signcryption Scheme. In Proceedings of the 2019 IEEE Globecom Workshops (GC Wkshps), Waikoloa, HI, USA, 9–13 December 2019.

37. Du, H.; Wen, Q.; Zhang, S. A Provably-Secure Outsourced Revocable Certificateless Signature Scheme Without Bilinear Pairings. *IEEE Access* **2018**, *6*, 73846–73855. [CrossRef]

38. Xiong, H.; Mei, Q.; Zhao, Y. Efficient and Provably Secure Certificateless Parallel Key-Insulated Signature Without Pairing for IIoT Environments. *IEEE Syst. J.* **2020**, *14*, 310–320. [CrossRef]

© 2020 by the authors. Licensee MDPI, Basel, Switzerland. This article is an open access article distributed under the terms and conditions of the Creative Commons Attribution (CC BY) license (http://creativecommons.org/licenses/by/4.0/).

Article

A Measurement-Based Frame-Level Error Model for Evaluation of Industrial Wireless Sensor Networks

Yun-Shuai Yu [1] and Yeong-Sheng Chen [2],*

[1] Department of Computer Science and Information Engineering, National Formosa University, Yunlin 632301, Taiwan; yys@nfu.edu.tw
[2] Department of Computer Science, National Taipei University of Education, Taipei 106320, Taiwan
* Correspondence: yschen@tea.ntue.edu.tw; Tel.: +886-2-2732-1104 (ext. 53459)

Received: 25 May 2020; Accepted: 15 July 2020; Published: 17 July 2020

Abstract: Industrial wireless sensor networks (IWSNs) are a key technology for smart manufacturing. To identify the performance bottlenecks in an IWSN before its real-world deployment, the IWSN must first be evaluated through simulations using an error model which accurately characterizes the wireless links in the industrial scenario within which it will be deployed. However, the traditional error models used in most IWSN simulators are not derived from the real traces observed in industrial environments. Accordingly, this study first measured the transmission quality of IEEE 802.15.4 in a one-day experiment in a manufacturing factory and then used the measurement records to construct a second-order Markov frame-level error model for simulating the performance of an IWSN. The proposed model was incorporated into the simulator of OpenWSN, which is an industrial WSN implementing the related IEEE and IETF standards. The simulation results showed that the proposed error model improved the accuracy of the estimated transmission reliability by up to 12% compared to the original error model. Moreover, the estimation accuracy improved with increasing burst losses.

Keywords: IWSNs; error models; IEEE 802.15.4; second-order Markov chain; OpenWSN; transmission reliability

1. Introduction

Industrial wireless sensor networks (IWSNs) play a key role in smart manufacturing by enabling a wide range of monitoring, control and optimization processes. In a typical IWSN, the operating parameters of a machine in the factory, e.g., the vibration frequency, pressure, flow rate, coolant level, and so on, are measured by one or more sensors and the sensor signals are then transmitted wirelessly to a gateway, from which they are passed to a remote server via a wired or wireless network for further processing and/or storage. Typically, the server performs online calculations on the sensed data using domain knowledge and transmits appropriate control commands back to the factory to control the corresponding actuators, provided that the calculation results satisfy certain predefined conditions. For example, the server may instruct the machine to close a valve or reduce the conveyor speed. The remote server may also analyze the sensed data offline, e.g., by inputting the manufacturing parameters and product yield data into a Deep Neural Network (DNN) in order to derive the optimal manufacturing parameters, for example.

There are two international standards for the IWSN protocol stack in industry at present, namely WirelessHART [1], established by the HART Communication Foundation (HCF), and ISA 100.11a [2], developed by the International Society of Automation (ISA). In both standards, the data transmissions in the physical layer are performed using IEEE 802.15.4-2006 [3], while media access is achieved using Time Division Multiple Access (TDMA), and routing at the network layer is conducted using Graph Routing. Academic researchers and industrial organizations, such as IEEE and IETF, have proposed many communication protocols and standards for IWSNs. Many of these protocols are implemented in OpenWSN [4], an open-source project using IEEE 802.15.4-2006 in the hardware layer (like WirelessHART and ISA 100.11a) and Time-Slotted Channel Hopping (TSCH) technology in IEEE 802.15.4e [5] in the lower part of the Media Access Control (MAC) layer. In addition, the higher part of the MAC layer adopts the Internet Protocol version 6 (IPv6) over the Time Slotted Channel Hopping mode of IEEE 802.15.4e (6TiSCH) Minimal Scheduling Function (MSF) and the 6TiSCH Operation Sublayer (6top) Protocol (6P) formulated by the 6TiSCH working group of the IETF. The network layer adopts protocols such as IPv6, Internet Control Message Protocol version 6 (ICMPv6), IPv6 over Low-Power Wireless Personal Area Networks (6LowPAN) and Routing Protocol for Low-Power and Lossy Networks (RPL) formulated by the IETF, while the transport layer uses User Datagram Protocol (UDP) and Constrained Application Protocol (CoAP), also formulated by the IETF. The application layer is left to the user for customization.

The transmission reliability, delay and lifetime of an IWSN are all critically dependent on its design. However, an IWSN is often composed of hundreds or even thousands of sensing nodes. Thus, optimizing the IWSN design in-situ using experimental methods is extremely challenging, if not impossible. Accordingly, before the actual deployment of the IWSN, it is desirable to evaluate the function and effectiveness of the network through systematic simulations using a simulator model as close as possible to that of the environment in which the network will be deployed. Among all the measures of the IWSN performance, the transmission reliability is one of the most important, and is defined as the probability of successful delivery of the sensed data from the sensing nodes to the gateway used to route the data to the remote server. In real-world environments, the transmission of frames via the wireless medium may fail for various reasons, including noise interference, low signal energy due to multipath attenuation, and so on. To simulate such transmission failures, it is necessary to employ an error model which accurately characterizes the errors associated with the real wireless environments. In particular, simulations on the cases with lots of transmission failures are important for applications requiring dependable operations, e.g., gas metering applications. The OpenSim simulator [6] in OpenWSN uses an independent error model (also known as the Bernoulli error model), which assumes that the frame transmission success has a fixed probability (generally referred to as the Frame Delivery Ratio (FDR)). However, the transmission quality of wireless links is affected by many factors in the field environment and frequently changes over time. Consequently, the simple Bernoulli error model limits the ability of the OpenWSN simulator to accurately evaluate the transmission performance of an IWSN.

To address this problem, the present study derives a frame error model which more accurately represents the wireless conditions in a real-world factory environment by measuring the transmissions in a machine-intensive factory over a period of almost 24 h and then using the measurement records to construct a frame-level second-order Markov model as an error model. Note that the proposed error model is mainly suitable for the industrial environments where there are only IEEE 802.15.4 devices operating in 2.4 GHz Industrial, Scientific and Medical (ISM) band (that is, it will be inappropriate to apply our model in the factory environment with devices supporting IEEE 802.11.). It is shown that the Cumulative Distribution Functions (CDFs) of the number of consecutively received correct frames (i.e., the correct-frame burst length) and the number of consecutively received error frames (i.e., the error-frame burst length) synthesized by the proposed model are very close to those of the original records. Moreover, the simulation results show that the proposed model improves the estimation accuracy of the transmission reliability by up to 12% compared to that achieved using the original independent error model in OpenWSN.

The remainder of this paper is organized as follows. Section 2 briefly reviews the related work in the literature. Section 3 describes the data collection methods employed in the present study and gives the preliminary analysis results. Section 4 describes the proposed error model and verifies its correctness. Section 5 analyzes and explains the overestimation errors of the independent error model in OpenWSN for the transmission reliability. Section 6 presents and discusses the simulation results obtained using the proposed error model. Finally, Section 7 provides some brief concluding remarks.

2. Related Work

In general, error models for IWSNs can be categorized into two main types, namely bit-level error models and frame-level error models. Models of the former type consider each bit of a frame to be either erroneous or correct, and assume that a frame is transmitted successfully if no bit is wrong. By contrast, models of the latter type judge the entire frame directly as either erroneous or correct. Generally speaking, bit-level error models provide subtler channel-state information than frame-level models, whereas the frame-level models offer better scalability due to their greater simplicity.

Nobre et al. [7–9] developed a WirelessHART communication module for Network Simulator 3 (NS-3) [10] based on a bit-level two-state Markov chain error model designated as the Gilbert/Elliot model. Remke et al. [11] modeled WirelessHART networks with bit failures using a Binary Symmetric Channel (BSC) model [12] and link failures using a two-state Markov chain. Barac et al. [13] analyzed the bit- and symbol-error nature of IEEE 802.15.4 transmissions in actual industrial sites and then employed the collected error traces to evaluate the performance of a lightweight Reed-Solomon (15, k) block code.

Gao et al. [14] adopted a frame-level two-state Markov error model to evaluate the Quality-of-Service (QoS) performance of IEEE 802.15.4 MAC under bursty channel errors. Petrova et al. [15] analyzed the performance of IEEE 802.15.4 based on Received Signal Strength Indicator (RSSI), Frame Error Rate (FER) and run length distribution measurements obtained in indoor and outdoor environments. It was shown that the independent and two-state Markov error models both provided an adequate modeling performance for short transmission distances, but the two-state Markov model slightly outperformed the independent model over longer distances. Wijetunge et al. [16] used a three-dimensional discrete-time Markov chain model to analyze the IEEE 802.15.4 MAC protocol with Acknowledgement (ACK) frame transmission.

Iqbal and Khayam [17] conducted transmission experiments in a two-story building for transmission distances of 5 to 12 m, a frame size of 20 bytes and a transmission rate of 10 frames per second. The transmission records were then used to construct a two-level error model consisting of a two-state Markov model in the first level for predicting the error probability of the frames, and a third-order Markov model in the second level for-predicting the error probability of each bit in any frames flagged in the first level. However, since the transmission records were not collected in factories, there is no guarantee that the two-level error model can be applied in IWSN environments. Ilyas and Radha [18] measured the transmission quality in office, home and outdoor environments over channel 26 (2479–2481 MHz) for a transmission power of 0 dBm and a transmitted frame size of 41 bytes. The measurement results obtained for partially lost frames and frames that failed the Cyclic Redundancy Check (CRC) test were then used to construct a a discretized exponential Probability Density Function (PDF) error model. However, as for the study of Iqbal and Khayam [17], the model is not directly applicable to factory environments. Moreover, the experimental measurements did not consider the case where the frames were not received at all.

Striccoli et al. [19] proposed a Markov error model with K states to account for the various conditions a lossy wireless channel may undergo. Each state contained two substates, ON and OFF, modeled with probability distributions of Error Free Bursts (EFBs) and Error Bursts (EBs) respectively, where an EFB represents a sequence of consecutive correct frames, while an EB represents a sequence of consecutive erroneous frames. The state of the error model was switched among the substates to generate EFB/EB from the probability distributions of the selected substate. However, the model was trained using data traces collected in laboratories rather than industrial environments. Furthermore, a frame was marked as correct only when the transmitter received acknowledgment of the frame's correct receipt. Hence, although the traces reflect the characteristics of bi-directional transmissions, they may not be suitable for unidirectional transmissions, such as beacon frame broadcasting.

Guntupalli et al. [20,21] proposed a frame-level error model for error-prone channels with one state in the loss macro-state and three states in the non-loss macro-state. However, the error model was configured according to field measurements [22] collected in an 802.11 Wireless Local Area Network

(WLAN) rather than IEEE 802.15.4 WLAN. Valle et al. [23,24] proposed a retransmission scheme for WSNs based on cooperative relays and network coding and evaluated its effectiveness using Objective Modular Network Testbed in C++ (OMNeT++) [25] with a semi-Markov error model [26].

3. Transmission Quality Measurement and Analysis

In this study, the wireless transmission quality in an industrial environment was recorded using the measurement system shown in Figure 1, consisting of a transmitter, a receiver and a recorder. The transmitter and receiver were constructed using Texas Instruments' CC2538dk development kit [27], incorporating a CC2538 chip with a built-in 2.4 GHz radio frequency (RF) transceiver compliant with IEEE 802.15.4 specifications. The transmitter was installed with Contiki OS 2.7 [28], together with a self-written program for continuously broadcasting signal frames with a frequency of 512 Hz and a length of 23 bytes. Similar to the previous studies in References [17,18], small frame size was adopted in our measurement experiment. However, the present study deliberately adopts a high traffic generation frequency from 10 frames per second to 512 frames per second for extracting the subtler features of the wireless link. In addition, the reasons for transmission using broadcasting technology are three-fold. First, it avoids retransmitting the same frame so that the receiver could better identify the frame loss. Assume that a unicast frame fails and one of its following retransmitted frames was successfully received. In such a case, the receiver cannot determine the number of transmission failures that actually happened. Second, the lack of retransmission mechanism guarantees that the high traffic generation frequency will not be affected. Third, the experimental architecture can be easily extended in the future to include more receivers and recorders without increasing the number of transmitters.

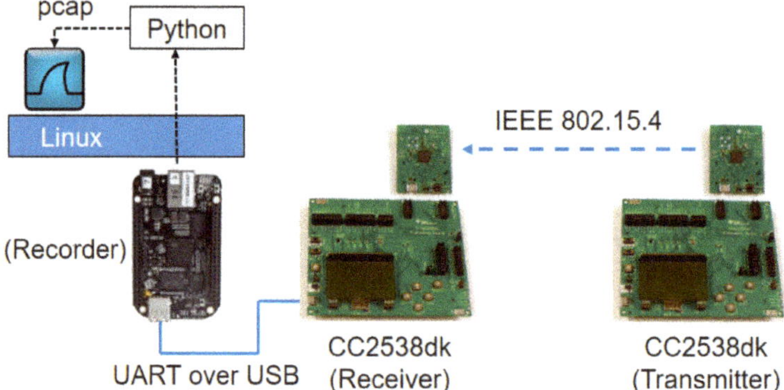

Figure 1. Devices for recording transmission quality.

The beacon frame format is shown in Figure 2, in which the beacon payload has the form of a 4-byte positive integer with an initial offset of 0 and an increment of one each time a frame is sent. The abbreviations MHR, MFR, GTS and FCS represent MAC header, MAC footer, guaranteed time slots and frame check sequence, respectively. The receiver was also installed with Contiki OS 2.7 and a self-written receiver program designed to send the received information (e.g., the frame contents at the MAC layer, the RSSI value and the CRC outcome, to the recorder with Universal Asynchronous Receiver-Transmitter (UART) signals over a Universal Serial Bus (USB)).

Octets: 2	1	2	8	2	1	1	4	2
Frame Control	Seq No	Source PAN	Source Address	Surperframe Spec	GTS	Pending Addresses	Beacon Payload	FCS
MHR				MAC Payload				MFR

Figure 2. Format of signal frame in trace experiments.

The recorder was built around a BeagleBone Black [29] single-board microcomputer. The USB signals received by the recorder were restored to UART signals by the Linux kernel embedded in the recorder and input to a Python program via a serial port. The experiment may be interrupted by some unpredictable events, such as power failure or collision caused by the workers. To reduce the impact of such interruptions, the Python program saved the information received each hour in the experimental measurement period to a pcap file (note that the pcap format is supported by many network packet monitoring systems, including Wireshark [30]).

To support the IWSN context considered in the present study, and in particular, to understand the effects of electromagnetic noise generated by the working machines on the transmission quality of IEEE 802.15.4 wireless communications, the measurement process was conducted in a machine-intensive factory in southern Taiwan. To protect trade secrets, the factory prohibits the use of any devices supporting IEEE 802.11 (e.g., access points and smartphones) in the factory. Consequently, the measured traces were free of interference from IEEE 802.11 or Bluetooth networks. Moreover, to avoid disruption of the factory operations, the trace experiment was performed only once, as described below.

Figure 3 illustrates the basic layout of the experimental sensing field with dimensions of approximately 17×30 m (length $\times$ width). Although the actual dimensions of the field are approximately 100 m in length and width, only the areas related to the experiment are drawn for ease of illustration. The factory had two ceiling heights, namely 8.4 m in the majority of the building (upper part of Figure 3) and 5 m in the remainder (bottom part of Figure 3). The transmitter was located in one corner of the building (see the red dot in Figure 3) and was placed on top of a cabinet with a height of 1.6 m. There were windows behind the cabinet, and hence the transmitter was exposed to sunlight during the afternoon period of the experimental process. The transmission power was set as −1 dBm throughout the entire measurement process and the transmissions were performed using channel 25 (2474–2476 MHz). The receiver was positioned together with the recorder on a shelf with a height of approximately 2 m in the diagonally opposite corner of the building (see the blue dot in the upper-left-hand corner of Figure 3). The receiver and transmitter were separated by a distance of approximately 35 m, but were not in direct line of sight of one another due to the presence of lathe machines on the factory floor with a height of around 2.5 m. In the path between the transmitter and the receiver, except for the static obstacles, there existed lots of dynamic obstacles. The raw and semi-finished materials on the machines in the path were often moved from one machine to another by the workers. Also, some workers might often walk across the path. Thus, the number of obstacles in the transmission path was dynamic. The measurement period commenced at 15:45 on one day and ended at 15:30 the following day. During the experimental period, the lathes operated independently in accordance with their particular work orders. That is, not all of the machines worked together, and none of the machines operated over the entire experimental period. Furthermore, in the factory, the daily work content is roughly the same, although there might be slight differences. That is, a day is just a cycle. Therefore, the data measured in one day could be used to characterize the factory environment. In Reference [18], the authors stated that "We collected error traces of approximately 10 million packets in a way that provides, to the authors' best knowledge, an unprecedented level of insight into the effects of the wireless channel state on the level of corruption of packets." By contrast, there were up to 43,728,940 transmission records collected in our experiment. According to our studies and the results presented in Reference [18], this amount of data could be able to provide valuable insights into the quality of the wireless link.

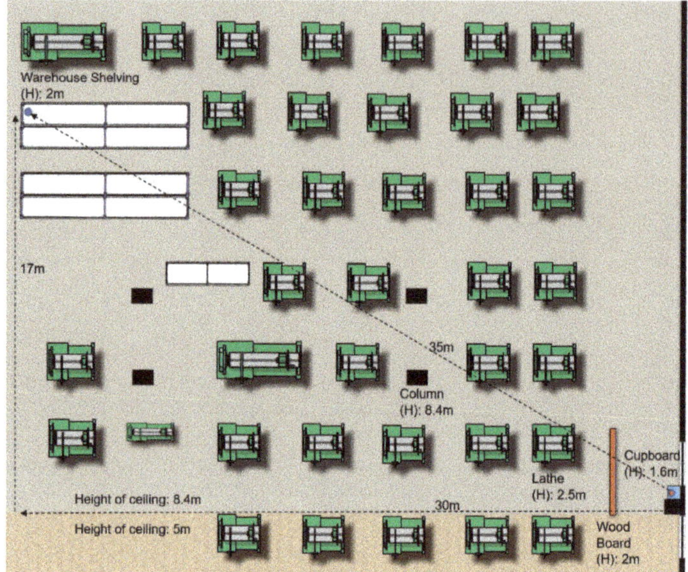

Figure 3. Layout of experimental environment.

Figure 4 shows the average RSSI values recorded in every hour of the experimental period, where the 0th hour runs from 15:45 to 16:44. It is seen that even though the distance between the transmitter and the receiver remained unchanged, the average RSSI varies significantly over time. In fact, a detailed inspection of Figure 4 shows that the difference between the largest and smallest average RSSI values is more than 8 dBm. The variation in the average RSSI is most likely the result of changes in the indoor temperature, noise interference produced by machines, the blocking or reflection effect of obstacles between the transmitter and the receiver, and so on. The measurement results in Reference [31] show that the RSSI decreases when the temperature increases. Notably, the factory faces the sun in the afternoon and the heat gets trapped inside until night. It might take hours to dissipate the heat after the sunset. Hence, the room temperature went down after 22:00. As a result, the average RSSI from afternoon to midnight is generally low, most probably because of the high room temperature caused by sunlight exposure.

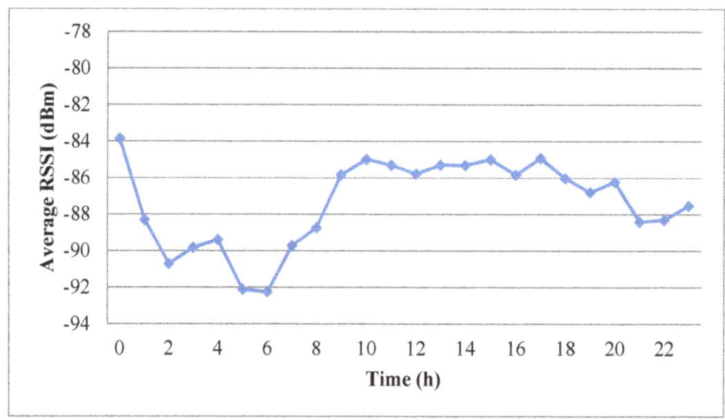

Figure 4. Average Received Signal Strength Indicator (RSSI) value in each hour of the experimental measurement period.

Figure 5 shows the variation in the FDR over the 24 h experimental period. As expected, the tendency of the FDR is very similar to that of the RSSI in Figure 4. The correlation coefficient between the RSSI and the FDR can be calculated as follows:

$$Correl(x, y) = \frac{\sum\limits_{i=0}^{23} (x_i - \bar{x})(y_i - \bar{y})}{\sqrt{\sum\limits_{i=0}^{23} (x_i - \bar{x})^2} \sqrt{\sum\limits_{i=0}^{23} (y_i - \bar{y})^2}}, \tag{1}$$

where x_i is the average RSSI value in a certain hour i, $\bar{x}$ is the average of x_i ($i = 0, 1, \ldots, 23$), y_i is the FDR in a certain hour i, and $\bar{y}$ is the average of y_i ($i = 0, 1, \ldots, 23$).

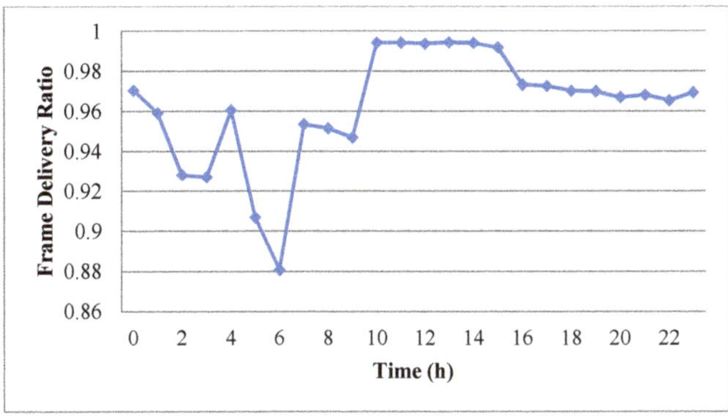

Figure 5. Frame Delivery Ratio (FDR) in every hour of the experimental measurement period.

The correlation coefficient was found from Equation (1) to have a value of 0.857. In other words, the FDR is extremely sensitive to the RSSI. Thus, when constructing the error model using the measured traces, the parameters in the model should be trained using the transmission records acquired during time intervals with a similar RSSI value. For example, the measurement data obtained in the fifth and sixth hours are suitable for training the model together since the average RSSI values are around −92 dBm in both hours. Similarly, the data from the fourth, seventh and eighth hours are also suitable for training the model together since they all have average RSSI values of around −89 dBm. Finally, the data from the ninth to eighteenth hours can also be used to train the model together since their average RSSI values are all around −85 to −86 dBm. In the subsequent simulations, the error model can then switch among the parameters trained using different sets of trace data in accordance with a switching policy formulated in advance by the IWSN investigator.

Referring to Figure 5, the FDR has its lowest value in the sixth hour of the experimental period (i.e., 21:45–22:44). In other words, the transmission quality within the factory is particularly poor during this period. As a result, the transmission records acquired in the sixth hour were scrutinized particularly carefully. In particular, the records in the corresponding pcap file were converted into binary records arranged in chronological order from left to right, as shown in Figure 6, in which values of 0 and 1 indicate that the frame transmitted at the particular time was successfully and unsuccessfully received, respectively. Note that failure cases were recorded if the corresponding frame was partially lost, or the length of the received frame was the same as that of the transmitted frame, but the received frame failed to pass the cyclic redundancy check.

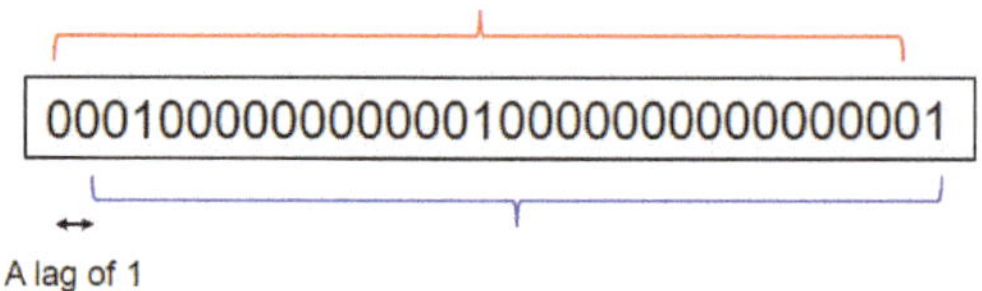

Figure 6. Illustrative example of binary record.

Equation (1) was then applied to determine the autocorrelation between the data corresponding to the binary record of the first minute of the sixth hour (21:45). Note that in implementing Equation (1), x and y were taken as two values in the selected trace with a certain time lag between them. For example, referring to Figure 6, the data enclosed within the lower curly bracket lag those within the upper curly bracket by a notional time value of 1. Figure 7 shows the autocorrelation coefficients computed for the data in the first minute of the selected trace. As described in Section 3, the transmitter used in the present experiments had a broadcast frequency of 512 Hz. In other words, the transmitter broadcasts 512 transmissions each second. As a result, the time lag in Figure 7 has a value of 1/512 s. In general, the results show that the data received at the receiver with a gap of 1 s have almost no correlation with each other. The autocorrelation of the data within the first 0.3 s of the time interval is around 0.1, or more. Figure 8 shows the autocorrelation of the data received over a period of 20/512 s within the first minute of the recorded trace. The above results show that the success or failure of the current transmission does depend on the previous transmission results and mainly depends on the last transmission results. Furthermore, this implies that the collected trace data could be used to train the proposed second-order Markov model, which will be elaborated in the following section.

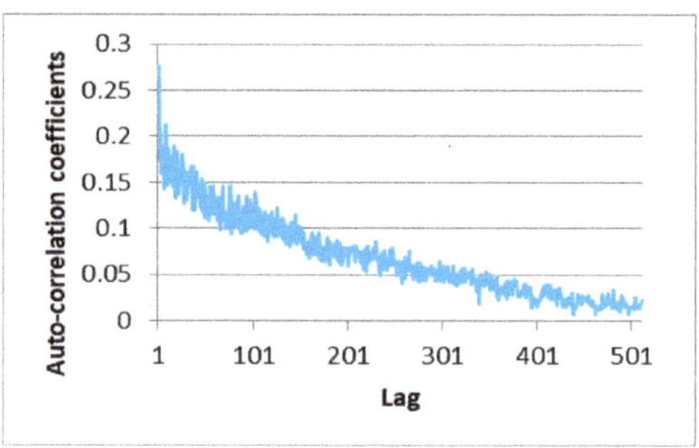

Figure 7. Autocorrelation coefficients of received data in first minute (21:45) of trace obtained in the sixth hour.

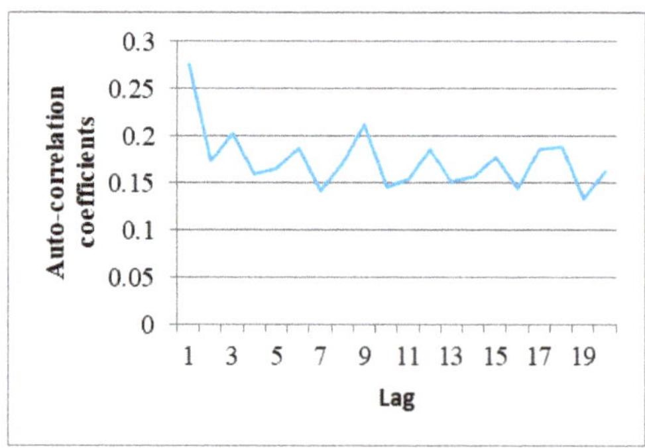

Figure 8. Autocorrelation coefficients of received data in time interval of 20/512 s within first minutes (21:45) of trace obtained in the sixth hour.

4. Proposed Error Model

The conventional two-state Markov model assumes that the current state is correlated only with the previous state. That is, it is independent of the states before the previous state. However, the autocorrelation coefficients presented above gradually decrease as the time lag increases. Thus, in general, a Markov model with higher orders provides a better representation of the true error situation within the factory environment than one with lower orders. It is noted, however, that the improvement offered by this higher-order model reduces as the order increases. In practice, increasing the order of the Markov model incurs a greater computation cost and larger memory requirement. Thus, determining the optimal order of the Markov model which provides the best tradeoff between the improved realism and the cost is essential. Consequently, the present study deliberately adopts a second-order Markov frame-level error model for evaluating the accuracy of IWSN simulators.

Since the transmission quality in the sixth hour is the worst among all the hours in the experimental period, while that in the tenth hour is the best (see Figure 5), the binary records of the first minutes in these two hours were used to train the independent model, the two-state Markov model and the second-order Markov model. The FDRs of the independent model derived from the two binary records were found to be 0.825 and 0.994, respectively. The training results for the two-state Markov model are shown in Figure 9, where the circles indicate the states and the digits in the circles represent the results of the previous transmission (i.e., 0: successful transmission, 1: failed transmission). In addition, the links between the circles indicate the change from the old state to the new state, where the end without an arrow is the old state and the end with an arrow is the new state. Finally, the numbers attached to the links denote the corresponding transition probabilities. Figure 10 shows the training results obtained for the second-order Markov model (note that the meanings of the symbols in Figure 10 are the same as those in Figure 9 except for the two digits in each circle, where the digit on the right indicates the result of the previous transmission, while that on the left indicates the result of the transmission before the previous one). For ease of discussion, the transition probability in the second-order Markov model is denoted as $pABC$, where p is the probability, AB is the old state and BC is the new state. For instance, the probability of transitioning from state 11 to state 10 is denoted as $p110$. From the results presented in Figures 9 and 10, it is clear that the transition probabilities trained by the two datasets are very different. In other words, the results support the inference above that the parameters in the error model should be trained based on the transmission records obtained during the time intervals with similar average RSSI values.

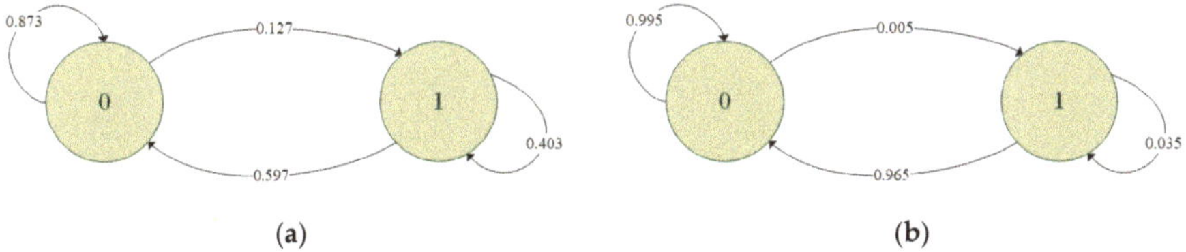

(a)   (b)

Figure 9. Training results for the two-state Markov model using: (**a**) records in the first minute of the sixth hour, and (**b**) records in the first minute of the tenth hour.

After training the three error models described above, one-minute binary records were synthesized using each of the three models (note that the starting state for the two-state Markov model was set as 0, while that for the second-order Markov model was set as 00). The CDFs of the correct-frame burst length derived from the original records and the three synthesized records are plotted in Figures 11 and 12 for the first minute of the sixth hour and the first minute of the tenth hour, respectively. Figure 11 shows that the CDF of the correct-frame burst length derived using the second-order Markov model is closer to that of the correct-frame burst length derived using the original records than those derived using the independent model or two-state Markov model. However, referring to Figure 12, the three error models have a similar performance. Figures 13 and 14 show the CDFs of the error-frame burst length derived from the original records and the synthesized records for the first minute of the sixth and tenth hour, respectively. It is seen in Figure 13 that the performance of the second-order Markov model is comparable to that of the two-state Markov model and is better than that of the independent model. However, for the records collected in the first minute of the tenth hour, all three models have a similar performance (see Figure 14). This finding is reasonable since the transmissions in the tenth hour have the best quality among all the traces collected in the experimental period. In other words, the trace data in the tenth hour contain only a small number of error-frame bursts for training purposes. Consequently, for all three models, the length of most error-frame bursts is equal to 1. Overall, however, the results presented in Figures 11–14 show that the proposed second-order Markov error model outperforms both the independent model and the two-state Markov model when the transmission quality is poor.

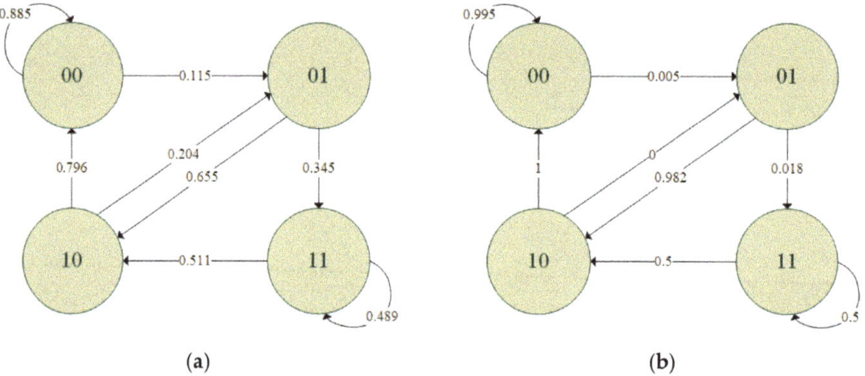

(a)   (b)

Figure 10. Training results for the second-order Markov model using: (**a**) records in the first minute of the sixth hour, and (**b**) records in the first minute of the tenth hour.

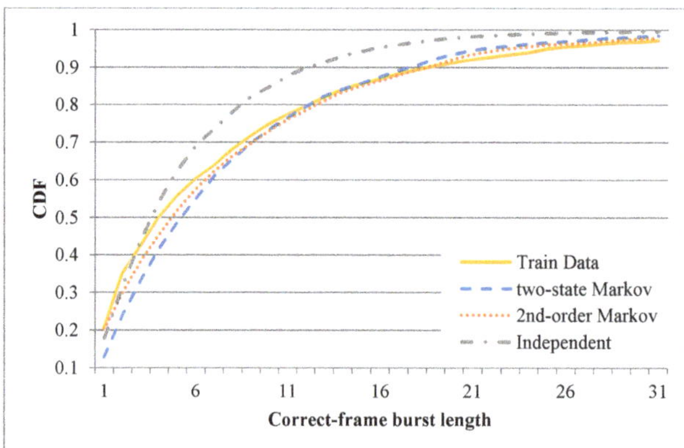

Figure 11. Cumulative Distribution Functions (CDFs) of correct-frame burst length derived using different error models based on recorded trace in the first minute of the sixth hour.

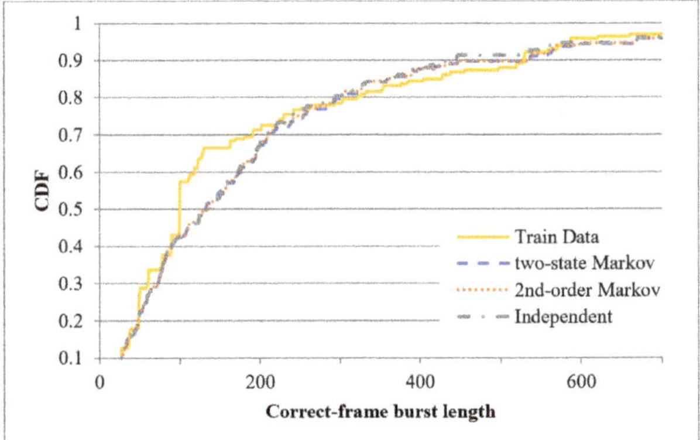

Figure 12. CDFs of correct-frame burst length derived using different error models based on recorded trace in the first minute of the tenth hour.

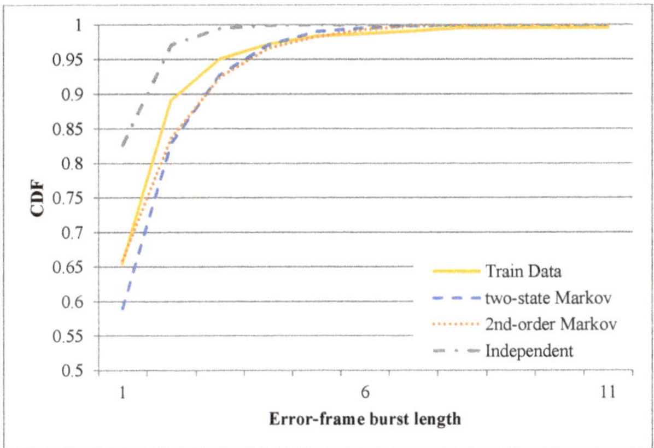

Figure 13. CDFs of error-frame burst length derived using different error models based on recorded trace in the first minute of the sixth hour.

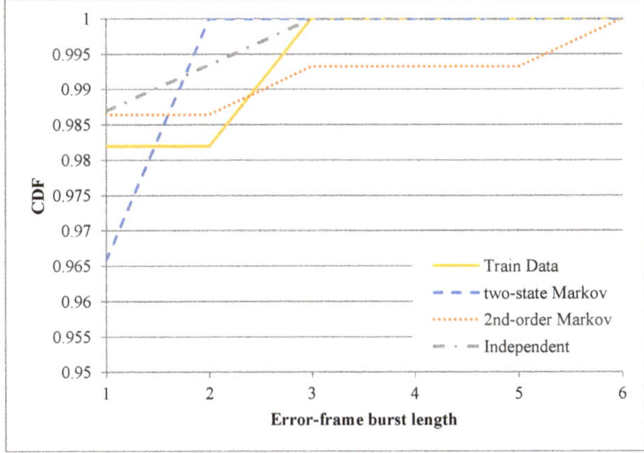

Figure 14. CDFs of error-frame burst length derived using different error models based on recorded trace in the first minute of the tenth hour.

As described above, the lack of error-frame bursts in the trace limits the ability of the trained error models to describe the error-frame bursts in the real environment. To address this problem, a further training process was performed using additional experimental traces. In particular, the trace data of the first 2^n minutes of the sixth hour were used to retrain the three models, where n is an integer ranging from 0 to 5. The authenticity of the resulting error models was quantified by computing the Kullback-Leibler (K-L) divergence [17] of the synthesized data relative to the original records, i.e.,

$$D(p \parallel q) = \sum_x p(x) \log \frac{p(x)}{q(x)}, \tag{2}$$

where x is the correct-frame (or error-frame) burst length, p is the PDF of the original records and q is the PDF of the synthetic data. The closer to zero the K-L divergence is, the more accurately the error model captures the burstiness of the original records.

Figures 15 and 16 show the results obtained for the K-L divergence between the correct-frame bursts and error-frame bursts of the original records and those of the synthetic data produced by the three error models, respectively (note that for both figures, the plotted data show the average values obtained over 100 computation processes performed with different synthetic data). As shown in Figure 15, the divergence of all three models reduces (i.e., the authenticity of the models increases) as a greater number of original records are employed in the training process for the correct-frame bursts. A similar tendency is noted for the two Markov models for the error-frame bursts, as shown in Figure 16. However, for both training processes, the divergence improves only very slightly as the length of the training data is increased beyond 4 min. It is additionally noted in Figure 16 that the independent model performs poorly in capturing error-frame bursts in lossy environments.

Based on the results shown in Figures 15 and 16, the three models were retrained using the trace data collected in the first 4 min of each hour in the experimental period. For each training process, three 4-min binary records were synthesized using the three error models, respectively. The authenticity of the resulting error models was then evaluated using a new performance metric R, defined as:

$$R = \left| \frac{D_m}{D_i} \right|, \tag{3}$$

where D_i is the K-L divergence of the independent model and D_m is the K-L divergence of the two-state Markov model or second-order Markov model. In other words, a value of R less than 1 indicates that the Markov model (two-state or second-order) outperforms the independent model, and vice versa. Figure 17 shows the value of R for the correct-frame bursts in each hour of the experimental period. For the fourth and sixth hours, both Markov models outperform the independent model. Furthermore, the two models provide a comparable performance to the independent model in all of the other hours. It is additionally noted that the second-order Markov model significantly outperforms the two-state Markov model in the sixth hour. Figure 18 presents the equivalent performance results for the error-frame bursts. For ease of presentation, the y-axis is plotted with a base 10 logarithmic scale. Hence, a value of R less than 0 indicates that the Markov models outperform the independent model, and vice versa. The results show that the second-order Markov model generally outperforms the other two models, particularly in the third and seventh hour.

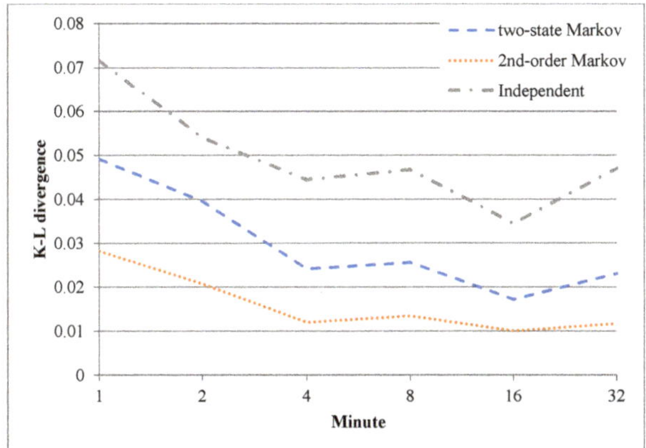

Figure 15. Kullback-Leibler (K-L) divergence between correct-frame bursts of original records and those of synthetic data.

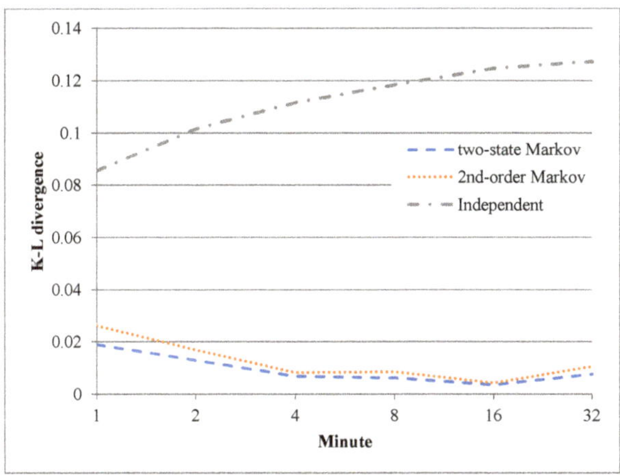

Figure 16. K-L divergence between error-frame bursts of original records and those of synthetic data.

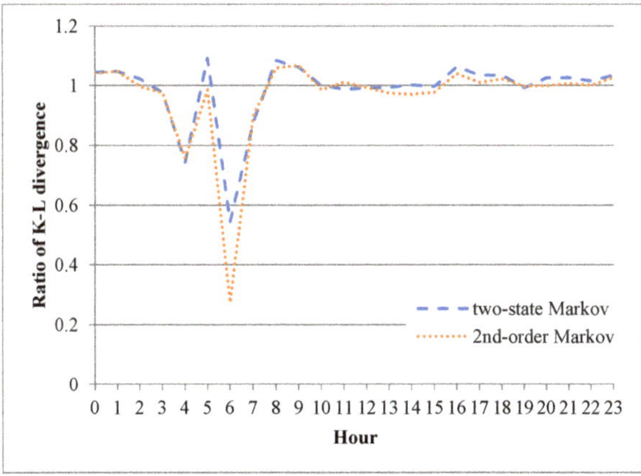

Figure 17. K-L divergence ratios of Markov error models for correct-frame bursts in each hour of the experimental period.

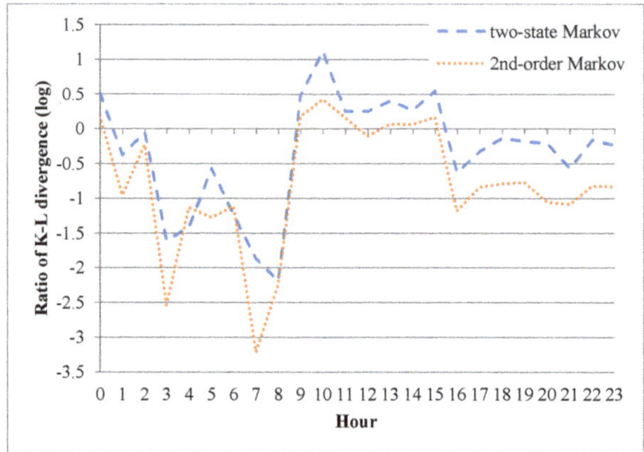

Figure 18. K-L divergence ratios of Markov error models for error-frame bursts in each hour of the experimental period (note that the K-L divergence ratio is plotted using a logarithm base 10 scale for ease of presentation).

5. Overestimation Errors of Independent Model

According to the IEEE 802.15.4e standard, a data frame may be retransmitted three times at most. In other words, each data frame has a maximum of four transmission opportunities. Furthermore, after a transmitting node sends a data frame to a receiving node, the receiving node must return an ACK frame within a certain time interval. If the transmitting node receives the correct ACK frame within this interval, it deems the transmission of the data frame to have been a success. Otherwise, it performs retransmission up until the maximum number of allowable retransmissions. Assuming that the transmission quality of the wireless links can be modeled by the FDR, the transmission errors can be described by the original independent error model in OpenWSN. Let P denote the FDR from a transmitting node to a receiving node. The probability that the receiving node fails to correctly receive the data frame in the first transmission can therefore be written as:

$$1 - P. \tag{4}$$

Accordingly, the probability that the receiving node fails to correctly receive the data frame for the first transmission and the three consecutive retransmissions can be written as:

$$(1 - P)^4. \tag{5}$$

Excluding the above probability, the remaining probability is given as follows:

$$1 - (1 - P)^4, \tag{6}$$

where this probability represents the probability that the receiving node successfully receives the data frame within the maximum number of retransmissions. This probability can be used to represent the transmission reliability of a single wireless link. Note that due to the dynamic nature of the transmission quality in wireless networks, the probability, P, in the above formulae is unlikely to be a constant value.

OpenWSN adopts the average FDR computed over an extended time interval as the FDR for each moment. In other words, the independent error model assumes that all the frame delivery failure events are evenly dispersed over the considered time interval. However, in practical wireless networks, the frame transmission failures are more likely to burst in a short period, T. In other words, the transmission reliability of the wireless links is likely to be relatively low in the time interval T, but relatively high in all the other time intervals. As a result, the transmission reliability in OpenWSN simulations tends to be overestimated since the error frame burst lengths generated by the independent error model are very likely to be less than four.

For illustration purposes, assume that a transmitting node sends one data frame to a receiving node every minute for a total of 10 min, and in one of these minutes, the frame cannot be successfully delivered due to physical obstacles in the transmission path or the presence of interference, for example. Assume further that in the other 9 min, the data frames (9 in total) are successfully delivered without the need for retransmission. In other words, the overall average FDR is $9/(4 + 9) = 0.692$. In this example, the transmitting node transmits a total of 10 data frames within 10 min, of which 1 data frame is not delivered due to poor wireless link conditions, but the other 9 data frames are successfully delivered. That is, the transmission reliability is equal to 0.9. By contrast, since the average FDR is 0.692, Equation (6) gives the transmission reliability as $1 - (1 - 0.692)^4 = 0.991$. In other words, the transmission reliability evaluated by the independent error model (0.991) overestimates the actual transmission reliability (0.9).

6. Simulation Results and Discussion

Simulations were performed to compare the performance of the second-order Markov error model proposed in the present study with the original independent error model in the OpenWSN simulator. The simulation environment consisted of two nodes, namely Node 1 serving as the gateway and Node 2 serving as the sensing device. The sensing device generated a 17-byte application-layer protocol data unit (APDU) every 3 s, where four of these bytes were serial numbers starting from the value of 1. In each experiment, Node 2 generated a total of 2000 distinct APDUs and sent them to Port Number 15001 of the gateway. Each time the gateway received a UDP packet with a destination port number of 15001, it added the packet to a text file for offline performance evaluation. The sensing node generated data for transmission only when it synchronized with at least one node in the OpenWSN network. It is noted that this approach is consistent with the general design of IWSN applications that repeatedly report the readings from a sensor. Since there is no datum generated during the period for which the sensing node is desynchronized from the network, the construction and maintenance of the IWSN has a trivial impact on the transmission reliability.

When the data frames carrying the above-mentioned APDU (or the corresponding ACK frames) were lost, the retransmission mechanism was automatically invoked. During retransmissions, the newly generated data frames cannot be sent immediately. In such a case, a buffer may increase the opportunities that the newly generated data frames be delivered later. However, in lossy environments, the buffer tends to be full, and hence the most-recently sensed data are simply dropped. Practical IWSNs generally prefer more recently-sensed data to older data since it is precisely the latest information on the plant state which is the most critical importance for most control applications [32]. However, the sensing nodes in IWSNs are usually resource-constrained and therefore cannot implement a buffering mechanism. As a result, the applications installed in IWSNs rarely allocate buffers for sensing data. In accordance with these observations, the testing application used in the present simulations also did not implement a buffering mechanism. That is, if the sensing node was still busy sending a previous data frame, the currently generated data frame was simply dropped. As a consequence, the error frame bursts not only affected the transmission reliability directly, but also impacted the transmission reliability indirectly through the frame drops produced during data retransmission.

Prior to performing the simulations, the trace acquired in the sixth hour of the experimental period was inspected to identify the 4 min segment having the poorest transmission quality. It was found that the records within the time interval from the 51st minute to the 54th minute had the lowest FDR (0.753). The transition probabilities $p000$, $p010$, $p100$ and $p110$ of the second-order Markov model trained using this segment were 0.860, 0.595, 0.746 and 0.379, respectively. In the subsequent simulation performed using the original independent error model in OpenWSN, Node 1 (the gateway) received a total of 1950 distinct APDUs. In other words, the transmission reliability was equal to 0.975 (i.e., $1950/2000 = 0.975$).

By contrast, in the simulation performed using the proposed second-order Markov error model, the gateway received a total of 1825 distinct APDUs. In other words, the transmission reliability was around 0.913. The level of overestimation of the transmission reliability was therefore reduced by $(0.975 - 0.913) \div 0.913 = 0.0679$, i.e., by around 6.8% in the considered scenario.

It is reasonable to assume that there exist industrial environments in which the transmission quality is poorer than that in the factory considered in the present study. For such environments, it is further reasonable to characterize the transmission reliability using the proposed second-order Markov model with transition probabilities $p000$, $p010$, $p100$ and $p110$, specified as fractions of the transition probabilities $p000$, $p010$, $p100$ and $p110$ trained by the records within the time interval extending from the 51st to 54th minute of the sixth hour in the measured experimental trace. Thus, four further simulations were performed with fractions of 0.9, 0.8, 0.7 and 0.6, respectively. The corresponding results are shown in Table 1, where the data in the first row correspond to the experiment described above, while the data in the second to fifth rows correspond to the simulations performed with fractions ranging from 0.9 to 0.6, respectively. For each simulation, the transmission reliability is denoted as *TR*. Furthermore, columns *Our TR* and *Original TR* show the transmission reliabilities derived by the simulator with the proposed second-order Markov model and the independent model, respectively. Finally, the column *Equation (6) TR* shows the transmission reliability computed using Equation (6).

Table 1. Accuracy Improvements.

Fraction	*p000*	*p010*	*p100*	*p110*	*Our TR* [1]	*FDR*	*Original TR*	*Equation (6) TR*	*AI* [2]
1	0.860	0.595	0.746	0.379	0.913	0.753	0.975	0.996	0.068
0.9	0.774	0.536	0.671	0.341	0.820	0.618	0.892	0.979	0.088
0.8	0.688	0.476	0.596	0.303	0.716	0.515	0.804	0.945	0.123
0.7	0.602	0.417	0.522	0.265	0.606	0.415	0.667	0.883	0.100
0.6	0.516	0.357	0.447	0.227	0.494	0.327	0.546	0.795	0.105

[1] TR denotes Transmission Reliability and [2] AI denotes Accuracy Improvement.

It could be seen that the differences between the reliability results obtained from the independent error model and those derived from Equation (6) increase with a decreasing fraction. To investigate this phenomenon, we devised a particular function in the simulator for recording the frame drops produced during retransmission, and once again performed a simulation with FDR being 0.327, which is the same as the FDR in the fifth row of Table 1. The experimental results showed that 887 data frames were not received, and, among them, 606 data frames were dropped by the sender during data retransmission. In other words, only 1394 (i.e., $2000 - 606 = 1394$) data frames are processed by the independent error model. Excluding the dropped frames, the transmission reliability is $(2000 - 887) \div 1394 = 0.798$, which is very close to the derived transmission reliability (i.e., 0.795) using Equation (6). Note that Equation (6) does not take into account the frame drops produced during data retransmission, which is the main reason for the inconsistency between the data in columns *Original TR* and *Equation (6) TR*.

In each simulation, a record was made of the total number of frames successfully received and the total number of transmissions. The corresponding FDR was then derived as the ratio of the former to the latter. Let the data in columns *Our TR* and *Original TR* be denoted as x and y, respectively. The accuracy improvement (AI) of the second-order Markov model over the original independent model can then be evaluated as:

$$AI = \frac{y - x}{x}. \tag{7}$$

It is seen in Table 1 that the accuracy improvement obtained by the second-order Markov model increases with an increasing burst rate. Overall, the results show that the proposed model improves the accuracy of the transmission reliability estimates by 0.123 (see the third row in Table 1), for which the FDR is equal to approximately 0.5.

7. Conclusions

This paper has proposed a second-order Markov model for estimating the frame-level transmission errors in an industrial WSN based on the experimental traces obtained in a real-world factory environment over a ~24 h period. A statistical analysis has shown that the proposed model provides a better description of the transmission quality in industrial WSNs than conventional methods. In addition, the simulation results have shown that the proposed error model improves the accuracy of the estimated transmission reliability in the OpenWSN simulator by up to 12% compared to that achieved using the original independent error model. The performance advantage of the proposed model is particularly apparent when the transmission failure events are non-uniformly dispersed in a certain time interval.

Author Contributions: Conceptualization, Y.-S.Y. and Y.-S.C.; methodology, Y.-S.Y.; software, Y.-S.Y.; validation, Y.-S.Y. and Y.-S.C.; formal analysis, Y.-S.Y.; investigation, Y.-S.Y.; resources, Y.-S.Y.; data curation, Y.-S.Y. and Y.-S.C.; writing—original draft preparation, Y.-S.Y. and Y.-S.C.; writing—review and editing, Y.-S.Y. and Y.-S.C.; visualization, Y.-S.Y.; supervision, Y.-S.Y. and Y.-S.C.; project administration, Y.-S.Y.; funding acquisition, Y.-S.Y. and Y.-S.C. All authors have read and agreed to the published version of the manuscript.

Funding: This work was partially supported by the Ministry of Science and Technology, Taiwan, under grant No. MOST 108-2221-E-150-008. Also, the APC was funded by the Ministry of Science and Technology, Taiwan, under the same grant No.

Conflicts of Interest: The authors declare no conflict of interest.

References

1. IEC 62591:2016. Industrial Networks—Wireless Communication Network and Communication Profiles—WirelessHART™. Available online: https://webstore.iec.ch/publication/24433 (accessed on 15 July 2020).

2. IEC 62734:2014/AMD1:2019. Industrial Network—Wireless Communication Network and Communication Profiles—ISA 100.11a. Available online: https://standards.iteh.ai/catalog/standards/iec/2d7dcecd-4731-4c9a-92ad-75e70f8f3318/iec-62734-2014-amd1-2019 (accessed on 15 July 2020).

3. *802.15.4-2006—IEEE Standard for Information Technology—Local and Metropolitan Area Networks—Specific Requirements—Part 15.4: Wireless Medium Access Control (MAC) and Physical Layer (PHY) Specifications for Low Rate Wireless Personal Area Networks (WPANs)*; IEEE: Piscataway, NJ, USA, 2006. [CrossRef]

4. Watteyne, T.; Vilajosana, X.; Kerkez, B.; Chraim, F.; Weekly, K.; Wang, Q.; Glaser, S.; Pister, K. OpenWSN: A standards-based low-power wireless development environment. *Trans. Emerg. Telecommun. Technol.* **2012**, *23*, 480–493. [CrossRef]

5. *802.15.4e-2012—IEEE Standard for Local and Metropolitan Area Networks—Part 15.4: Amendment 1: MAC Sublayer*; IEEE: Piscataway, NJ, USA, 2012. [CrossRef]

6. OpenSim. Available online: https://openwsn.atlassian.net/wiki/spaces/OW/pages/13434892/OpenSim (accessed on 20 May 2020).

7. Nobre, M.; Silva, I.; Guedes, L.A.; Portugal, P. Towards a WirelessHART module for the ns-3 simulator. In Proceedings of the 2010 IEEE 15th Conference on Emerging Technologies Factory Automation (ETFA 2010), Bilbao, Spain, 13–16 September 2010; pp. 1–4.

8. Nobre, M.; Silva, I.; Guedes, L.A. Reliability evaluation of wirelesshart under faulty link scenarios. In Proceedings of the 2014 12th IEEE International Conference on Industrial Informatics (INDIN), Porto Alegre, Brazil, 27–30 July 2014; pp. 676–682.

9. Nobre, M.; Silva, I.; Guedes, L.A. Performance evaluation of WirelessHART networks using a new network simulator 3 module. *Comput. Electr. Eng.* **2015**, *41*, 325–341. [CrossRef]

10. Nsnam ns-3. Available online: https://www.nsnam.org/ (accessed on 20 May 2020).

11. Remke, A.; Wu, X. WirelessHART modeling and performance evaluation. In Proceedings of the 2013 43rd Annual IEEE/IFIP International Conference on Dependable Systems and Networks (DSN), Budapest, Hungary, 24–27 June 2013; pp. 1–12.

12. Shannon, C.E. A mathematical theory of communication. *SIGMOBILE Mob. Comput. Commun. Rev.* **2001**, *5*, 3–55. [CrossRef]

13. Barac, F.; Gidlund, M.; Zhang, T. Scrutinizing Bit-and Symbol-Errors of IEEE 802.15.4 Communication in Industrial Environments. *IEEE Trans. Instrum. Meas.* **2014**, *63*, 1783–1794. [CrossRef]

14. Gao, J.; Hu, J.; Min, G.; Xu, L. QoS Performance Analysis of IEEE 802.15.4 MAC in LR-WPAN with Bursty Error Channels. In Proceedings of the 2009 Fifth International Conference on Mobile Ad-hoc and Sensor Networks, Wuyishan, China, 14–16 December 2009; pp. 252–256.

15. Petrova, M.; Riihijarvi, J.; Mahonen, P.; Labella, S. Performance study of IEEE 802.15.4 using measurements and simulations. In Proceedings of the IEEE Wireless Communications and Networking Conference, WCNC 2006, Las Vegas, NV, USA, 3–6 April 2006; Volume 1, pp. 487–492.

16. Wijetunge, S.; Gunawardana, U.; Liyanapathirana, R. Performance analysis of IEEE 802.15.4 MAC protocol for WSNs in burst error channels. In Proceedings of the 2011 11th International Symposium on Communications Information Technologies (ISCIT), Hangzhou, China, 12–14 October 2011; pp. 286–291.

17. Iqbal, A.; Khayam, S.A. Improving WSN Simulation and Analysis Accuracy Using Two-Tier Channel Models. In Proceedings of the 2008 IEEE International Conference on Communications, Beijing, China, 19–23 May 2008; pp. 349–353.

18. Ilyas, M.U.; Radha, H. Measurement Based Analysis and Modeling of the Error Process in IEEE 802.15.4 LR-WPANs. In Proceedings of the IEEE INFOCOM 2008—The 27th Conference on Computer Communications, Phoenix, AZ, USA, 13–18 April 2008; pp. 1274–1282.

19. Striccoli, D.; Boggia, G.; Grieco, L.A. A Markov Model for Characterizing IEEE 802.15.4 MAC Layer in Noisy Environments. *IEEE Trans. Ind. Electron.* **2015**, *62*, 5133–5142. [CrossRef]

20. Guntupalli, L.; Martinez-Bauset, J.; Li, F.Y.; Weitnauer, M.A. Aggregated Packet Transmission in Duty-Cycled WSNs: Modeling and Performance Evaluation. *IEEE Trans. Veh. Technol.* **2017**, *66*, 563–579. [CrossRef]

21. Guntupalli, L.; Martinez-Bauset, J.; Li, F.Y. Performance of frame transmissions and event-triggered sleeping in duty-cycled WSNs with error-prone wireless links. *Comput. Netw.* **2018**, *134*, 215–227. [CrossRef]

22. Boggia, G.; Camarda, P.; D'Alconzo, A. Performance of Markov models for frame-level errors in IEEE 802.11 wireless LANs. *Int. J. Commun. Syst.* **2009**, *22*, 695–718. [CrossRef]

23. Valle, O.T.; Montez, C.; Araujo, G.M.; Moraes, R.; Vasques, F. A WSN data retransmission mechanism based on network coding and cooperative relayers. In Proceedings of the 2015 IEEE World Conference on Factory Communication Systems (WFCS), Palma de Mallorca, Spain, 27–29 May 2015; pp. 1–4.

24. Valle, O.T.; Montez, C.; Araujo, G.M.; Vasques, F.; Moraes, R. NetCoDer: A Retransmission Mechanism for WSNs Based on Cooperative Relays and Network Coding. *Sensors* **2016**, *16*, 799. [CrossRef] [PubMed]

25. OMNeT++ Discrete Event Simulator. Available online: https://omnetpp.org/ (accessed on 20 May 2020).

26. Willig, A. Antenna redundancy for increasing transmission reliability in wireless industrial LANs. In Proceedings of the EFTA 2003. 2003 IEEE Conference on Emerging Technologies and Factory Automation. Proceedings (Cat. No.03TH8696), Lisbon, Portugal, 16–19 September 2003; Volume 1, pp. 7–14.

27. CC2538DK CC2538 Development Kit. TI.com. Available online: http://www.ti.com/tool/CC2538DK (accessed on 20 May 2020).

28. Contiki, O.S. Contiki 2.7. Available online: https://github.com/contiki-os/contiki/tree/release-2-7 (accessed on 20 May 2020).

29. BeagleBoard.org-black. Available online: https://beagleboard.org/black (accessed on 20 May 2020).

30. Wireshark: Go Deep. Available online: https://www.wireshark.org/ (accessed on 18 June 2020).

31. Guidara, A.; Fersi, G.; Derbel, F.; Jemaa, M.B. Impacts of temperature and humidity variations on RSSI in indoor wireless sensor networks. *Procedia Comput. Sci.* **2018**, *126*, 1072–1081. [CrossRef]

32. Sadi, Y.; Ergen, S.C. Joint optimization of wireless network energy consumption and control system performance in wireless networked control systems. *IEEE Trans. Wirel. Commun.* **2017**, *16*, 2235–2248. [CrossRef]

© 2020 by the authors. Licensee MDPI, Basel, Switzerland. This article is an open access article distributed under the terms and conditions of the Creative Commons Attribution (CC BY) license (http://creativecommons.org/licenses/by/4.0/).

Article

Hierarchical Anomaly Detection Model for In-Vehicle Networks Using Machine Learning Algorithms

Seunghyun Park and Jin-Young Choi *

School of Cybersecurity, Korea University, Seoul 02841, Korea; cloakingmode@korea.edu
* Correspondence: choi@formal.korea.ac.kr

Received: 13 June 2020; Accepted: 13 July 2020; Published: 15 July 2020

Abstract: The communication and connectivity functions of vehicles increase their vulnerability to hackers. The unintended failure and malfunction of in-vehicle systems caused by external factors threaten the security and safety of passengers. As the controller area network alone cannot protect vehicles from external attacks, techniques to analyze and detect external attacks are required. Therefore, we propose a multi-labeled hierarchical classification (MLHC) intrusion detection model that analyzes and detects external attacks caused by message injection. This model quickly determines the occurrence of attacks and classifies the attack using only existing classified attack data. We evaluated the performance of the model by analyzing its learning space. We further verified the model by comparing its accuracy, F1 score and data learning and evaluation times with the two layers multi-class detection (TLMD) and single-layer multi-class classification (SLMC) models. The simulation results show that the MLHC model has the highest F1 score of 0.9995 and is 87.30% and 99.92% faster than the SLMC and TLMD models in terms of detection time, respectively. Consequently, the proposed model can classify both the type and existence or absence of attacks with high accuracy and can be used in interior communication environments of high-speed vehicles with a high throughput.

Keywords: controller area network; intrusion detection system; in-vehicle network security; machine learning; hierarchical approach; anomaly detection; MLHC

---

1. Introduction

High connectivity and automotive electronics are two major developments in modern vehicles, which are evolving to provide various convenience features to drivers. Vehicle connectivity using smart devices and cellular network has enabled the consumption of various contents in the vehicle through an infotainment platform. Particularly, vehicle-to-vehicle communication has enabled the sharing of driving information and dangerous situations on the road. Likewise, vehicle-to-infrastructure communication has broadened the prospects of autonomous vehicles, which have depended on existing sensors only, through the exchange of traffic signals and flows. Furthermore, vehicles are evolving to giant smart devices by being equipped with safety devices, such as forward collision-avoidance and lane-keeping assists, as well as convenience devices, such as telematics and power supply electric devices.

However, such diverse connectivity of vehicles increases their points of attack and exposure to external attacks. As the current controller area network (CAN) message frame lacks authentication or access control mechanisms, in-vehicle data transfer is performed without the use of security techniques. Furthermore, as the in-vehicle controllers are interconnected, the complexity of the architecture increases. The interferences or mutual effects between controllers may cause unintended motions or failures, thus posing further threats to the cybersecurity of vehicles or the safety of passengers.

Existing connected vehicles attain security by configuring a separate dedicated network for in-vehicle Internet services, such as telematics, and separating the connectivity services of the vehicle from the Internet. However, the dedicated network is costly to construct and operate, and it has limitations in opening the platform to expand connectivity-related services. Hence, a more fundamental solution to protect the devices without depending on the traditional communication network security is now required because dedicated Internet services and local area network system have been combined.

To design the cybersecurity of a mission-critical environment, such as vehicles, the characteristics of the external network environment, such as vehicle domain and machine-to-machine (M2M) communication, should be considered. Particularly, intrusion detection or prevention systems of in-vehicle network protection require high accuracy. If important messages in the vehicle are mistaken for an attack and blocked, the vehicle may malfunction and develop safety problems. Therefore, false alarms must be prevented in the intrusion prevention of in-vehicle networks.

Additionally, real-time response is critical for the cybersecurity of vehicles. Malicious attacks on moving vehicles are directly linked to the safety of passengers, pedestrians and other vehicles. Therefore, when external attack messages are identified, the vehicle must be able to implement response measures in real time. However, due to the nature of embedded environments, such as vehicles, there are constraints in temporal and spatial resources. As the available resources for learning and classifying intrusion data are limited, a real-time intrusion detection system (IDS) having high accuracy should be constructed, and it should be able to function with the minimum available computing power of the vehicle.

In 2015, a Jeep Cherokee was remotely hacked and reported to raise awareness of the cybersecurity of vehicles [?]. In a recent article [?], the author suggested that we should not only depend on defending against attacks because it is impossible to produce vehicles with perfect security system to disable hacking, but we should also design the security system to detect attacks and respond appropriately.

Therefore, in this study, we developed a model for detecting anomalous behaviors and attacks caused by message injection on vehicles in real time with high accuracy. We applied a hierarchical data analysis technique for detecting and classifying attack data. Furthermore, to train the intrusion detection model, we minimized misdetections and no-detections using a machine learning algorithm. An appropriate algorithm for the dataset was selected to detect the attack data, and a simulation environment was set up to derive the optimal hyperparameters. Particularly, we propose a method to quickly detect the existence or absence of attacks hierarchically by learning the behaviors of the CAN data. The accuracy of the model was increased to make it applicable to an actual vehicle environment, and a model with real-time responsiveness and using limited resources was implemented. Accuracy, F1 score and detection time were applied as valid metrics to evaluate the proposed model. Using these metrics, we obtained an improved model to detect attacks and anomaly behaviors that flowed into vehicles. The contributions of this study are as follows.

- This is the first study that presents a hierarchical data analysis model for simultaneously classifying the presence or absence of an attack, an attack type and a vehicle type to detect anomaly behaviors in vehicles.
- We present a detection model that includes hyperparameters and an optimal classification algorithm for detection.

The rest of this paper is organized as follows. Section **??** introduces existing related studies. Section **??** details the CAN message frame and topology for an understanding of vehicle cybersecurity. Section **??** describes the dataset we used, as well as the concrete data analysis method and analysis model proposed in this paper. This includes the algorithm for vehicle data analysis, performance measurement metrics and hypothesis space comparison of models for in-vehicle data analysis. Section **??** interprets the simulation results and verifies the effectiveness of the proposed method by comparing it with existing results. In Section **??**, we present the conclusion and future research direction.

2. Related Work

This section highlights existing works related to this study. The problems in each domain, existing methods to solve them, advantages and disadvantages of the solutions and constraints are stated.

Song et al. [?] proposed an intrusion detection model that learns the sequential pattern of in-vehicle network traffic and detects message insertion attacks according to traffic changes. The structure of the inception-ResNet model designed for large-scale images was used, and the deep convolutional neural network was redesigned by reducing the architecture complexity. Particularly, the authors experimented with a dataset extracted from actual vehicle environment and suggested that detecting complex, irregular random attacks has an advantage. The experiment compared long short-term memory (LSTM), artificial neural network, support vector machine, k-nearest neighbors (kNN) [?], naïve Bayes (NB) and decision tree (DT) [?] algorithms. Zhang et al. [?] proposed a vehicle intrusion detection model based on the neural network algorithm. They compared detection performances using gradient descent with momentum and adaptive gain, and they performed verification and evaluation by applying data collected from actual vehicles. Further, the authors proposed a host-type intrusion detection model for in-vehicle intrusion detection. However, host-type IDS may be inefficient in a broadcast-type communication environment, such as CAN. This architecture is impractical in an embedded environment using limited resources as duplicate detections are performed because every controller receives the same message, and each controller must secure separate resources for intrusion detection. Kang et al. [?] proposed a deep neural network (DNN)-based IDS to monitor the CAN message frame. The DNN model was pre-trained using a deep-belief network. The authors used probability-based feature vectors extracted from packets in learning and training to classify messages as normal or attack. The experiment demonstrated that an accurate detection ratio of approximately 0.98 can be provided in real-time response.

Hoppe et al. [?] placed an anomaly-based IDS in the CAN bus to monitor network traffic. The IDS detects randomly manipulated messages by comparing them with normal patterns. Four attack scenarios related to the CAN bus were presented and classified using the established computer emergency response team taxonomy. It includes technical and managerial considerations to protect the in-vehicle network in comparison with the traditional information technology system, and the countermeasures are discussed by analyzing security vulnerability and potential safety implications. Taylor et al. [?] suggested an anomaly detection method based on the LSTM neural network to detect attacks on the CAN bus. The authors analyzed data by manipulating the identifiers (IDs) of the message frame in a dataset extracted from vehicles rather than infusing attack traffic into the in-vehicle network. By assuming that the CAN traffic was regular, they detected traffic outside the normal sequence in five dataset manipulation scenarios. The result of detecting the known attacks of the CAN bus showed potential for development and provided follow-up tasks to improve the experimental method and detection model. Wang et al. [?] proposed a distributed anomaly detection framework using hierarchical temporal memory (HMM) to strengthen the security of the in-vehicle CAN bus. This method evaluates the output using an abnormal score mechanism that learns the prior state of the CAN network and predicts the flow data. The authors extracted CAN traffic and modified the data fields manually. In addition, they created attack data by replaying the captured traffic on the dataset. They claimed that the area under the curve score was higher than those of the recurrent neural network and HMM, but a method of efficiently detecting attacks where multiple IDs interact without relying on a single message ID should also be considered. Furthermore, experiments are required on indices related to time or resource utilization to examine the applicability of the proposed model to an actual vehicle environment.

The common limitation of the studies mentioned above is that the existing models only determine whether the attack, which is injected in the in-vehicle network, has occurred. In an actual vehicle environment, merely distinguishing between an attack and benign status is insufficient. It is highly important to provide additional information for immediately determining the target affected by the type of attack. It may be easy to inject the attack data in a network and track the sign of occurrence.

However, a large amount of computation, which is proportional to the number of target labels, is required to extensively determine the semantics of the attack injected into the vehicle. To address these limitations and satisfy the requirements of an IDS in an actual vehicle environment, we propose a learning model that can not only determine whether an attack occurred, but also classify the attack type and target vehicle.

3. In-Vehicle Network Security

To define the proposed multi-labeled hierarchical classification (MLHC) model, this section describes the vehicle CAN message frame, CAN bus structure and attack vector for the vehicle.

3.1. Controller Area Network Message Frame and Topology

The CAN is the most representative in-vehicle network technology developed by Robert Bosch GmbH [?] in the early 1980s. Its specifications are still being expanded as a major protocol was used in On-Board Diagnostics II standard. The International Organization for Standardization (ISO) standardized the CAN by ISO 11898 [?] and is still expanding it. This standard was designed to enable communication between in-vehicle microcontrollers and devices and is used for information exchange between electronic control units (ECUs). The CAN device transfers data in packets in message frame units on the CAN network. The message frame does not contain the source or target addresses but only the IDs related to priorities. The real-time priority-based message transfer system follows IDs composed of an 11- or 29-bit string, and a lower ID has a higher priority. First, whether the CAN bus is in use is determined before sending a message to the CAN node, and then collision between messages is detected. When two nodes send a message simultaneously, the message with a higher priority is first sent, and then the message with a lower priority is delayed.

The CAN message frame is divided into base and extended formats depending on the length of the arbitration field, as shown in Figure **??**. The base format supports the CAN 2.0A protocol, whereas the extended format supports the CAN 2.0B protocol, and it also accepts the CAN 2.0A protocol. We describe the fields used in the present paper, and the abbreviations for the remaining fields are presented in the Abbreviation Section.

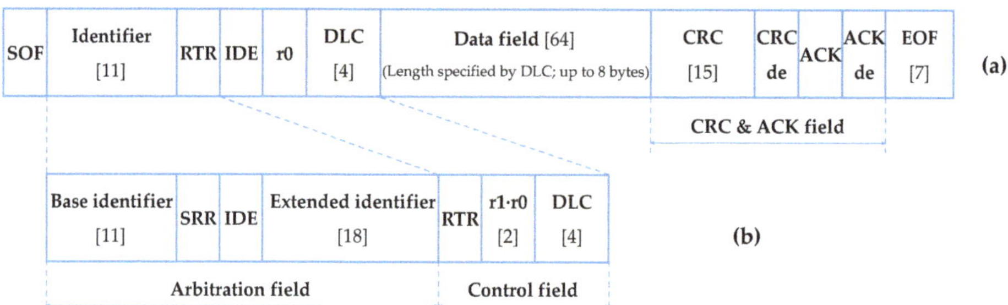

Figure 1. Controller area network (CAN) message structures: (a) base format; and (b) extended format.

- Base identifier (11 bits): This is the first part of the identifier that indicates the priority of message frames and commonly exists in the standard and extended frames.
- Data length code (DLC, 4 bits): DLC expresses the byte length of the data field in the message frame.
- Data field (64 bits): This is a payload for loading actual data to be sent from one node to the other; a maximum of 8 bytes can be used.

The ECU is a component of the in-vehicle network. It is an embedded device that controls other in-vehicle controllers or devices. The ECU contains input and output interfaces for interconnecting the microcontroller unit, memories (such as read-only and random-access memories), sensors and actuators. The ECU collects and analyzes data from sensors, and it generates control signals and sends them to actuators.

Figure **??** illustrates the CAN topology composed of the in-vehicle network and controllers. The ECUs are grouped as the domain controller for logically distinguishing vehicle functions by use, and the CAN bus enables mutual cooperation or control between the ECUs by interconnecting them. Vehicle ethernet may be used for interconnecting controllers that require high-speed communication, and the media-oriented systems transport network is often used for multimedia communication. A gateway may be installed to control diagnostic communication or external interfaces and installing an IDS function for monitoring the CAN traffic inside this gateway may be effective. As shown in Figure **??**, external attacks may be injected through a diagnostic bus connected to the CAN bus or an external interface, and this can aid hacking by dominating the CAN bus or ECU.

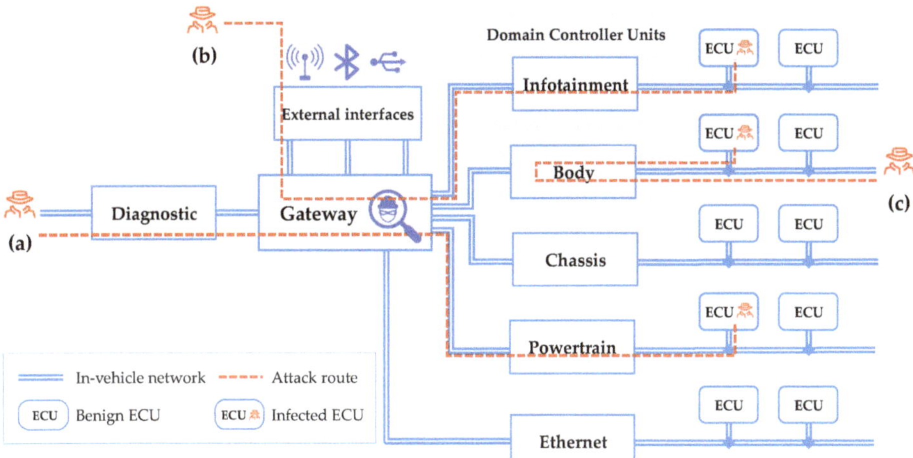

Figure 2. CAN topology and attack vectors: **(a)** external interfaces; **(b)** diagnostic bus; and **(c)** occupation of CAN bus.

3.2. Attack Vectors on In-Vehicle Network

Attack vectors of confidentiality, integrity and availability aspects need to be considered for defense against vehicle cyberattacks. Attackers can seize the rights for a vehicle or the systems connected to a vehicle and randomly tap major traffic in the vehicle or peek into sensitive information, such as the location of the vehicle. They can also attempt to launch a denial-of-service attack to manipulate the ECU software by reprogramming it. Additionally, they can generate large-scale traffic inside the vehicle to disable normal messages. By entering the in-vehicle network and injecting random messages, hackers can threaten the confidentiality, integrity and availability of the vehicle. Threats of compromising the security objectives of in-vehicle systems are outlined in Table **??**.

Table 1. Summary of security objectives and corresponding threats on in-vehicle network.

Security Objectives	Threats	Related Work	Attack Vectors
Availability	Damage of the internal systems by denial-of-service attack (flooding)	[?]	CAN bus, gateway, external interface
	Interference with short-range communication or sensor recognition	[? ? ?]	External interface, sensor
	Unintended service interruption (fuzzing)	[?]	CAN bus, ECU
	Blockage of normal message flow	[?]	CAN bus, gateway
Confidentiality	Illegal upgrade or acquisition of rights	[? ?]	ECU, memory
	Access to unauthorized information	[?]	External interface, ECU, memory
	Information leakage by damaged applications (malfunction)	[?]	ECU
	Acquisition of the encryption key by sniffing	[?]	External interface
Integrity	Forging and falsification of control messages	[?]	CAN bus, ECU
	Injection of malicious messages and forced operation of the controller (fuzzing)	[?]	CAN bus, gateway, ECU
	Manipulation of the firmware and update with a tampered firmware	[? ?]	ECU
	Installation of backdoor	[?]	ECU

A monumental event in vehicle cybersecurity occurred in 2015 when Miller and Valasek [?] hacked Jeep Cherokee and opened it to the media and at a hacking conference. They demonstrated a hacking attack targeted at a real moving vehicle by using the vulnerabilities of the cellular network and external interface of the connected service. They accessed the CAN bus through the head unit of a remote vehicle and successfully updated a tampered firmware by acquiring the rights of the controller. After acquiring the control rights of the vehicle, they could remotely operate not only the audio and wiper of the moving vehicle, but also the brakes and steering wheel. Consequently, Fiat Chrysler Automobiles recalled 1.4 million vehicles that could be attacked and was fined $105 million. Furthermore, Tencent's Keen Security Lab [?] recently seized the rights of a Lexus NX300 using the vulnerability of the audio-video navigation system in the vehicle. They informed the manufacturer that they invaded the CAN bus and successfully injected a malicious message that can cause the vehicle to malfunction and warned of the vulnerability on their blog.

Various attack vectors that may damage the security objectives of vehicles in an in-vehicle network topology are shown in Figure ??. Various remote-connection external interfaces such as Wi-Fi hotspot and Bluetooth are used, as well as the Internet and cellular networks. It is also possible to form sessions with remote vehicles by scanning the M2M network of a specific communication service provider for connectivity services and searching the Internet protocol address and open service ports of the vehicle. In addition, the controller can be operated by force or reprogrammed using diagnostic communication that bypasses the authentication system of the gateway in an in-vehicle network. Once a specific controller is seized, it is possible to launch an attack to occupy the network and stop services by sending many CAN messages with manipulated priorities to the CAN bus.

4. Materials and Methods

4.1. Multi-Labeled Hierarchical Classification (MLHC) Process

The overall process of the proposed model is illustrated in Figure ??. The CAN traffic extracted from vehicles is preprocessed to enable the classifier to learn and evaluate it. The data analysis model uses a classification algorithm, preconfigured hyperparameters and performance evaluation metrics.

The analysis model is trained by injecting training data, and the performance of the trained model is evaluated using test data. The intrusion detection module, including the trained model in an actual application environment, is used to detect follow-up information, such as attack or benign, vehicle type and attack type, after receiving the CAN message frame as input.

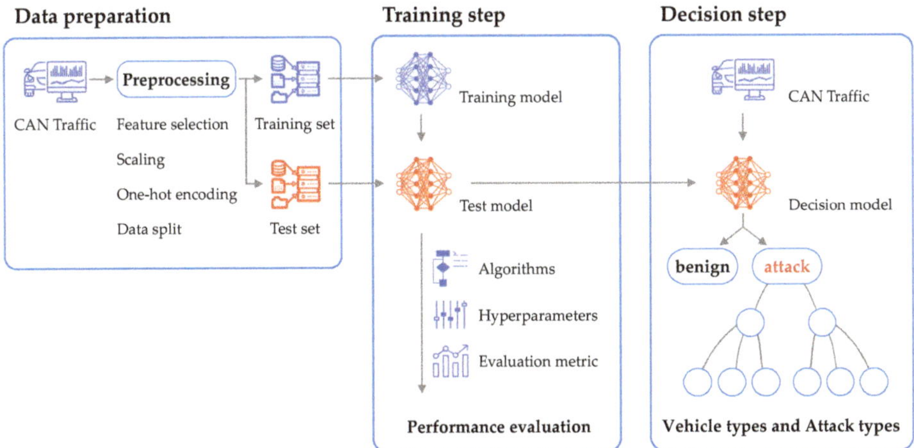

Figure 3. Overall multi-labeled hierarchical classification (MLHC) process.

4.1.1. Dataset

The scheme of the in-vehicle network intrusion detection challenge dataset released by Han et al. [?] included CAN ID, DLC and data payload, reflecting the CAN message structure; the timestamp when each data sample was recorded was added into this dataset. They also added a binary label to indicate whether it corresponds to an attack or benign status, whether the data sample is that of an attack or a normal state. We selected this dataset because it includes data extracted from an actual vehicle environment and allows a hierarchical structure of detailed data in the lower layers, such as attack type and vehicle type, for training the vehicle IDS model. The dataset comprises a total of 12 files, with three types of attack data and three vehicle types in normal and message-injected states. This dataset was constructed using data from vehicle models from three vehicle manufacturers. Furthermore, a group of vehicles using the same CAN database formed a vehicle type, and this depended on the vehicle manufacturer that designs the CAN databases. The distributions of the data in each data type are outlined in Table ??.

Table 2. Statistics of CAN intrusion dataset.

Classification			Total	Vehicle Types		
				Vehicle_A	Vehicle_B	Vehicle_C
Total			1,735,840	402,956	535,041	797,843
		Benign	1,552,526	366,510	468,527	717,489
Attack types	Attack	Flooding	88,150	22,587	32,422	33,141
		Fuzzing	63,742	5812	18,118	39,812
		Malfunction	31,422	8047	15,974	7401

The message injection into the in-vehicle network was attempted in three attack types as follows. For the flooding attack, several messages were injected with a high-priority CAN ID to induce service delay. For the fuzzing attack, random CAN IDs were injected in brute force until the pre-defined valid CAN ID in the vehicle reacted. For the malfunction attack, valid CAN IDs for each vehicle type were collected in advance, random data fields were configured using the IDs and tampered values were injected. The dataset can be expanded without limitation when additional information is required, such as attack type and vehicle type.

4.1.2. Data Preprocessing

For the classifier to learn the CAN traffic for data analysis, the data preprocessing step illustrated in Figure **??** is required. The CAN IDS dataset used in this model consists of 12 files, which are separated by vehicle type and attack type, and only attack or benign is expressed by binary classification. However, as the vehicle type or attack type is not classified in advance in an actual environment, the intrusion detection module should be able to detect anomalies, even in an environment of random combinations of vehicle types or attack types. Therefore, in this model, to enable the classification of vehicle type and attack type from the incoming data, each unit dataset was integrated into one data frame as shown in Equation (**??**):

$$S = \sum_{v_{type}} \sum_{a_{type}} S_{v_{type}, a_{type}} \tag{1}$$

where S is the total dataset required for data analysis, v_{type} is the vehicle type and a_{type} is the attack type. The unit dataset $S_{v_{type}, a_{type}}$ is subdivided by attack type and vehicle type, and the existing binary codes are encoded in multiple sub-labels to express additional information, such as vehicle type or attack type.

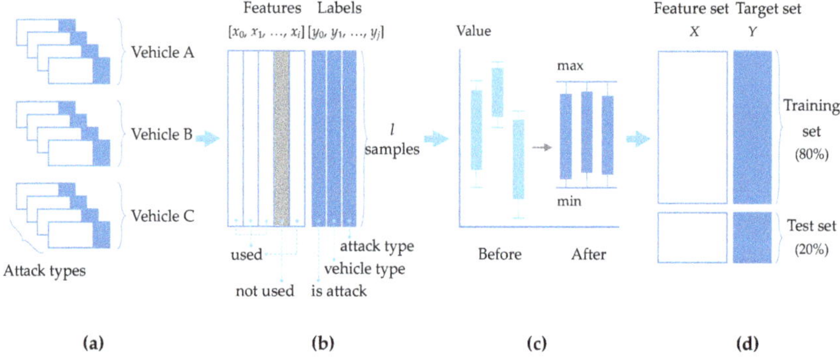

Figure 4. MLHC data preprocessing: (**a**) initial dataset; (**b**) merging and feature selection; (**c**) scaling; and (**d**) data split.

The features of this dataset include timestamp, time interval, CAN ID, DLC and eight data bytecodes for payload. The feature set of the input data is extracted using the improved feature selection (IFS) method proposed by Park and Choi [?]. This method uses correlations and cross-entropy between the features to combine the high values derived from correlation and information gain. It finds both greedy features as well as the ones with the highest correlation. These two vectors are combined to determine the final features from the dataset that are highly correlated and have a strong impact on the classes. Consequently, timestamp is excluded from the original feature set, and the selected features are as follows: time interval, CAN ID, DLC and data payload. Particularly, the data payload is composed of 64-bit strings at the maximum and can be converted to a byte code string of a length specified by the DLC field. Normalization is applied to prevent underflow or overflow that may occur in the learning process and to evenly distribute the impact on each data string of the payload. The eight independent byte strings having the same values of sections from 0 to 255 are converted to eight floating point variables having a value between 0 and 1 using the min-max normalizer with minimum and maximum values as follows:

$$x_i' = \frac{x_i - \min(x_i)}{\max(x_i) - \min(x_i)} = \frac{x_i}{2^8 - 1} \tag{2}$$

where x_i' is a normalized value and x_i is an original vector of feature i.

The dataset S used as input contains a feature set X and target set Y. This is split into training, validation and target sets, which are used for learning. For the feature and target sets, S is divided into columns, whereas for the training, validation and test sets, S is divided into rows. $x_i^{(l)}$ and $y_j^{(l)}$ denote data elements at feature i and labels in the classification group j for sample l, respectively. In this study, the training and test sets were divided at the ratio 8:2. The model was trained using 80% of the total data, and the performance of the final model was evaluated using the remaining 20% samples. The test set was separated to prevent overfitting and to accurately predict the model performance in a new actual data environment. Notably, the test set was used only for evaluating the model and not for learning. Instead, part of the training data was divided and used for verification to measure the model performance in the learning stage and to obtain hyperparameters yielding excellent performance. This process is illustrated in Figure **??**.

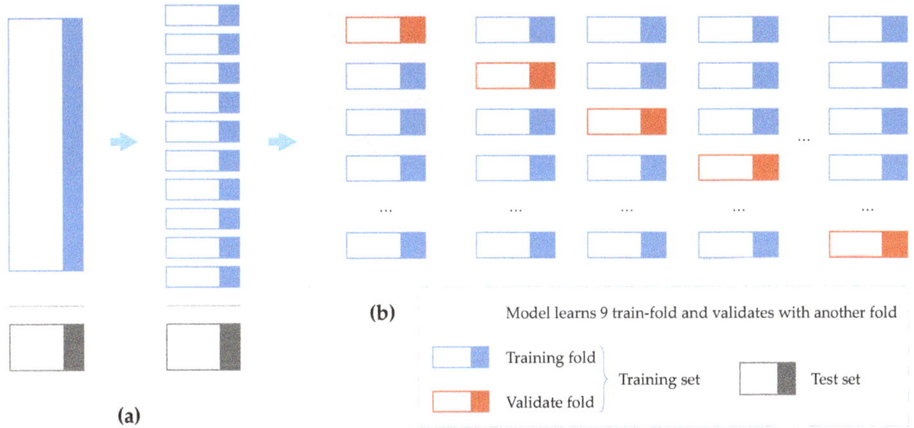

Figure 5. MLHC 10-fold cross validation: (**a**) dividing the training set into 10 folds; and (**b**) learning training-fold and validating with the other fold.

After dividing the training set into 10 folds, the model was trained with nine different folds, and the model performance was verified with the remaining fold. The learning was performed 10 times; nine folds were used for training, and the remaining one fold was used for validation.

Additional information must be present in the target data, for example, vehicle information and attack type, as well as the attack or benign of the CAN message. The label was excluded from the feature set for training because it was used to evaluate the learning result in supervised learning. Rather, the label was included in the target data and reorganized to express the additional information, such as vehicle information and attack type, as well as the attack or benign of the CAN message. To hierarchically classify data traffic as suggested in this study, the target data must also form a similar data structure. As shown in Figure **??**, the first row of the target data classifies attack or benign, and the lower rows include a hierarchical structure to distinguish the vehicle information or attack type only for attack data. Furthermore, the target data were designed to have a multi-labeled form so that the additional information can be included. Finally, the output data become a vector set including sub-vectors.

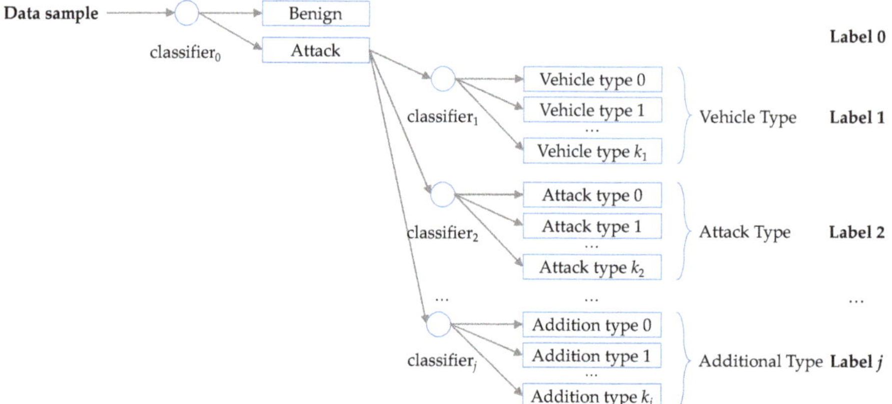

Figure 6. MLHC target labeling.

4.2. MLHC Model

The objective of this study was to effectively detect anomaly behaviors, such as message injection attack, in the CAN traffic of vehicles. To detect intrusion or anomaly behaviors external to the vehicle, an intrusion detection module is required in the CAN bus. Prior studies have detected anomaly behaviors by training normal CAN traffic and analyzing the time interval between messages, or by using machine learning algorithms. In this present study, we adopted a hierarchical approach using multi-label and multi-class classifiers. Hence, we propose a machine-learning-based multi-labeled method for detecting intrusions into the CAN and classifying attack techniques in a hierarchical manner. The multi-class classifier can identify more various categories of data with one classifier as compared to binary classification, and the multi-labeled classifier can contain various types of information simultaneously in a single classifier. This section explains the learning process and algorithm of the hierarchical intrusion detection method using the multi-labeled technique proposed in this study. This subsection describes the MLHC algorithm and compares the space of hypothesis and accuracy according to the classification model.

4.2.1. MLHC Algorithm

The MLHC algorithm and its deployment (see Algorithm **??**). The data preprocessing process described in Section **??** is described on Lines 1–4, and the model learning process is described on Lines 5–17.

In the preprocessing stage, we use the IFS method to select the features for the model (Line 1). Then, we normalize the features using min-max normalization, as described in Equation (**??**) (Line 2). The training and test sets are split (Line 3); the training set is divided into k folds using k-fold cross-validation (Line 4).

In the learning stage, the algorithm searches through the training data of each training dataset S_{train}, determines whether the data sample $x^{(l)}$ is benign or attack using the first classifier c_0 and records the result in $\hat{y}_0^{(l)}$ (Line 7). If the data sample indicates a benign state, it is not classified further, and the learning of the corresponding sample is terminated (Lines 8–9). Otherwise, $\hat{y}_j^{(l)}$ (Line 10), the result of additional classification using the sub-classifier c_j is obtained and stored in the detailed information vector $\hat{V}^{(l)}$ (Lines 12–13). $\hat{Y}$, which is returned as the result of the model, is composed of a set comprising $\hat{y}^{(l)}$ as its elements, as shown in Equation (**??**):

$$\hat{Y}^{(l)} = \left\{ \hat{y}^{(l)} \mid l \in \{0, 1, \ldots, n(S)\} \right\} \tag{3}$$

where l is an index of a sample of dataset S. Regarding dataset S, S_{train} is the training set and S_{test} is the test set. This is generally expressed as S. The result for each sample l can be expressed as a concatenation of $\hat{y}_0^{(l)}$ and $\hat{V}^{(l)}$ (Line 16), as expressed in Equation (??):

$$\hat{y}^{(l)} = \begin{bmatrix} \hat{y}_0^{(l)} & \hat{V}^{(l)} \end{bmatrix} \tag{4}$$

where $\hat{y}_0^{(l)}$ is a binary classification result to determine whether sample l is a benign or an attack case. $\hat{V}^{(l)}$ is a vector set that expresses additional information if $\hat{y}^{(l)}$ is an attack, and it can be expressed in detail as Equation (??):

$$\hat{V}^{(l)} = \begin{cases} \varnothing, & \text{for } \hat{y}_0^{(l)} \text{ is } benign \\ \begin{bmatrix} \hat{y}_1^{(l)} & \hat{y}_2^{(l)} & \cdots & \hat{y}_j^{(l)} \end{bmatrix}, & \text{otherwise} \end{cases} \tag{5}$$

where $\hat{V}^{(l)}$ is an empty matrix if $\hat{y}_0^{(l)}$ is a benign case. On the other hand, $\hat{V}^{(l)}$ has a matrix of elements $\hat{y}_1^{(l)}, \hat{y}_2^{(l)}, \ldots, \hat{y}_j^{(l)}$ representing additional learning results by each classifier if $\hat{y}_0^{(l)}$ is an attack.

Algorithm 1 Multi-labeled hierarchical anomaly detection.

Input: S is a universal dataset including a feature set X and a target set Y.

Output: $\hat{Y}$ is a set of learning results including $\hat{y}_0^{(l)}$ and $\hat{V}^{(l)}$ for all samples l.

        $\hat{y}_0^{(l)}$ is a result of determining whether sample l is a benign or an attack.

        $\hat{V}^{(l)}$ is a combination of additional classification results for the attack sample $(j \neq 0)$.

1: Select features $x_i \in X$ using *Improved Feature Selection*.

2: Normalize x_i for all features using *min_max normalizer*.

3: Split the dataset S into a training set S_{train} and a test set S_{test} for $S_{train} \cap S_{test} = \varnothing$.

4: Split S_{train} into k folds and designate a fold $S_{validation}$ as the validation set excluded from the

    training set. $n(S) = n(S_{train}) + n(S_{validation}) + n(S_{test})$

5: **for** $x^{(l)} \in X \wedge X \subset S_{train}$ **do**

6:     Initialize $\hat{V}^{(l)} \leftarrow \varnothing$

7:     $\hat{y}_0^{(l)} \leftarrow c_0(x^{(l)})$

8:     **if** $\hat{y}_0^{(l)} = benign$ **then**

9:        // *do nothing*

10:    **else**

11:       **for** $c_j \in C \wedge j \neq 0$ **do**

12:         $\hat{y}_0^{(l)} \leftarrow c_j(x^{(l)})$

13:         $\hat{V}^{(l)} \leftarrow \text{add}(y_j^{(l)})$

14:       **end for**

15:    **end if**

16:    $\hat{y}^{(l)} \leftarrow \hat{y}_0^{(l)} \text{ concat } \hat{V}^{(l)}$

17: **end for**

4.2.2. Confusion Matrix and Evaluation Metric for MLHC

A confusion matrix is used to evaluate the classification results. In general, when the training results of the model are returned only in binary classification, the results are expressed in only two types, positive and negative, so they have a simple matrix, as presented in Table **??**.

Table 3. Confusion matrix for binary classification.

		Prediction	
		Positive	Negative
Actual	Positive	True Positive	False Negative
	Negative	False Positive	True Negative

However, the proposed MLHC method contains more information than the typical confusion matrix because it is a multi-class method that processes data of various categories and contains various classification results simultaneously. Similar to the existing confusion matrix, the confusion matrix indicates true negative (TN) or true positive (TP) if the benign sample is classified accurately as benign, or the sub-classification information of the attack sample, such as vehicle type and attack type, is accurately detected. Furthermore, the matrix classifies it as false negative (FN) if attack detection is missed because the sample containing sub attack information is misclassified as normal and as false positive (FP) if normal data are erroneously detected as attack; a sub attack classification result is then returned. The difference from the existing confusion matrix is that if the model classifies a data sample as attack, classification results of various categories are included in the layers below the attack. If the first classifier accurately detected an attack but erroneously classified additional information, such as vehicle type and attack type in the lower layers, it is classified as partial true positive (PTP). The hierarchical confusion matrix that contains PTP in the MLHC model is shown in Table **??**.

Table 4. Hierarchical confusion matrix.

				Prediction							
				Attack							
				VT_0			VT_1			···	Benign
				AT_0	AT_1	AT_2	AT_0	AT_1	AT_2	···	
		VT_0	AT_0	TP	PTP	PTP	PTP	PTP	PTP	···	FN
			AT_1	PTP	TP	PTP	PTP	PTP	PTP	···	FN
			AT_2	PTP	PTP	TP	PTP	PTP	PTP	···	FN
Actual	**Attack**	VT_1	AT_0	PTP	PTP	PTP	TP	PTP	PTP	···	FN
			AT_1	PTP	PTP	PTP	PTP	TP	PTP	···	FN
			AT_2	PTP	PTP	PTP	PTP	PTP	TP	···	FN
		···	···	···	···	···	···	···	···	···	···
	Benign			FP	FP	FP	FP	FP	FP	···	TN

For the model's performance, among the accuracy classification indies, accuracy and F1 score are used as shown in Equations (**??**) and (**??**), respectively.

$$Accuracy = \frac{TN + \sum TP}{TN + \sum TP + \sum FN + \sum FP + \sum PTP} \tag{6}$$

where accuracy represents the ratio of accurate classification of attack cases as attack and benign cases as benign among all cases. For attack cases, only TP cases where even the additional information type is correct are counted as follows. The precision, which represents the probability that the actual correct answer is included among the values predicted as attack (i.e., $P_{predict}$) by the classifier, is expressed as follows:

$$Precision = \frac{positive\ detections}{whole\ detections\ of\ an\ algorithm} = \frac{\sum TP}{P_{Predict}} = \frac{\sum TP}{\sum TP + \sum FP + \sum PTP} \qquad (7)$$

However, precision does not include the PTP cases where the vehicle type or attack type is not accurately detected.

The recall, which represents the probability that the actual attack cases noted as P are accurately predicted as attack by the classifier, is expressed as follows:

$$Recall = \frac{positive\ detections}{total\ number\ of\ existing\ positives} = \frac{\sum TP}{P} = \frac{\sum TP}{\sum TP + \sum FN + \sum PTP} \qquad (8)$$

As with the precision, PTP cases are not included in recall. Precision and recall have a trade-off relationship with each other. When the recall is raised by adjusting the parameters of the algorithm, false alarms increase; if the conditions are strengthened to reduce false alarms, the recall drops. Therefore, recall and precision should be considered together. Hence, in this study, we used F1 score, which is the harmonic mean of these two items, as follows:

$$F1score = 2 \times \frac{Precision \times Recall}{Precision + Recall} \qquad (9)$$

4.2.3. Space of Hypothesis

The space of hypothesis $H(S,C)$, which represents the space set of the model, product of the number of samples and number of classifiers, increases in proportion to the quotient of the data depth. It can be expressed as Equation (??):

$$H(S,C) = (n(S) \times n(C))^{depth} \qquad (10)$$

where S is the set of all samples, C is the set of classifiers for distinguishing the type of each target and depth is the number of layers of each classifier. The related notations are outlined in Table ??.

Table 5. Summary of notations.

Notation	Description
S	Set of full datasets which containing benign, attack and attack types extracted from several vehicle models.
S_α, S_β	Subsets of S, each composed of attack and benign samples, respectively. $S = S_\alpha \cup S_\beta$, $S_\alpha \cap S_\beta = \varnothing$, $S_\beta = \{(X,Y) \mid y_0 = 0\}$.
l	Index of sample at the line of l in S. $(0 \leq l \leq n(S))$.
x_i	Features are elements of feature set X, and i is an index of the feature. $(x_i \in X$ and $i = 0, 1, \ldots, I)$
y_j	Types of target are elements of target set Y, and j is the number of classifiers. Especially if the index j of y is zero, (i.e., y_0) is an indicator of whether the target is benign or attack. $(y_j \in Y$ and $j = 0, 1, \ldots, J)$
$\hat{y}_j$	Predicted target classes are elements of prediction result set $\hat{Y}$. $(\hat{y}_j \in \hat{Y})$
$\hat{V}$	Subset of $\hat{Y}$ excluding y_0 represents multiple predicted types from classifiers. $(\hat{y}_0 \notin \hat{V}, \hat{V} \subset \hat{Y}$ and $\hat{y}_0 = 0)$
c_j	Classifiers are elements of classifier set C, and j is an index of classifier. $(c_j \in C)$
k_j	The number of types on each classifier c_j.
$H(S,C)$	Space of hypothesis.

In this section, the existing two models, two-layer multi-class detection (TLMD) and single-layer based multi-class classification (SLMC), are compared in terms of space set with our proposed data learning model MLHC. The TLMD model proposed by Yuan et al. [?] performs multi-class classification independently in each layer by two independent classifiers using the C5.0 algorithm and

NB algorithm, respectively. By contrast, the method proposed by Aburomman and Reaz [?] is an SLMC model that contains a multi-class classifier using a support vector machine that has a weight in one layer.

Figure ??a illustrates the traditional model TLMD, which repeats the learning of the total dataset for the number of classifiers, and the computation of TLMD is shown in Equation (??):

$$H_{TLMD}(S,C) = \frac{n(S)}{c_0} \times \frac{n(S)}{c_1} \times \cdots \times \frac{n(S)}{c_j} = (n(S))^{j+1} \cdot \prod_{i=0}^{j} \left(\frac{1}{c_i} \right) \tag{11}$$

where the number of sample data to be learned in each classifier is $n(S)/c_j$, and training is repeated for the number of classifiers c_j.

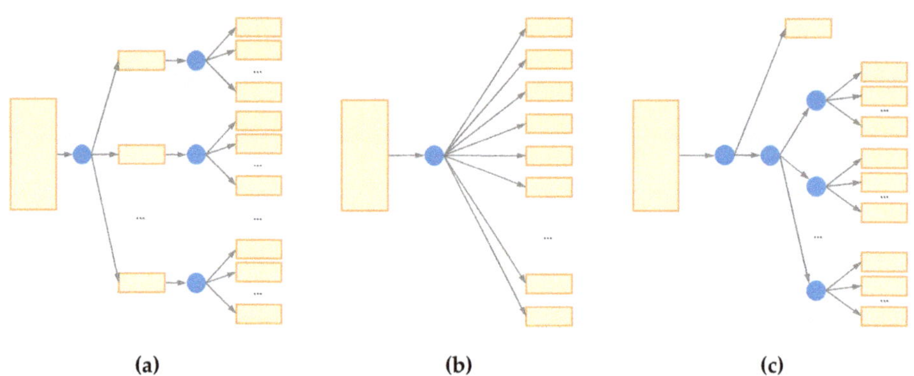

Figure 7. Comparison classification methods: (**a**) two layers multi-class detection model; (**b**) single-layer based multi-class classification model; and (**c**) multi-labeled hierarchical classification model (our approach).

Figure ??b illustrates the SLMC for classifying all the target data using one classifier. The multi-class classification method is used because the number of classes k_j classified by every classifier C must be expressed. The computation of SLMC is expressed as Equation (??):

$$H_{SLMC}(S,C) = \left\{ n(S) \times (c_0 \times c_1 \times \cdots \times c_j) \right\}^1 = n(S) \cdot \prod_{i=0}^{j} c_i \tag{12}$$

where the target data are expressed as a combination of all data types that can be expressed by each classifier. Therefore, classifier C is $c_0 \times c_1 \times \cdots \times c_j$, and the depth is one.

By contrast, our proposed MLHC method in Figure ??c forms one classifier by combining multi-class classification and multi-labeled classification. Therefore, the computation of the MLHC is expressed as Equation (??):

$$H_{MLHC}(S,C) = \left\{ n(S) \times (1 + c_1 \times \cdots \times c_j) \right\}^1 = n(S) \cdot \left(1 + \prod_{i=1}^{j} c_i \right) \tag{13}$$

Compared to Equation (??), Equation (??) can reduce the amount of computation for benign data because it does not perform a separate classification process if the result of classifier c_0 of the first layer is benign. To compare them with each other, the two equations are rearranged after replacing $n(S) \cdot \prod_{i=1}^{j} c_i$ with δ. $H_{SLMC}(S,C)$ and $H_{MLHC}(S,C)$ are expressed as Equations (??) and (??), respectively:

$$H_{SLMC}(S,C) = n(S) \cdot c_0 \cdot \prod_{i=1}^{j} c_i = \delta \cdot c_0 \tag{14}$$

$$H_{MLHC}(S,C) = n(S) + n(S) \cdot \prod_{i=1}^{j} c_i = \delta + n(S) \tag{15}$$

In the SLMC model, an increase in data types to be classified means that the space of hypothesis increases according to the multiplicative function. By contrast, in the MLHC model, classifier c_0 of the first layer determines benign or attack; if it is benign, classification stops. Therefore, the amount of computation can be reduced for the amount of benign data. When the present dataset, where 89.39% of the total data is benign, is applied, only 10.61% of the attack data is used to classify the vehicle type and attack type. Hence, the space of hypothesis is reduced for the ratio of attack data.

5. Results and Discussion

5.1. Simulation Environments

In the simulation, the data were learned using the learning model described in Section **??**, and the performance was compared by measuring accuracy and time. For the intrusion detection model of the in-vehicle network, we used the dataset [?] released from the challenge of in-vehicle intrusion detection. The model was trained and verified by randomly extracting 80% of the data samples from a total of 1.73 million data samples, and the model performance was evaluated using the remaining 20% of the data samples. To classify attack or benign, vehicle type and attack type of CAN traffic, the data samples were learned as multi-labels, and the targets were classified as multi-classes to accommodate various vehicle types and attack techniques.

We used four machine learning algorithms to compare the performance of the proposed method. The stochastic gradient descent (SGD) algorithm [?] is an iterative algorithm used for optimizing objective functions such that they have suitable smoothness properties. We used SGD in our study to compare the performance of the machine learning algorithms, as it reduces the computational burden associated with high-dimensional optimization problems, thereby achieving faster iterations, although the convergence rate obtained is low. In the kNN classification algorithm, the input consists of the k-closest training examples in the feature space. An object is classified by a plurality vote of its neighbors, with the object being assigned to the class most common among its k nearest neighbors. We used this algorithm in our study, as it is basic and capable of performing multi-class classification for performance evaluation.

The DT algorithm constructs a tree structure where each non-leaf node represents an attribute evaluation and each leaf node represents a class label. This algorithm can effectively analyze and classify the data to identify the attributes with information gain. We also used DT in our study as it is a classification algorithm and can achieve good performance depending on the type of dataset used. Furthermore, the random forest (RF) algorithm [?] is a kind of ensemble learning that is used for classification and regression. It returns the classification and average prediction results from the DTs and is therefore an extension of DT. We used the RF algorithm as well, to address the problem of overfitting on the training data and for obtaining a high accuracy.

To evaluate the performance of the classification model, detection rate and training time were selected as evaluation metrics. Accuracy, recall, precision and F1 score were calculated to evaluate the accuracy of the model in a reliable manner, and the elapsed time for training and evaluation of the model were measured. For the reference to evaluate whether the data samples were accurately classified, we used the hierarchical confusion matrix illustrated in Table **??**. This matrix does not include PTPs in TPs where the vehicle type or attack type is incorrect even if the attack or benign is accurately detected. We implemented classifiers using our novel method specified in Algorithm **??** and measured the accuracy.

5.2. Simulation Results

Table **??** compares and outlines the simulation results based on the four machine learning algorithms, namely, SGD, kNN, DT and RF, in terms of the detection rate; these models are described in Section **??**. The results are rounded from the fifth decimal place. Among the three models described, the RF algorithm shows a high positive detection rate of 0.99 or higher. Particularly, the MLHC model proposed in this study showed the highest detection rates evenly in the other three algorithms. The algorithm having the highest F1 score in each model and a graph of F1 score are shown in Figure **??**. All three models showed the highest performance with RF. If the training time is not considered, it can be seen that the F1 score of the model is the highest in MLHC, followed by TLMD and SLMC. The reason for the higher detection rate of MLHC as compared to the other models can be explained as follows.

Table 6. Simulation results for detection rate.

Evaluation Metric	Algorithm	TLMD	SLMC	MLHC
Accuracy	SGD	0.9111	0.9055	0.9336
	kNN	0.9431	0.9924	0.9950
	DT	0.9617	0.9984	0.9997
	RF	0.9989	0.9992	0.9999
Precision	SGD	0.9203	0.2877	0.9001
	kNN	0.9228	0.9399	0.9526
	DT	0.9398	0.9915	0.9974
	RF	0.9983	0.9934	0.9993
Recall	SGD	0.9065	0.6193	0.9171
	kNN	0.9216	0.9312	0.9573
	DT	0.9416	0.9872	0.9986
	RF	0.9986	0.9934	0.9998
F1 score	SGD	0.9133	0.3929	0.9085
	kNN	0.9222	0.9355	0.9550
	DT	0.9407	0.9893	0.9980
	RF	0.9984	0.9934	0.9995

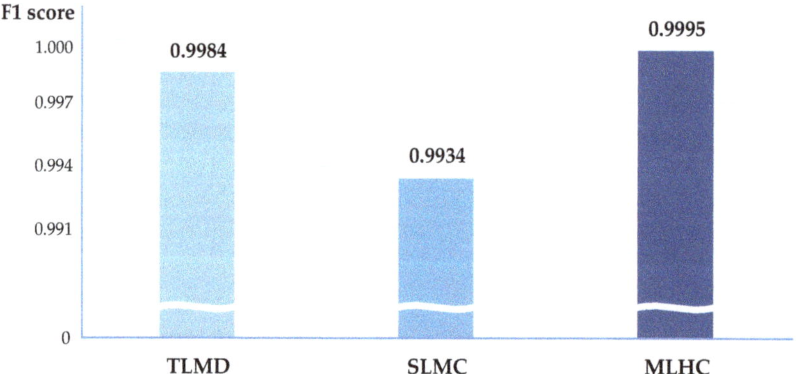

Figure 8. Simulation results for the best F1 score of each model.

MLHC determines whether an attack has occurred and then classifies the attack information in a hierarchical manner. Therefore, benign and attack data are separated for each data sample in the first stage itself. Subsequently, the model uses only the attack data when classifying specific attack information such as the attack type and vehicle type. Therefore, in this model, the benign data do not contribute to any errors. Consequently, it can be seen that the MLHC model shows a higher detection rate than the TLMD model, which contains two layers and the SLMC model, which comprises a single layer.

Table **??** illustrates the measurement result of the time elapsed for training and model evaluation in each model. For the training data, 1,388,672 data samples corresponding to 80% of all data samples were extracted randomly. Each model was evaluated using the remaining 20% (347,168) of the data samples. The first method TLMD uses independent classifiers in each layer to classify the attack type and vehicle type from the CAN traffic data. For this, $\prod_i^{j=9}(n(S)/c_i)$ needs to be computed in the three layers using Equation (**??**) during training. Consequently, TLMD took the largest amount of time for training and evaluating all the algorithms. In addition, in the case of the SLMC model, the hypothesis space is proportional to the product of the number of sample and the number of classifiers. The hypothesis space is represented by $n(S) \cdot \prod_i^{j=24} c_i$ as shown in Equation (**??**), and the number of classifiers is 24.

Table 7. Simulation results for elapsed time.

Evauation Metric	Algorithm	TLMD	SLMC	MLHC
Training Time (s)	SGD	320.871	17.793	2.110
	kNN	4495.222	715.893	4.609
	DT	272.642	11.138	0.618
	RF	1445.077	236.428	13.033
Test time (s)	SGD	14.048	0.078	0.042
	kNN	659.283	117.097	4.707
	DT	11.208	0.063	0.008
	RF	180.634	6.170	0.753

On the contrary, the MLHC model uses a classifier to learn the entire data and then determines if a data sample represents a an attack or benign state. In this method, the benign data that do not require additional analysis, such as vehicle type or attack type, are excluded from the sub-classification targets. Therefore, Equation (**??**) is used to reduce the amount of calculation as many as the number of benign data compared SLMC of Equation (**??**). Therefore, since in an MLHC model using a single classifier, the benign data (89.4% of the total data) need not be reclassified, 99.92% of the learning time is reduced on average, as compared to the TLMD model.

Figure **??** shows the number of CAN messages that can be processed per unit time for each algorithm of each model. The kNN and RF of the TLMD model processed 528 and 1927 test messages per second, respectively, whereas the kNN of the SLMC model processed 2973 messages per second. Considering that 1 Mbps of CAN has 50% of channel utilization, 5000 or more messages must be processed per second. Therefore, the three types of models are not suitable for processing the flooding messages in real time. If high-speed CAN communication in the future is considered, the DT algorithm of the MLHC model that can process 43.5 million messages per second should be used to prevent the bottleneck of the intrusion detection module.

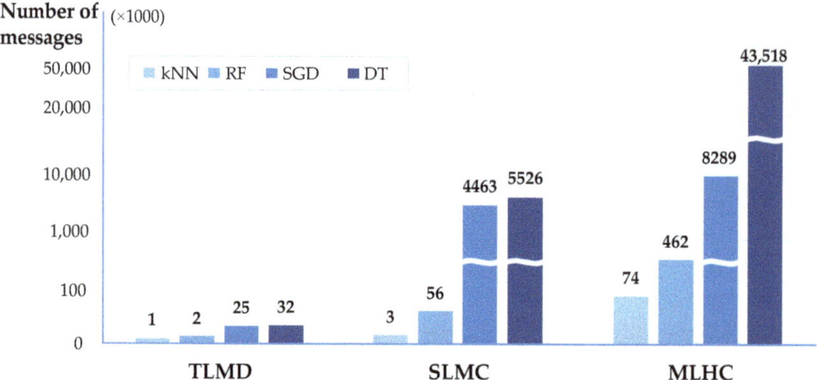

Figure 9. Simulation results: Number of CAN messages that can be processed per second.

6. Conclusions

This paper proposes the MLHC learning model that hierarchically classifies attacks using a machine learning algorithm to detect anomaly behaviors of the in-vehicle network accurately and rapidly. The MLHC method can make quick judgements about attack or benign cases for in-vehicle networks by learning the CAN traffic, and it can classify additional detailed information when an attack is detected. A learning model that accommodates multi-labeled multi-class schemas was designed to include various attributes simultaneously while classifying various types of attack data. To evaluate the performance of our model, we applied four machine learning algorithms to existing models and compared accuracy, precision, recall, F1 score and elapsed times for training step and test step.

The simulation results show that the proposed MLHC model achieved high accuracy when based on the RF algorithm and rapid detection when based on the DT algorithm. Both algorithms derived F1 scores higher than 0.998. Thus, we conclude that the DT and RF algorithms are applicable to high-speed internal communication environments, as well as in CAN for analyzing 43 million and 46 million CAN message frames per second, respectively.

In the future, we plan to train and verify intrusion detection models based on traffic injected into vehicles after directly generating messages of various attack types in addition to fuzzing, flooding and malfunction. Furthermore, we will additionally analyze the vehicle ethernet traffic beyond the CAN for target networks to investigate methods of applying the traditional intrusion detection and prevention patterns to the in-vehicle network. In addition, in the future, we intend to investigate the parallel processing method [?] for fast data processing in real time against sequential message injection attacks.

Author Contributions: Conceptualization, S.P. and J.-Y.C.; methodology, S.P.; software, S.P.; validation, S.P. and J.-Y.C.; data curation, S.P.; writing—original draft preparation, S.P.; writing—review and editing, J.-Y.C.; visualization, S.P.; supervision, J.-Y.C.; project administration, S.P. and J.-Y.C.; and funding acquisition, J.-Y.C. All authors have read and agreed to the published version of the manuscript.

Funding: This work was supported by Institute for Information and communications Technology Promotion (IITP) grant funded by the Korea government (MSIT) [No. 2018-0-00532, Development of High-Assurance ($\geq$EAL6) Secure Microkernel].

Conflicts of Interest: The authors declare no conflict of interest.

Abbreviations

The following abbreviations are used in this manuscript:

MLHC	Multi-labeled hierarchical classification
TLMD	Two layers multi-class detection
SLMC	Single-layer multi-class classification
CAN	Controller area network
M2M	Machine-to-machine
IDS	Intrusion detection system
LSTM	long short-term memory
kNN	k-nearest neighbors
NB	Naïve Bayes
DT	Decision tree
DNN	Deep neural network
ID	Identifier
HTM	Hierarchical temporal memory
ISO	International Organization for Standardization
ECU	Electronic control unit
SOF	Start of frame
RTR	Remote transmission request
SRR	Substitute remote request
IDE	Identifier extension
DLC	Data length code
CRC	Cyclic redundancy check
ACK	Acknowledgement
EOF	End of frame
IFS	Improved feature selection
TN	True negative
TP	True positive
FN	False negative
FP	False positive
PTP	Partial true positive
SGD	Stochastic gradient descent
RF	Random forest

References

. Miller, C.; Valasek, C. Remote Exploitation of an Unaltered Passenger Vehicle. In Proceedings of the Black Hat USA 2015, Las Vegas, NV, USA, 1–6 August 2015; pp. 1–91.

. Miller, C. Lessons learned from hacking a car. *IEEE Des. Test* **2019**, *36*, 7–9. [CrossRef]

. Song, H.M.; Woo, J.; Kim, H.K. In-vehicle network intrusion detection using deep convolutional neural network. *Veh. Commun.* **2020**, *21*, 100198. [CrossRef]

. Cover, T.M.; Hart, P. Nearest Neighbor Pattern Classification. *IEEE Trans. Inf. Theory* **1967**, *13*, 21–27. [CrossRef]

. Quinlan, J.R. Induction of Decision Trees. *Mach. Learn.* **1986**, *1*, 81–106. [CrossRef]

. Zhang, Y.; Chen, X.; Jin, L.; Wang, X.; Guo, D. Network Intrusion Detection: Based on Deep Hierarchical Network and Original Flow Data. *IEEE Access* **2019**, *7*, 37004–37016. [CrossRef]

. Kang, M.J.; Kang, J.W. Intrusion detection system using deep neural network for in-vehicle network security. *PLoS ONE* **2016**, *11*, 1–17. [CrossRef] [PubMed]

. Hoppe, T.; Kiltz, S.; Dittmann, J. Security threats to automotive CAN networks Practical examples and selected short-term countermeasures. *Reliab. Eng. Syst. Saf.* **2011**, *96*, 11–25. [CrossRef]

. Taylor, A.; Leblanc, S.; Japkowicz, N. Anomaly detection in automobile control network data with long short-term memory networks. In Proceedings of the IEEE International Conference on Data Science and Advanced Analytics (DSAA 2016), Montreal, QC, Canada, 17–19 October 2016; pp. 130–139. [CrossRef]

Wang, C.; Zhao, Z.; Gong, L.; Zhu, L.; Liu, Z.; Cheng, X. A Distributed Anomaly Detection System for In-Vehicle Network Using HTM. *IEEE Access* **2018**, *6*, 9091–9098. [CrossRef]

Kiencke, U.; Dais, S.; Litschel, M. Automotive Serial Controller Area Network. *SAE Trans.* **1986**, *95*, 823–828.

International Organization for Standardization (ISO). ISO 11898-1: 2015 Controller Area Network (CAN), 2015. Available online: https://www.iso.org/standard/63648.html (accessed on 15 July 2020).

Cho, K.T.; Shin, K.G. Error Handling of In-vehicle Networks Makes Them Vulnerable. In Proceedings of the 2016 ACM SIGSAC Conference on Computer and Communications Security, Vienna, Austria, 24–28 October 2016; pp. 1044–1055. [CrossRef]

Checkoway, S.; McCoy, D.; Kantor, B.; Anderson, D.; Shacham, H.; Savage, S.; Koscher, K.; Czeskis, A.; Roesner, F.; Kohno, T. Comprehensive experimental analyses of automotive attack surfaces. In Proceedings of the 20th USENIX Security Symposium, San Francisco, CA, USA, 10–12 August 2011; pp. 77–92.

Woo, S.; Jo, H.J.; Lee, D.H. A Practical Wireless Attack on the Connected Car and Security Protocol for In-Vehicle CAN. *IEEE Trans. Intell. Transp. Syst.* **2015**, *16*, 993–1006. [CrossRef]

Bharati, S.; Podder, P.; Mondal, M.R.H.; Robel, R.A. Threats and Countermeasures of Cyber Security in Direct and Remote Vehicle Communication Systems. *J. Inf. Assur. Secur.* **2020**, *15*, 153–164.

Koscher, K.; Czeskis, A.; Roesner, F.; Patel, S.; Kohno, T.; Checkoway, S.; McCoy, D.; Kantor, B.; Anderson, D.; Shacham, H.; et al. Experimental security analysis of a modern automobile. In Proceedings of the 2010 IEEE Symposium on Security and Privacy, Berkeley/Oakland, CA, USA, 16–19 May 2010; pp. 447–462. [CrossRef]

Vasenev, A.; Stahl, F.; Hamazaryan, H.; Ma, Z.; Shan, L.; Kemmerich, J.; Loiseaux, C. Practical Security and Privacy Threat Analysis in the Automotive Domain: Long Term Support Scenario for Over-the-Air Updates. In Proceedings of the 5th International Conference on Vehicle Technology and Intelligent Transport Systems (VEHITS 2019), Heraklion, Crete, Greece, 3–5 May 2019; pp. 550–555. [CrossRef]

Steger, M.; Dorri, A.; Kanhere, S.S.; Römer, K.; Jurdak, R.; Karner, M. Secure Wireless Automotive Software Updates Using Blockchains: A Proof of Concept. In Proceedings of the Advanced Microsystems for Automotive Applications (AMAA 2017), Berlin, Germany, 25–26 September 2017; pp. 137–149. [CrossRef]

Van Bulck, J.; Mühlberg, J.T.; Piessens, F. VulCAN: Efficient component authentication and software isolation for automotive control networks. In Proceedings of the 33rd Annual Computer Security Applications Conference (ACSAC 2017), Orlando, FL, USA, 4–8 December 2017; pp. 225–237. [CrossRef]

Choi, W.; Joo, K.; Jo, H.J.; Park, M.C.; Lee, D.H. VoltageIDS: Low-level communication characteristics for automotive intrusion detection system. *IEEE Trans. Inf. Forensics Secur.* **2018**, *13*, 2114–2129. [CrossRef]

Wang, E.; Xu, W.; Sastry, S.; Liu, S.; Zeng, K. Hardware Module-based Message Authentication in Intra-Vehicle Networks. In Proceedings of the 2017 ACM/IEEE 8th International Conference on Cyber-Physical Systems (ICCPS), Pittsburgh, PA, USA, 18–21 April 2017; pp. 207–216. [CrossRef]

Narayanan, S.N.; Mittal, S.; Joshi, A. OBD-SecureAlert: An Anomaly Detection System for Vehicles. In Proceedings of the IEEE International Conference on Smart Computing (SMARTCOMP 2016), St. Louis, MO, USA, 18–20 May 2016; pp. 5–10. [CrossRef]

Martinelli, F.; Mercaldo, F.; Nardone, V.; Santone, A. Car Hacking Identification through Fuzzy Logic Algorithms. In Proceedings of the IEEE International Conference on Fuzzy Systems (FUZZ-IEEE), Naples, Italy, 9–12 July 2017; pp. 1–7. [CrossRef]

Tencent Keen Security Lab. Experimental Security Assessment on Lexus Cars. Available online: https://keenlab.tencent.com/en/2020/03/30/Tencent-Keen-Security-Lab-Experimental-Security-Assessment-on-Lexus-Cars/ (accessed on 10 July 2020).

Woo, S.; Moon, D.; Youn, T.Y.; Lee, Y.; Kim, Y. CAN ID Shuffling Technique (CIST): Moving Target Defense Strategy for Protecting In-Vehicle CAN. *IEEE Access* **2019**, *7*, 15521–15536. [CrossRef]

Han, M.L.; Kwak, B.I.; Kim, H.K. Anomaly intrusion detection method for vehicular networks based on survival analysis. *Veh. Commun.* **2018**, *14*, 52–63. [CrossRef]

Park, S.; Choi, J.-Y. Malware Detection in Self-Driving Vehicles Using Machine Learning Algorithms. *J. Adv. Transp.* **2020**, *2020*, 3035741. [CrossRef]

Yuan, Y.; Huo, L.; Hogrefe, D. Two Layers Multi-class Detection Method for Network Intrusion Detection System. In Proceedings of the 2017 IEEE Symposium on Computers and Communications (ISCC), Heraklion, Crete, Greece, 3–6 July 2017; pp. 767–772. [CrossRef]

Aburomman, A.A.; Reaz, M.B.I. A novel weighted support vector machines multiclass classifier based on differential evolution for intrusion detection systems. *Inf. Sci.* **2017**, *414*, 225–246. [CrossRef]

Robbins, H.; Monro, S. A Stochastic Approximation Method. *Ann. Math. Stat.* **1951**, *22*, 400–407. [CrossRef]

Breiman, L. Random forests. *Mach. Learn.* **2001**, *45*, 5–32. [CrossRef]

Gao, R.; Zhao, M.; Ye, T.; Ye, F.; Wang, Y.; Luo, G. Smartphone-based real time vehicle tracking in indoor parking structures. *IEEE Trans. Mob. Comput.* **2017**, *16*, 2023–2036. [CrossRef]

© 2020 by the authors. Licensee MDPI, Basel, Switzerland. This article is an open access article distributed under the terms and conditions of the Creative Commons Attribution (CC BY) license (http://creativecommons.org/licenses/by/4.0/).

Article

A Maze Matrix-Based Secret Image Sharing Scheme with Cheater Detection

Ching-Chun Chang [1], Ji-Hwei Horng [2,*], Chia-Shou Shih [3] and Chin-Chen Chang [3,4]

1   Department of Electronic Engineering, Tsinghua University, Beijing 100084, China; c.c.chang.phd@gmail.com
2   Department of Electronic Engineering, National Quemoy University, Kinmen 89250, Taiwan
3   Department of Information Engineering and Computer Science, Feng Chia University, Taichung 40724, Taiwan; jsfcu1129@gmail.com (C.-S.S.); ccc@o365.fcu.edu.tw (C.-C.C.)
4   School of Computer Science and Technology, Hangzhou Dianzi University, Hangzhou 310018, China
*   Correspondence: horng@email.nqu.edu.tw; Tel.: +886-(82)-313553

Received: 19 May 2020; Accepted: 6 July 2020; Published: 7 July 2020

Abstract: Secret image sharing is a technique for sharing a secret message in such a fashion that stego image shadows are generated and distributed to individual participants. Without the complete set of shadows shared among all participants, the secret could not be deciphered. This technique may serve as a crucial means for protecting private data in massive Internet of things applications. This can be realized by distributing the stego image shadows to different devices on the Internet so that only the ones who are authorized to access these devices can extract the secret message. In this paper, we proposed a secret image sharing scheme based on a novel maze matrix. A pair of image shadows were produced by hiding secret data into two distinct cover images under the guidance of the maze matrix. A two-layered cheat detection mechanism was devised based on the special characteristics of the proposed maze matrix. In addition to the conventional joint cheating detection, the proposed scheme was able to identify the tampered shadow presented by a cheater without the information from other shadows. Furthermore, in order to improve time efficiency, we derived a pair of Lagrange polynomials to compute the exact pixel values of the shadow images instead of resorting to time-consuming and computationally expensive conventional searching strategies. Experimental results demonstrated the effectiveness and efficiency of the proposed secret sharing scheme and cheat detection mechanism.

Keywords: secret image sharing; maze matrix; cheat detection; cheater identification

1. Introduction

Massive Internet of things (Massive IoT) involves an immense number of devices that require to be connected reliably and gigantic loads of data that need to travel safely through the Internet. With the growing public concerns over Internet privacy and security, there is an urgent appeal for research into secure communications in massive IoT. Pioneering works include the aggregate-signcryption [1], decentralized blockchain [2], FORGE system [3], and chaotic maps [4]. In this paper, we address this issue with a novel approach based on secret image sharing.

We propose to conceal the private data into a pair of image shadows and transmit them to separate devices over public networks. An authorized recipient should be able to access the image shadows stored on the separate devices and retrieve the private data via low-cost computations. The core component of the proposed secret image sharing scheme is the maze matrix, which belongs to a group of reference matrices originating from steganographic methods.

Steganography is the art and science of hiding information. It can be used to protect secret information by concealing it into cover images. These techniques can be broadly categorized into the transformed domain [5–9] and the spatial domain [10–17] methodologies. For the former class of methods, some commonly used transformations are the discrete cosine transform (DCT) [5,6], vector quantization (VQ) [7], and absolute moment block truncation coding (AMBTC) [8,9]. As for the latter class of methods, reference matrix-based algorithms have proved to be efficient in terms of the distortion versus capacity tradeoff. Common magic matrix-based steganographic schemes include the exploiting modified direction (EMD) [10,11], the turtle shell [12], the octagon-shaped shell [13], and the Sudoku [14–17] schemes.

Another closely related research stream is visual cryptography, which was first proposed by Naor and Shamir [18]. Typical visual cryptography schemes encrypt a secret image by breaking it up into n shares of obfuscated meaningless images, which are then printed onto separate transparencies. When k out of n transparencies are stacked and overlaid, the secret image will appear and become recognizable, where k is a pre-defined threshold. Methods of visual cryptography has constantly evolved, and the later developments contrived to produce shares in such a form that they themselves are images with meaningful contents [19–22].

A significant visual cryptography (SVC) [23] was recently proposed to securely transfer real-time images without compromising the visual quality. In the author's scheme, random share values are hidden in a cover image by LSB embedding. The signific secret image with induced errors can be revealed using a (k, n) SVC scheme, while the exact secret image can be revealed using an (n, n) scheme. However, this scheme is not capable of detecting cheaters.

As a notable improvement, a verifiable secret sharing scheme with combiner verification and cheater identification [24] was recently developed. Its share generation and secret reconstruction mechanisms were based on the polynomial interpolation technique invented by Shamir [25]. Its combiner verification and cheater identification were realized via a pre-shared key and a verifier code generated from the combiner's ID and password.

A recent development by Liu et al. [26] demonstrated that it is possible to identify the tampered shadows by restricting the use of elements at certain locations of the reference matrix and checking justness of the mapped elements in the secret extracting process. Through this mechanism, dishonest behaviors can be detected without the help of a pre-shared secret key or a password system.

In this paper, we proposed a novel secret image sharing scheme for massive IoT applications. The image shadows were generated under the guidance of the maze matrix. By leveraging the special characteristics of the maze matrix, we were able to inspect whether cheating behaviors took place. A two-layered cheat detection mechanism was devised. A joint cheat detection can discover cheating behaviors and a blind cheater identification can trace which shadow is inauthentic.

The proposed scheme shares the same merits as Liu et al.'s scheme, as that no pre-shared secret key or password system is required. In addition to this, the proposed maze matrix was explicitly designed to enable the scheme to detect cheats under the paradigm of secret sharing. Moreover, we formulated a pair of Lagrange polynomials to compute the exact pixel values of the shadow images rather than adopting time-consuming and computationally expensive conventional searching strategies. As a consequence, the time efficiency of the proposed share construction algorithm can be dramatically improved.

This remainder of this paper is organized as follows. Section 2 reviews a state-of-the-art secret image sharing scheme. Section 3 presents the proposed secret image sharing scheme based on maze matrix and the two-layered cheat detection mechanism. Experimental results and performance comparisons are shown in Section 4. This paper is concluded in Section 5.

2. Related Work

In this section, we briefly review the secret image sharing scheme proposed by Liu et al. [26] with a discussion of its merits and demerits. Our proposed scheme was based on the similar framework and is introduced in the next section.

The secret image sharing scheme proposed by Liu et al. [26] allows a dealer to share secret message into two different meaningful images. It adopts the turtle shell matrix $M(p_{1i}, p_{2i})$, proposed by Chang et al. [12], to guide the embedding of secret message, as shown in Figure 1. Before constructing secret shares of shadow images, the binary stream of secret message is converted to 8-ary secret set $S = \{sg_k | k = 1, 2, \ldots, n\}$. The pixels of two distinct grayscale cover images with size $H \times W$ are rearranged into $C_1 = \{p_{1i} | i = 1, 2, \ldots, H \times W\}$ and $C_2 = \{p_{2i} | i = 1, 2, \ldots, H \times W\}$. Each pair of pixels (p_{1i}, p_{2i}) is used to embed a secret digit sg_k in a way like conventional reference matrix-based data hiding scheme.

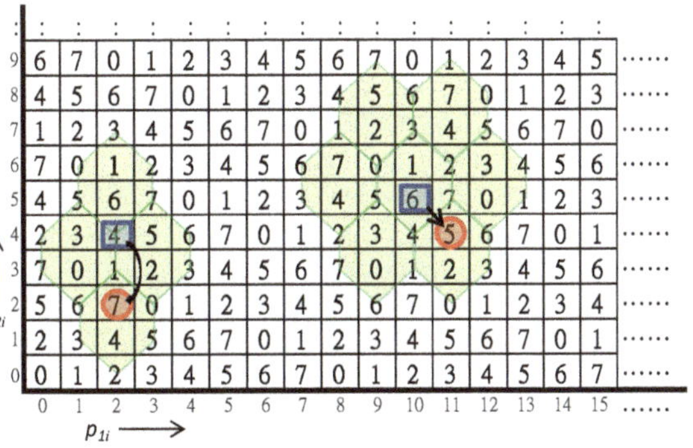

p_{2i}	0	1	2	3	4	5	6	7	8	9	10	11	12	13	14	15	
9	6	7	0	1	2	3	4	5	6	7	0	1	2	3	4	5	······
8	4	5	6	7	0	1	2	3	4	5	6	7	0	1	2	3	······
7	1	2	3	4	5	6	7	0	1	2	3	4	5	6	7	0	······
6	7	0	1	2	3	4	5	6	7	0	1	2	3	4	5	6	······
5	4	5	6	7	0	1	2	3	4	5	6	7	0	1	2	3	······
4	2	3	4	5	6	7	0	1	2	3	4	5	6	7	0	1	······
3	7	0	1	2	3	4	5	6	7	0	1	2	3	4	5	6	······
2	5	6	7	0	1	2	3	4	5	6	7	0	1	2	3	4	······
1	2	3	4	5	6	7	0	1	2	3	4	5	6	7	0	1	······
0	0	1	2	3	4	5	6	7	0	1	2	3	4	5	6	7	······
	0	1	2	3	4	5	6	7	8	9	10	11	12	13	14	15	······ $p_{1i} \rightarrow$

Figure 1. The turtle shell matrix for secret image sharing scheme.

For the purpose of cheating detection, the elements in the reference matrix are classified into back elements and edge elements. As implied by the name, an edge element is an element located on the common edges of adjacent hexagons. On the contrary, a back element is located inside a single hexagon. The embedding rules are as follows. The cover pixel pair (p_{1i}, p_{2i}) is applied to locate a reference element in the matrix first. For an edge reference element, the rocket-shaped turtle shells as shown in the figure are the candidates of embedding, while the flower-shaped turtle shells are the candidates for a back-reference element. By searching the candidates to find the nearest back element that $M(p'_{1i}, p'_{2i}) = sg_k$, the obtained pixels (p'_{1i}, p'_{2i}) are recorded to the image shadows. After all secret digits are embedded, the shadow images $S_1 = \{p'_{1i} | i = 1, 2, \ldots, H \times W\}$ and $S_2 = \{p'_{2i} | i = 1, 2, \ldots, H \times W\}$ are constructed. By restricting the embedding candidates to the back elements only, the cheating event can be detected while the shadow pixel pair (p'_{1i}, p'_{2i}) is mapped to an edge element $M(p'_{1i}, p'_{2i})$.

Two typical examples of data hiding are illustrated in Figure 1. In the first example, the cover pixel pair is $(p_{1i}, p_{2i}) = (2, 4)$ and the secret digit is $sg_k = 7$. First, the cover pixel pair $(2, 4)$ is mapped to the edge reference element $M(2, 4)$. By searching its associated rocket-shaped candidate turtle shells, the only matched back element is $M(2, 2) = 7 = sg_k$. The recorded shadow pixels are therefore $(p'_{1i}, p'_{2i}) = (2, 2)$. Although $M(3, 5)$ and $M(0, 3)$ are also matched with the secret digit, they are not back elements and thus conflict with the embedding rule.

The second example uses $(p_{1i}, p_{2i}) = (10, 5)$ and $sg_k = 5$ as inputs. The reference element $M(10, 5)$ is a back element, therefore the candidates of embedding are the flower-shaped turtle shells shown in the figure. The matched candidates $M(9, 5)$ and $M(12, 7)$ are edge elements and excluded. Two legal candidates are $M(11, 4)$ and $M(9, 8)$. The nearest matched back element $M(11, 4)$ is the final solution and the shadow pixels are given by $(p'_{1i}, p'_{2i}) = (11, 4)$.

To extract secret data, both shares of the image shadows should be obtained from the participants. The corresponding pair of pixels from the two shadows is mapped to the secret digit through the guidance of the turtle shell matrix. In case an edge element is mapped, we can conclude someone is cheating. The exact cheater can only be identified by a faithful participant. To overcome this weak point, we propose a new scheme in the following section.

3. The Proposed Secret Image Sharing Scheme

The proposed secret image sharing scheme was to convert two distinct cover images into a pair of shadow images through the guidance of a new proposed maze matrix. By cooperating the pair of shadow images occupied by two different participants, the embedded secret data could be extracted. In addition, a cheater detection mechanism was devised such that any cheating share of shadow images could be detected without help of the other share.

3.1. The Maze Matrix

The maze matrix was constructed using a basic structure matrix of size 6×6 as enclosed by the red square shown in Figure 2. Distinct numbers in the radix-16 number system were arranged by circulating the outmost boundary of the region except for a horizontal and a vertical gap. Other elements were marked with 'x'. By repeated mirroring operations, the rest of a 256×256 maze matrix was constructed. The first mirror matrix of the red basic structure to the p_α direction of axis was enclosed by a blue square in the figure. The resulting matrix $M(p_\alpha, p_\beta)$ looks like a big maze map and was named the maze matrix.

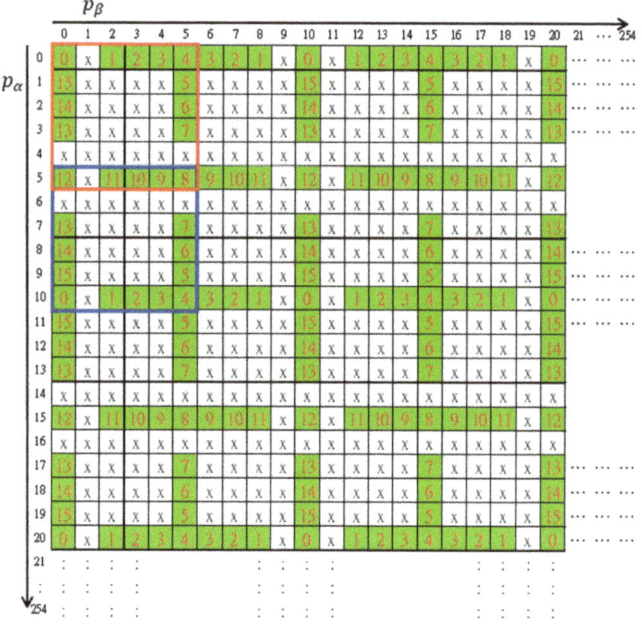

Figure 2. The maze matrix for secret image sharing scheme.

3.2. The Data Embedding and Extraction Scheme

Following the same problem formulation as the turtle shells matrix-based secret image sharing scheme [26], we constructed a pair of image shadows using a pair of distinct cover images through the guidance of the proposed maze matrix.

Before constructing secret shares of shadow images, the binary stream of secret message was converted to 16-ary secret set $S = \{sg_k | k = 1, 2, \dots, n\}$. Pixels of two distinct grayscale cover images with size $H \times W$ are rearranged into $C_1 = \{p_{1i} | i = 1, 2, \dots, H \times W\}$ and $C_2 = \{p_{2i} | i = 1, 2, \dots, H \times W\}$.

Each pair of pixels (p_{1i}, p_{2i}) was used to embed a secret digit sg_k through the guidance of the maze matrix. For a cover pixel pair (p_{1i}, p_{2i}), it was mapped to the maze matrix $M(p_{1i}, p_{2i})$ first. Then, we searched the neighboring elements to find the nearest matched element $M(p'_{1i}, p'_{2i}) = sg_k$ and record the shadow pixels $\left(p'_{1i}, p'_{2i}\right)$ to the shadow images. After all secret digits were embedded, the shadow images $S_1 = \left\{p'_{1i} \middle| i = 1, 2, \ldots, H \times W\right\}$ and $S_2 = \left\{p'_{2i} \middle| i = 1, 2, \ldots, H \times W\right\}$ were constructed. Note that the elements marked with 'x' can never be the target element. This property will be applied to devise the cheating detection mechanism.

The data extraction process is rather simple: Collect the pair of image shadows provided by different participants and construct the same maze matrix as embedding. Then, consecutively extract secret digits by $sg_k = M\left(p'_{1i}, p'_{2i}\right)$ until all secrets are extracted.

To detect cheating events, the 'x'-marked forbidden zone is the key. Any pair of shadow pixels which maps to an 'x'-marked element indicates someone is cheating. In addition, while the mapped element lies at a horizontal or vertical gap of the maze matrix, the exact cheater can be identified.

The detailed algorithms of the data embedding and extraction processes are discussed in the following subsections. In the last subsection, we discuss the cheater detection mechanism of the proposed secret image sharing scheme.

3.3. The Sshare Construction Algorithm

As described in the previous subsection, a secret digit is embedded by modifying the cover pixel pair to the target element through the guidance of maze matrix. However, the searching process is time-consuming. To improve the embedding efficiency, we devised a Lagrange polynomial to determine the target element of modification. Let $\left(p_x, p_y\right)$ be the cover pixel pair in the range $0 \leq p_x \leq 4$, $0 \leq p_y \leq 4$. According to the maze matrix as shown in Figure 2, the target elements of modification for embedding different secret digits are listed in Table 1.

Table 1. Target elements of modification for different secret digits.

sg_j	(p'_x, p'_y)	sg_j	(p'_x, p'_y)	sg_j	(p'_x, p'_y)	sg_j	(p'_x, p'_y)
0	(0, 0)	4	(0, 5)	8	(5, 5)	12	(5, 0)
1	(0, 2)	5	(1, 5)	9	(5, 4)	13	(3, 0)
2	(0, 3)	6	(2, 5)	10	(5, 3)	14	(2, 0)
3	(0, 4)	7	(3, 5)	11	(5, 2)	15	(1, 0)

Let

$$X = \{x_0, x_1, x_2, \cdots, x_{15}\} = \{0, 0, 0, 0, 0, 1, 2, 3, 5, 5, 5, 5, 5, 3, 2, 1\}, \tag{1}$$

$$Y = \{y_0, y_1, y_2, \cdots, y_{15}\} = \{0, 2, 3, 4, 5, 5, 5, 5, 5, 4, 3, 2, 0, 0, 0, 0\}. \tag{2}$$

The modified pixel pair $\left(p'_x, p'_y\right)$ can be represented by the Lagrange polynomial functions of s_j as shown below:

$$p'_x(sg_j) = \sum_{r=0}^{15} x_r \prod_{\substack{k \neq r \\ k = 0}}^{15} \frac{(sg_j - k)}{(r - k)}, \tag{3}$$

$$p'_y(sg_j) = \sum_{r=0}^{15} y_r \prod_{\substack{k \neq r \\ k = 0}}^{15} \frac{(sg_j - k)}{(r - k)}. \tag{4}$$

By leveraging the periodic property of the maze matrix, we modulated a reference element $M(p_\alpha, p_\beta)$ to the fundamental period of $M(p_x, p_y)$, $0 \leq p_x \leq 9$, $0 \leq p_y \leq 9$. Then, the fundamental period was further divided into four reflective symmetric parts. According to the secret digit s_j to be embedded, a quasi-target element $M\left(p'_x(sg_j), p'_y(sg_j)\right)$ can be obtained. By reflection and backward modulation, the target element $M\left(p'_\alpha, p'_\beta\right)$ can be determined. The detailed algorithm is summarized as follows.

The construction of image shadows:

Input: Cover images C_1 and C_2, secret message S

Output: Image shadows S_1 and S_2

Step 1. Arrange the cover images into two separate pixel streams and convert the secret message to 16-ary secret digits.

$$C_1 = \{p_{1i}|i = 1, 2, \ldots, H \times W\}, \tag{5}$$

$$C_2 = \{p_{2i}|i = 1, 2, \ldots, H \times W\}, \tag{6}$$

where $H \times W$ is the image size.

$$S = \{sg_k|k = 1, 2, \ldots, n\}, \tag{7}$$

where n is the total number of digits.

Step 2. Retrieve a cover pixel pair (p_{1i}, p_{2i}) and let

$$\left(p_\alpha, p_\beta\right) = \begin{cases} (p_{1i}, p_{2i}), & \text{for } i \text{ is odd,} \\ (p_{2i}, p_{1i}), & \text{for } i \text{ is even.} \end{cases} \tag{8}$$

Step 3. Modulate the pixel values to the fundamental period.

$$p_x = mod(p_\alpha, 10), \tag{9}$$

$$p_y = mod\left(p_\beta, 10\right). \tag{10}$$

$$M = \left\lfloor \frac{p_\alpha}{10} \right\rfloor, \tag{11}$$

$$N = \left\lfloor \frac{p_\beta}{10} \right\rfloor. \tag{12}$$

Step 4. Using the Lagrange polynomial defined as Equations (1) to (4), determine the target element of modification.

For $0 \le p_x \le 4$ and $0 \le p_y \le 4$,

$$p'_\alpha = p'_x\left(sg_j\right) + 10 \times M; \tag{13}$$

$$p'_\beta = p'_y\left(sg_j\right) + 10 \times N. \tag{14}$$

For $5 \le p_x \le 9$ and $0 \le p_y \le 4$,

$$p'_\alpha = \left[10 - p'_x\left(sg_j\right)\right] + 10 \times M; \tag{15}$$

$$p'_\beta = p'_y\left(sg_j\right) + 10 \times N. \tag{16}$$

For $0 \le p_x \le 4$ and $5 \le p_y \le 9$,

$$p'_\alpha = p'_x\left(sg_j\right) + 10 \times M; \tag{17}$$

$$p'_\beta = \left[10 - p'_y\left(sg_j\right)\right] + 10 \times N. \tag{18}$$

For $5 \le p_x \le 9$ and $5 \le p_y \le 9$,

$$p'_\alpha = \left[10 - p'_x\left(sg_j\right)\right] + 10 \times M; \tag{19}$$

$$p'_\beta = \left[10 - p'_y\left(sg_j\right)\right] + 10 \times N. \tag{20}$$

Step 5. Record the shadow pixels.

$$\left(p'_{1i}, p'_{2i}\right) = \begin{cases} \left(p'_\alpha, p'_\beta\right), & \text{for } i \text{ is odd,} \\ \left(p'_\beta, p'_\alpha\right), & \text{for } i \text{ is even.} \end{cases} \tag{21}$$

Step 6. Repeat Step 2 to 5 until all secret digits are embedded.
Step 7. Output the pair of image shadows.

$$S_1 = \left\{ p'_{1i} \middle| i = 1, 2, \ldots, H \times W \right\}; \tag{22}$$

$$S_2 = \left\{ p'_{2i} \middle| i = 1, 2, \ldots, H \times W \right\}. \tag{23}$$

Note that there are many gaps at $\mathrm{mod}(p_\alpha, 10) = 4, 6$ and $\mathrm{mod}(p_\beta, 10) = 1, 9$ of the maze matrix. Using the conventional fixed assignment of $\left(p_\alpha, p_\beta\right) = (p_{1i}, p_{2i})$, the resulting $\left(p'_\alpha, p'_\beta\right)$ will lack the gapped pixel values. This may draw the eavesdropper's attention. To prevent the vacuums of pixel value, we alternatively assigned $\left(p_\alpha, p_\beta\right)$ with (p_{1i}, p_{2i}) and (p_{2i}, p_{1i}) in Step 2 and switched back in Step 5 coordinately. The asymmetric gapping of maze matrix in the p_α and p_β directions made it possible to cover the gaps by leveraging the alternating assignment.

We provide two examples to demonstrate the operation of embedding process. Assume the first cover pixel pair is $(p_{11}, p_{21}) = (83, 61)$ and the secret digit to be embedded is $sg_1 = 5$. Following the steps of embedding algorithm gives $\left(p_\alpha, p_\beta\right) = (83, 61)$, $\left(p_x, p_y\right) = (3, 1)$, $(M, N) = (8, 6)$, $(p'_x(5), p'_y(5)) = (1, 5)$, and $\left(p'_{11}, p'_{21}\right) = \left(p'_\alpha, p'_\beta\right) = (1 + 8 \times 10, 5 + 6 \times 10) = (81, 65)$. Let the second cover pixel pair and the second secret digit be $(p_{12}, p_{22}) = (83, 66)$ and $sg_2 = 14$. Following the same calculation gives $\left(p_\alpha, p_\beta\right) = (p_{22}, p_{12}) = (66, 83)$, $\left(p_x, p_y\right) = (6, 3)$, $(M, N) = (6, 8)$, $(p'_x(14), p'_y(14)) = (2, 0)$, and $\left(p'_\alpha, p'_\beta\right) = ((10 - 2) + 6 \times 10, 0 + 8 \times 10) = (68, 80)$, and $\left(p'_{12}, p'_{22}\right) = \left(p'_\beta, p'_\alpha\right) = (80, 68)$.

3.4. The Data Extraction Algorithm

The secret message can be extracted only through cooperation of the two shadow image owners. The secret data can be extracted by pairing the pixels from the two image shadows and applying each pixel pair to retrieve a 16-ary secret digit through the guidance of maze matrix. The 16-ary secret digits can be converted back to the binary secret stream if necessary. The data extraction algorithm is provided as follows.

The data extraction algorithm:
Input: image shadows S_1 and S_2
Output: secret message S
Step 1. Arrange the image shadows into two separate pixel streams.

$$S_1 = \left\{ p'_{1i} \middle| i = 1, 2, \ldots, H \times W \right\}; \tag{24}$$

$$S_2 = \left\{ p'_{2i} \middle| i = 1, 2, \ldots, H \times W \right\}, \tag{25}$$

where $H \times W$ is the image size.

Step 2. Construct the fundamental period of maze matrix $M\left(p_x, p_y\right)$, $0 \leq p_x \leq 9$, $0 \leq p_y \leq 9$ as shown in Figure 2.

Step 3. Retrieve a shadow pixel pair $\left(p'_{1i}, p'_{2i}\right)$ and let

$$\left(p_\alpha, p_\beta\right) = \begin{cases} \left(p'_{1i}, p'_{2i}\right), & \text{for } i \text{ is odd,} \\ \left(p'_{2i}, p'_{1i}\right), & \text{for } i \text{ is even.} \end{cases} \tag{26}$$

Step 4. Extract the secret digit sg_j and record to S.

$$sg_j = M\big(mod(p_\alpha,\ 10), mod(p_\beta,\ 10)\big). \tag{27}$$

Step 5. Repeat Step 3 and 4 until all secret digits are extracted.

Step 6. Convert the 16-ary secret digits back to the binary secret stream.

Now, we apply the embedding results in the previous subsection $(p'_{11}, p'_{21}) = (81, 65)$ and $(p'_{12}, p'_{22}) = (80, 68)$ as examples. For the first shadow pixel pair, the secret digit can be retrieved by directly calculating Equation (27), i.e., $sg_1 = M(mod(81,\ 10),\ mod(65,\ 10)) = M(1,5) = 5$. For the second pixel pair, the pixels should be swapped according to Equation (26), i.e., $(p_\alpha, p_\beta) = (68, 80)$. Then, calculate Equation (27),, i.e., $sg_2 = M(mod(68,\ 10),\ mod(80,\ 10)) = M(8,0) = 14$. Both secret digits coincided with the embedded ones.

3.5. The Cheat Event Detection and Cheater Detection Mechanism

The most creative part of our secret sharing scheme was the cheater detection mechanism. Referring to Figure 2, the 'x'-marked elements in the maze matrix were the traps. Any pair of shadow pixels which maps to an 'x'-marked element was illegal and served as key information for cheat event detection. The algorithm is given as follows.

The cheat detection algorithm:

Input: image shadows S_1 and S_2

Output: cheating pixel pairs F, cheating pixels F_1 and F_2

Step 1. Arrange the image shadows into two separate pixel streams.

$$S_1 = \{p'_{1i} | i = 1, 2, \ldots, H \times W\}; \tag{28}$$

$$S_2 = \{p'_{2i} | i = 1, 2, \ldots, H \times W\}, \tag{29}$$

where $H \times W$ is the image size.

Step 2. Construct the fundamental period of maze matrix $M(p_x, p_y)$, $0 \le p_x \le 9$, $0 \le p_y \le 9$ as shown in Figure 2.

Step 3. Retrieve a shadow pixel pair (p'_{1i}, p'_{2i}) and let

$$(p_\alpha, p_\beta) = \begin{cases} (p'_{1i}, p'_{2i}), & \text{for } i \text{ is odd,} \\ (p'_{2i}, p'_{1i}), & \text{for } i \text{ is even.} \end{cases} \tag{30}$$

Step 4. Detect cheating pixel pairs and individual cheating pixels.

if $M\big(mod(p_\alpha,\ 10), mod(p_\beta,\ 10)\big) = $ 'x',

    record i to F;

    if $mod(p_\alpha,\ 10) = 4$ or 6,

        record i to F_1 for i is odd; record i to F_2 for i is even.

    end

    if $mod(p_\beta,\ 10) = 1$ or 9,

        record i to F_2 for i is odd; record i to F_1 for i is even.

    end

end

Step 5. Repeat Step 3 and 4, until all pixel pairs are checked.

The cheat detection included two layers. The outer layer was a joint cheat event detection. The shadow pixel pair was mapped to the maze matrix and check the legality. If an 'x'-marked element was mapped, the index i of the pixel pair was recorded to F. Under such circumstances, we could conclude that a cheat event was detected. The exact cheater could only be determined by a faithful participant. The inner layer was a blind cheater detection. We checked whether the mapped element was located at a gap. If it was located at a horizontal gap, p_α was a tampered pixel no matter what value p_β is, because it was impossible to find a p_β to make the pixel pair $\left(p_\alpha, p_\beta\right)$ legal. For the same reason, a p_β trapped in a vertical gap was a tampered pixel, and the participant who shared this shadow pixel was the cheater. The output sets F_1 and F_2 recorded the indices of tampered pixels from image shadows S_1 and S_2, respectively.

4. Experimental Results

In this section, we give some experimental results to show the performance of the proposed secret image sharing scheme. Figure 3 shows six pairs of 512×512 grayscale cover images, including (a) Lena and baboon, (b) Tiffany and Barbara, (c) airplane and peppers, (d) boat and Goldhill, (e) toys and girl, and (f) Elaine and sailboat. According to the embedding capacity of the proposed scheme, we used a 362×362 grayscale secret image "office," as shown in Figure 4. The embedding capacity of a cover image pair was $512 \times 512 \times 4 = 1,048,576$ bits, while the secret image contained $362 \times 362 \times 8 = 1,048,352$ bits of data. The whole secret image can be embedded into a cover image pair. The remaining capacity was filled with random generated data.

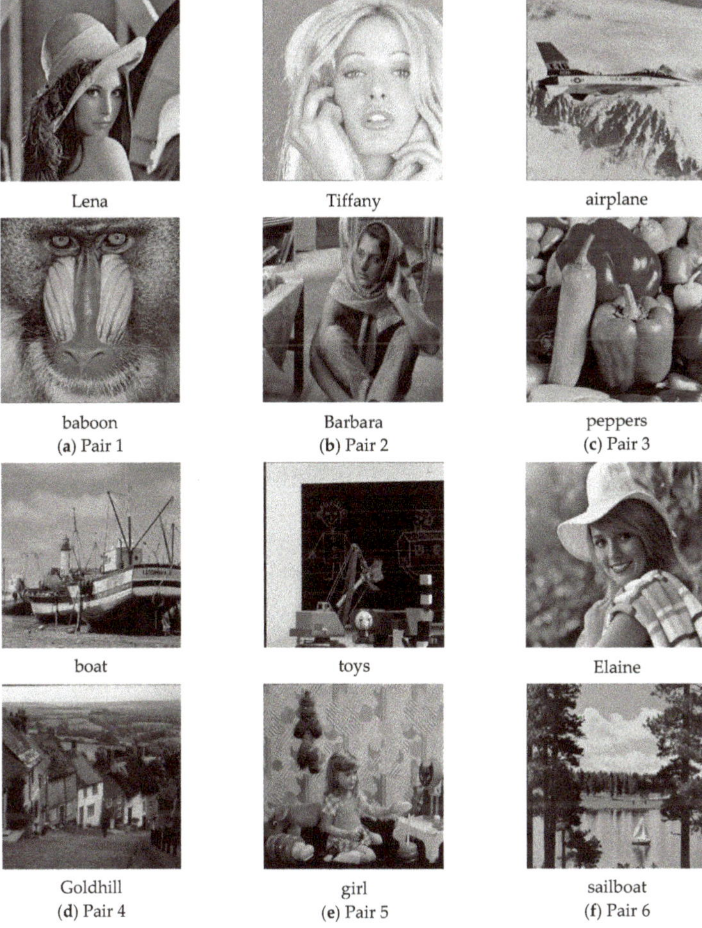

Lena	Tiffany	airplane
baboon	Barbara	peppers
(a) Pair 1	(b) Pair 2	(c) Pair 3
boat	toys	Elaine
Goldhill	girl	sailboat
(d) Pair 4	(e) Pair 5	(f) Pair 6

Figure 3. Six pairs of grayscale cover images.

office

Figure 4. Secret image.

This section includes four subsections. In the first subsection, we demonstrate the applicability of the proposed share construction and data extraction scheme. The visual quality of the secret image shadows is also assessed. In the second subsection, we measure the detection ratio of tampered image regions. The effectiveness of cheat event detection and cheater detection are discussed. In the third subsection, the performance, including visual quality, hiding capacity, and cheat detection effectiveness, is compared with the Liu et al.'s scheme, which shares the same framework of secret image sharing scheme. Finally, the time efficiency of the new proposed share construction scheme is compared with conventional version in the last subsection.

4.1. Share Construction and Data Extraction

To demonstrate the applicability of the proposed secret image sharing scheme, all six pairs of cover images were tested. Two examples of the experimental results are shown in Figures 5 and 6, where (a) and (b) are the cover images, (c) and (d) are the shadow images, and (e) is the recovered secret image. As shown in the figures, the difference between a cover image and its corresponding shadow image cannot be distinguished by human eyes.

(**a**) Cover image 1

(**b**) Cover image 2

(**c**) Shadow 1
(PSNR = 39.88 dB)

(**d**) Shadow 2
(PSNR = 39.89 dB)

Figure 5. *Cont.*

(**e**) Recovered secret image

Figure 5. Experimental results of cover image pair 1.

(**a**) Cover image 1

(**b**) Cover image 2

(**c**) Shadow 1
(PSNR = 39.86 dB)

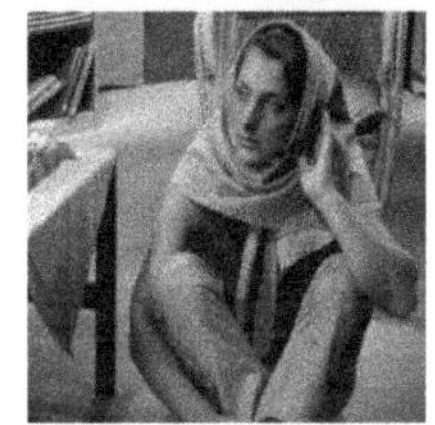

(**d**) Shadow 2
(PSNR = 39.90 dB)

(**e**) Recovered secret image

Figure 6. Experimental results of cover image pair 2.

To evaluate the visual quality of the shadow images, we applied the peak-signal-to-noise ratio (PSNR), defined by

$$PSNR = 10 \log_{10} \frac{255^2}{MSE} (\mathrm{dB}),$$ (31)

where MSE is the mean square error between the cover image C_k and its corresponding shadow image S_k, defined by

$$MSE = \frac{1}{H \times W} \sum_{i=1}^{H} \sum_{j=1}^{W} (C_k(i,j) - S_k(i,j))^2.$$ (32)

The visual quality and embedding capacity for the six cover image pairs are listed in Table 2.

Table 2. Experimental values of the proposed scheme.

	Cover Image 1	Cover Image 2	PSNR (dB)		Embedding Capacity (bits)
			Shadow 1	Shadow 2	
Pair 1	Lena	baboon	39.88	39.89	1,048,576
Pair 2	Tiffany	Barbara	39.86	39.90	1,048,576
Pair 3	airplane	peppers	39.87	39.88	1,048,576
Pair 4	boat	Goldhill	39.88	39.88	1,048,576
Pair 5	toys	girl	39.90	39.88	1,048,576
Pair 6	Elaine	sailboat	39.88	39.88	1,048,576

4.2. Cheat Event Detection and Cheater Detection

The six pairs of image shadows were then applied to test the cheating detection mechanism. In each pair of shadows, shadow 1 was tampered by inserting a small image into a local region while shadow 2 was kept faithful. Four results of the six experiments are provided in Figures 7–10, where (a) is the tampered shadow image 1, (b) is the faithful shadow image 2, and (c) is the result of joint detection. The detected cheat pixel pairs are illustrated by black pixels on the tampered shadow. The joint cheat detection ratio for the six test shadow pairs are listed in Table 3. In each test pair, the detection ratio was calculated by

$$DR_J = \frac{N(F)}{N},\qquad(33)$$

where $N(F)$ is the number of total detected cheat pixel pairs and N is the number of tampered pixels, i.e., the total number of pixels in the inserted small image. As shown in the table, DR_J of the joint cheat detection was around 0.42 and independent of the image features.

(a) Tampered shadow 1

(b) Faithful shadow 2

(c) Detection result

Figure 7. Joint cheat detection result 1.

(a) Tampered shadow 1

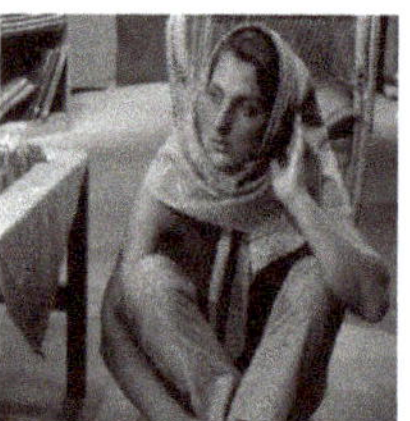

(b) Faithful shadow 2

(c) Detection result

Figure 8. Joint cheat detection result 2.

(**a**) Tampered shadow 1

(**b**) Faithful shadow 2

(**c**) Detection result

Figure 9. Joint cheat detection result 3.

(**a**) Tampered shadow 1

(**b**) Faithful shadow 2

(**c**) Detection result

Figure 10. Joint cheat detection result 4.

Table 3. Joint cheat detection ratio for the six shadow pairs.

Tampered Shadow	DR_J
Lena	0.42
Tiffany	0.42
airplane	0.42
boat	0.42
toys	0.42
Elaine	0.42

The blind cheater detection results for the six tampered shadows are listed in Table 4. The detection ratio for blind cheater detection is defined by

$$DR_{B1} = \frac{N(F_1)}{N},\tag{34}$$

where $N(F_1)$ is the number of total detected pixel in shadow 1 by blind cheater detection and N is the number of tampered pixels, i.e., the total number of pixels in the inserted small image. As shown in the table, DR_B of the blind cheater detection is around 0.20 and independent of the image features. Since the image shadow 2 was not tampered, the number of detected pixels $N(F_2)$ and thus DR_{B2} are both zeros.

Table 4. Blind cheater detection ratio for the six tampered shadows.

Tampered Shadow	DR_{B1}
Lena	0.20
Tiffany	0.20
airplane	0.20
boat	0.20
toys	0.20
Elaine	0.20

To investigate the effect of combinatorial tampering, we further designed an experiment in which both image shadows were tampered with dis-aligned regions. Example results are given in Figures 11 and 12, where (a) and (b) are the cover image pair, (c) and (d) are the detection results of joint cheat detection, and (e) illustrates the overview of total detected pixels. The experimental data for all six test shadow image pairs are listed in Table 5, where DR_1/DR_2 is the joint cheating detection ratio (DR_J) of the region that shadow 1/shadow 2 is tampered only; $DR_{1\cap2}$ is the DR_J of the region that both shadow1 and shadow 2 are tampered; $DR_{1\cup2}$ is the DR_J of the union tampered region. The joint cheating detection ratio (DR_1/DR_2) was around 43% for single tampered pixel pairs, while it was increased to 72% for combinatorial tampered pixel pairs ($DR_{1\cap2}$). Both of the percentage numbers were independent of the image features since the proposed data hiding scheme was a uniform embedding scheme [27]. The detection ratio of the union region ($DR_{1\cup2}$) depended on the percentage of overlapped region and was not an intrinsic characteristic of the proposed scheme.

(a) Tampered shadow 1

(b) Tampered shadow 2

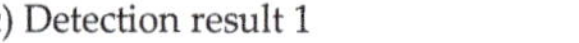

(c) Detection result 1

(d) Detection result 2

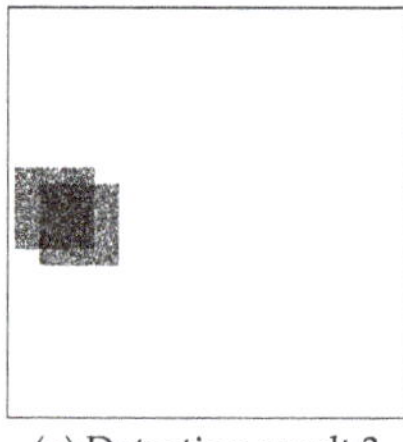

(e) Detection result 3

Figure 11. Joint cheating detection for combinatorial tampered shadows: Results for shadow pair 1.

(**a**) Tampered shadow 1

(**b**) Tampered shadow 2

(**c**) Detection result 1

(**d**) Detection result 2

(**e**) Detection result 3

Figure 12. Joint cheating detection for combinatorial tampered shadows: Results for shadow pair 3.

Table 5. DR values for the six combinatorial tampered shadow pairs.

	Shadow 1	Shadow 2	DR_1	DR_2	$DR_{1 \cap 2}$	$DR_{1 \cup 2}$
Pair 1	Lana	baboon	0.43	0.43	0.72	0.52
Pair 2	Tiffany	Barbara	0.42	0.42	0.73	0.52
Pair 3	airplane	peppers	0.44	0.42	0.73	0.53
Pair 4	boat	Goldhill	0.43	0.43	0.71	0.52
Pair 5	toys	girl	0.43	0.43	0.71	0.53
Pair 6	Elaine	sailboat	0.43	0.43	0.72	0.53

4.3. Comparison with Liu et al.'s Scheme [26]

The comparison of the proposed maze matrix-based data hiding scheme with the turtle shell matrix-based scheme [26] is provided in Table 6. The new proposed scheme can hide four bits of secret data for each pair of cover pixels, while the turtle shell matrix-based scheme can hide only three bits for each pair. The EC given in the table was measured by bits per pixel pair, one from cover image 1 and the other from cover image 2. Due to different embedding capacity, the PSNR of the proposed scheme was slightly lower than the turtle shell scheme. However, the degradation of visual quality could not be recognized by human eyes.

Table 6. Comparison of the proposed maze matrix-based scheme with the turtle shell-based scheme.

Hiding Scheme	PSNR	EC	DR_{JS}	DR_{JC}	DR_B
Maze matrix	39.88	4	0.43	0.72	0.20
Turtle shell [26]	41.71	3	0.50	0.50	—

The joint cheat detection ratio of the turtle shell scheme was 50% in both single tampered or combinatorial tampered cases. Although only the single tampered data was provided by the authors, the combinatorial tampered detection ratio can be analyzed easily. Since legal hiding locations are the back elements of turtle shells and such elements occupy 50% of the entire matrix, the theoretic cheating detection ratio was 50%. Our cheating detection mechanism outperformed the turtle shell scheme in combinatorial tampering, while the detection ratio was lower in single tampering.

The most creative part of the proposed scheme is the function of blind cheater detection. Without information of the other shadow, we detected 20% of tampered pixels in the shadow shared by a cheater. Meanwhile, the turtle shell scheme can only identify a cheater by a faithful participant.

4.4. Time Efficiency Evaluation

To assess the time efficiency of the proposed secret image sharing scheme, we listed the execution time required for the share construction program in Table 7 and the execution time for secret data extraction program in Table 8. The conventional reference matrix-based data hiding scheme and share construction scheme usually embed secret data by searching the nearest element that matches the intended secret digit and modify the pixel values accordingly. This type of searching procedures is often time-consuming. In this paper, a pair of Lagrange polynomials was derived to compute the coordinates of the matched element. Thus, the running time for share construction was drastically reduced. Referring to Table 7, up to 39% of execution time can be saved by leveraging the proposed approach. The execution time required for data extraction grogram is relatively short in comparison with the share construction program as shown in Table 8.

Table 7. Efficiency comparison of the proposed embedding scheme with conventional scheme.

Cover Images	Execution Time (sec)	
	Conventional Scheme	Proposed Scheme
Pair 1	0.1297	0.0692
Pair 2	0.1425	0.0747
Pair 3	0.1074	0.0737
Pair 4	0.1030	0.0703
Pair 5	0.1110	0.0708
Pair 6	0.1055	0.0709
Average	0.1165	0.0716

Table 8. Efficiency of the extraction scheme.

Stego Images	Execution Time (sec)
Pair 1	0.0366
Pair 2	0.0361
Pair 3	0.0382
Pair 4	0.0366
Pair 5	0.0411
Pair 6	0.0348
Average	0.0372

5. Conclusions

In this paper, we proposed a secret image sharing scheme based on a novel maze matrix. A pair of distinct cover images was used to carry secret data and a pair of shadow images was constructed under the guidance of the maze matrix. The secret data is extracted only if both authentic shadows are presented.

A two-layered cheat detection mechanism was devised to examine cheating behaviors as well as to ascertain the inauthentic shadow. In the outer cheat detection layer, the corresponding pair of pixels retrieved from the two shares was jointly used for detecting cheat events. The detection ratio was 43% for the cases in which single shadow was tampered and was 72% for the cases in which both shadows were tampered. In the inner blind cheater identification layer, the cheater's image share could be spotted without the information from the other share. The detection ratio of tampered pixels was 20% for the blind cheater identification.

An additional merit of the proposed scheme is time efficiency. By computing the pixel values of the image shadows with Lagrange polynomials instead of conventional searching algorithms, the proposed approach can save up to 39% of program execution time. In view of the effectiveness and low power consumption of the proposed scheme, the outlook for integrating it with massive IoT systems as a data security module shall be positive.

In the future world where massive IoT environment is fully established, secret image sharing will no longer be restricted to share secrets among human participants. The image shadows produced by the dealer can be transmitted via different routes to devices located at different sites. The shadow production and secret extraction will be executed via APPs installed on smartphones of the dealer and receiver. Uploading and downloading image shadows through IoT links will permit secret data to be communicated securely without the use of a preshared key or password system.

Author Contributions: Conceptualization, C.-C.C. (Ching-Chun Chang) and J.-H.H.; Data curation, C.-S.S.; Formal analysis, J.-H.H.; Funding acquisition, J.-H.H.; Investigation, C.-S.S.; Methodology, C.-C.C. (Ching-Chun Chang) and J.-H.H.; Project administration, C.-C.C. (Chin-Chen Chang); Resources, C.-S.S.; Software, C.-S.S.; Supervision, C.-C.C. (Chin-Chen Chang); Validation, J.-H.H.; Visualization, J.-H.H., C.-C.C. (Ching-Chun Chang) and C.-C.C. (Chin-Chen Chang); Writing: original draft, J.-H.H.; Writing: review & editing, J.-H.H., C.-C.C. (Ching-Chun Chang) and C.-C.C. (Chin-Chen Chang). All authors have read and agreed to the published version of the manuscript.

Funding: This research received no external funding.

Conflicts of Interest: The authors declare no competing interests.

References

1. Ullah, S.; Marcenaro, L.; Rinner, B. Secure smart cameras by aggregate-signcryption with decryption fairness for multi-receiver IoT applications. *Sensors* **2019**, *19*, 327. [CrossRef]
2. Li, Y.; Tu, Y.; Lu, J.; Wang, Y. A security transmission and storage solution about sensing image for blockchain in the Internet of Things. *Sensors* **2020**, *20*, 916. [CrossRef] [PubMed]
3. Chakraborty, T.; Jajodia, S.; Katz, J.; Picariello, A.; Sperli, G.; Subrahmanian, V.S. FORGE: A fake online repository generation engine for cyber deception. *IEEE Trans. Dependable Secure Comput.* **2019**. [CrossRef]
4. García-Guerrero, E.E.; Inzunza-González, E.; López-Bonilla, O.R.; Cárdenas-Valdez, J.R.; Tlelo-Cuautle, E. Randomness improvement of chaotic maps for image encryption in a wireless communication scheme using PIC-microcontroller via Zigbee channels. *Chaos Solitons Fractals* **2020**, *133*, 109646. [CrossRef]
5. Chang, C.; Lin, C.; Tseng, C.; Tai, W. Reversible hiding in DCT-based compressed images. *Inf. Sci.* **2007**, *177*, 2768–2786. [CrossRef]
6. Huang, F.; Qu, X.; Kim, H.; Huang, J. Reversible data hiding in JPEG images. *IEEE Trans. Circuits Syst. Video Technol.* **2016**, *26*, 1610–1621. [CrossRef]
7. Hu, Y. High capacity image hiding scheme based on vector quantization. *Pattern Recogn.* **2006**, *39*, 1715–1724. [CrossRef]
8. Lin, Y.; Hsia, C.; Chen, B.; Chen, Y. Visual IoT security: Data hiding in AMBTC images using block-wise embedding strategy. *Sensors* **2019**, *19*, 1974. [CrossRef]
9. Chang, C.; Wang, X.; Horng, J. A hybrid data hiding method for strict AMBTC format images with high-fidelity. *Symmetry* **2019**, *11*, 1314. [CrossRef]
10. Zhang, X.; Wang, S. Efficient steganographic embedding by exploiting modification direction. *IEEE Commun. Lett.* **2006**, *10*, 781–783. [CrossRef]
11. Kim, H.; Kim, C.; Choi, Y.; Wang, S.; Zhang, X. Improved modification direction schemes. *Comput. Math. Appl.* **2010**, *60*, 319–325. [CrossRef]

12. Chang, C.C.; Liu, Y.; Nguyen, T.S. A novel turtle shell based scheme for data hiding. In Proceedings of the 2014 tenth international conference on intelligent information hiding and multimedia signal processing, New York, NY, USA, 27–29 August 2014; pp. 89–93.

13. Leng, H. Generalized scheme based on octagon-shaped shell for data hiding in steganographic applications. *Symmetry* **2019**, *11*, 760. [CrossRef]

14. Chang, C.; Chou, Y.; Kieu, T. An information hiding scheme using Sudoku. In Proceedings of the 2008 3rd international conference on innovative computing information and control, Dalian, China, 18 June 2008; pp. 17–22.

15. Xia, B.; Wang, H.; Chang, C.; Liu, L. An image steganography scheme using 3D-Sudoku. *J. Info. Hiding Multimed. Sign Proc.* **2016**, *7*, 836–845.

16. He, M.; Liu, Y.; Chang, C. A mini-Sudoku matrix-based data embedding scheme with high payload. *IEEE Access* **2019**, *7*, 141414–141425. [CrossRef]

17. Horng, J.; Xu, S.; Chang, C.; Chang, C. An efficient data-hiding scheme based on multidimensional mini-SuDoKu. *Sensors* **2020**, *20*, 2739. [CrossRef]

18. Naor, M.; Shamir, A. Visual cryptography. In Proceedings of the Workshop on the Theory and Application of Cryptographic Techniques, Perugia, Italy, 9–12 May 1994; pp. 1–12.

19. Nakajima, M.; Yamaguchi, Y. Extended visual cryptography for natural images. *J. WSCG* **2002**, *10*, 2.

20. Kang, I.; Arce, G.; Lee, H. Color extended visual cryptography using error diffusion. *IEEE Trans. Image Process.* **2011**, *20*, 132–145. [CrossRef]

21. Patil, S.; Rao, J. Extended visual cryptography for color shares using random number generators. *IEEE Trans. Image Process.* **2012**, *1*, 399–410.

22. Jainthi, K.; Prabhu, P. A novel cryptographic technique that emphasis visual quality and efficiency by Floyd Steinberg error diffusion method. *Int. J. Res. Eng. Technol.* **2015**, *4*, 428–439.

23. Mary, G.S.; Kumar, S.M. Secure grayscale image communication using significant visual cryptography scheme in real time applications. *Multimed. Tools Appl.* **2019**, *79*, 1–20.

24. Kandar, S.; Dhara, B.C. A verifiable secret sharing scheme with combiner verification and cheater identification. *J. Inf. Secur. Appl.* **2020**, *51*, 102430. [CrossRef]

25. Shamir, A. How to share a secret. *Commun. ACM* **1979**, *22*, 612–613. [CrossRef]

26. Liu, Y.; Chang, C.C.; Huang, P.C. Security protection using two different image shadows with authentication. *Math. Biosci. Eng.* **2019**, *16*, 1914–1932. [CrossRef]

27. Liao, X.; Qin, Z.; Ding, L. Data embedding in digital images using critical functions. *Signal Process. Image Commun.* **2017**, *58*, 146–156. [CrossRef]

 © 2020 by the authors. Licensee MDPI, Basel, Switzerland. This article is an open access article distributed under the terms and conditions of the Creative Commons Attribution (CC BY) license (http://creativecommons.org/licenses/by/4.0/).

Article

Compression-Assisted Adaptive ECC and RAID Scattering for NAND Flash Storage Devices

Seung-Ho Lim [1] and Ki-Woong Park [2,*]

[1] Division of Computer and Electronic Systems Engineering, Hankuk University of Foreign Studies, Yongin 17035, Korea; slim@hufs.ac.kr

[2] Department of Computer and Information Security, Sejong University, Seoul 05006, Korea

* Correspondence: woongbak@sejong.ac.kr

Received: 27 March 2020; Accepted: 20 May 2020; Published: 22 May 2020

Abstract: NAND flash memory-based storage devices are vulnerable to errors induced by NAND flash memory cells. Error-correction codes (ECCs) are integrated into the flash memory controller to correct errors in flash memory. However, since ECCs show inherent limits in checking the excessive increase in errors, a complementary method should be considered for the reliability of flash storage devices. In this paper, we propose a scheme based on lossless data compression that enhances the error recovery ability of flash storage devices, which applies to improve recovery capability both of inside and outside the page. Within a page, ECC encoding is realized on compressed data by the adaptive ECC module, which results in a reduced code rate. From the perspective of outside the page, the compressed data are not placed at the beginning of the page, but rather is placed at a specific location within the page, which makes it possible to skip certain pages during the recovery phase. As a result, the proposed scheme improves the uncorrectable bit error rate (UBER) of the legacy system.

Keywords: NAND flash memory; P/E cycle; compression; adaptive ECC; RAID scattering; stripe log; parity

---

1. Introduction

With advances in process geometry scale-down of flash memory, a rapid increase in the capacity of NAND flash memory chips has been achieved. The main reason for this rapid increase in capacity is the increase in the number of bits stored per cell. However, as this number increases, the inter-electron interference degree increases, resulting in an increase in the error rate of the flash memory [1,2]. In addition, as the program/erase cycle (P/E cycle) increases; that is, as the usage time increases, the physical characteristics of the cell deteriorate, and the error rate increases rapidly. This is a serious drawback of computer systems based on NAND flash memory storage devices.

The typical approach to rectify this short endurance and increase in errors is using error-correction codes (ECCs) [3,4]. NAND flash memory devices use the ECC module to create additional parity for data-error correction. The parity generated by the ECC module is stored in the out-of-bound (OOB) region within a page; a page is the unit of read and write operations. Therefore, ECC is used for error correction in page units. Although ECC parities can achieve error-correction gain, if there are more errors than an ECC can correct, they may have no choice but to return decoding failures, if used without any additional error-recovery scheme for the failed data. However, creating more ECC parity is a complex and difficult process because the ECC module is a type of hardware block within the flash memory controller. One useful approach for creating more ECC parities for specific data is to use lossless compression [5,6], where additional ECC parity is created and stored in the free area after compression of the data.

In contrast, as a complement to in-page ECC decoding failures, an error management method using another parity technique between outer-pages, such as RAID5, has often been applied.

Many previous works show that parity with the RAID5 technique can reduce the page error rate [7–10]; however, it has also been shown to result in increased additional write operations and low space utilization owing to RAID parity management. That is, the RAID parity operation creates an additional write operation in the parity update process. From a space management point of view, the smaller the size of a cluster that has RAID parity, the larger the number of parities, resulting in a space overhead.

In this paper, we propose an enhancement scheme for an in-page ECC parity ability and outer-page RAID parity management with the help of lossless data compression. In the proposed system, incoming data are compressed through a compression module, and then, so-called adaptive ECC and RAID-scattering schemes are applied to this compressed data step by step. For in-page ECC improvement, ECC parity is created by applying ECC encoding only to the compressed data. Compression reduces the source length applied to ECC encoding, which results in lowering of the code rate.

The code rate is defined as a rate occupied by data in a codeword, so the lower the code rate, the higher the parity rate; thus, the error correction capability is improved by compressing data and applying ECC only to these compressed data.

After compression, data are then placed in each offset position within the page, in which the placement position is determined by its position in the RAID stripe.

We refer this placement scheme as RAID-scattering scheme. Since the data are compressed and occupy a smaller area than the page, each datum occupies only a specific part of the page, and the rest becomes a non-use area, and the value of the non-use can be set a known value. As a result, the specific positioning of each compressed data creates a region where non-use regions overlap each other in a stripe. Owing to the overlapping of the non-use area, the overhead of restoring data in RAID is reduced, because when data are restored with RAID parity algorithm for the error-occurred page in the RAID stripe, the non-use area of the error page does not need to be restored.

As a result, some pages can be skipped during the restoration process.

In other words, it results in more pages being recovered by each RAID parity, and parity overhead is reduced at the same level as the RAID reliability. In addition, this paper describes a FTL architecture, a FTL management scheme and the associated metadata for parity logging that can efficiently update and manage RAID parity on NAND flash devices. The scheme proposed in this paper is an extension of our previous work [11] that only considered in-page ECC management. This paper extends and highlights a combination of in-page ECC management and outer-page parity management, as well as detailed metadata structure of RAID parity management.

The organization of this paper is as follows. In Section 2, background and related work are presented. The proposed adaptive ECC and RAID-scattering scheme is described in Section 3, and experimental results are shown in Section 4. Finally, Section 5 concludes this paper.

2. Background and Related Work

In this section, the background of this research area is first described, which includes basics of flash memory and flash memory-based systems. Then, the related work is described.

2.1. NAND Flash Memory-Based Storage Devices

Figure 1 describes typical NAND flash-based storage devices. The basic hardware architecture of NAND flash memory-based storage devices consists of a high-performance controller, DRAM main memory, and an array of NAND flash memory chips. NAND flash memory is a nonvolatile semiconductor storage device that is formed by integrating a floating gate-based semiconductor device. The main components include a page, which is a read and write unit, and a block, which is an erase unit. The page sizes are mainly about 4 KB, 8 KB, and 16 KB, and dozens of pages are collected to form a block. There are three internal commands used for NAND flash memory: read, program, and erase. The read and program commands transfer data to and from the flash chip, and a data transfer unit is a page. Actually, the erase command does not transfer data and works in block units. A client device can make only two types of requests, namely, a read and a write request. The read request is related to the read command and the write request is related to the program command.

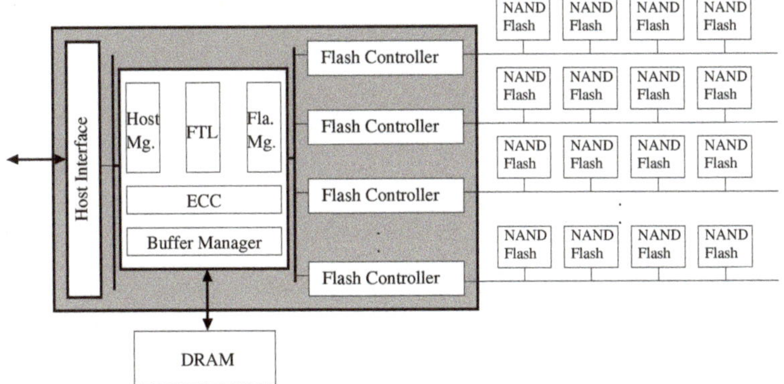

Figure 1. Basic architecture for NAND flash-based storage devices.

Currently, triple-level cells (TLC) are the most commonly used NAND flash memory. It stores three bits per cell, where each of the three bits of data in a TLC flash cell are either programmed or erased, which means the cell has a total of eight different states from 000 to 111. In contrast, a single-level cell (SLC) and a multi-level cell (MLC) store one bit and two bits per cell, respectively. With several states, the quantized voltage level of memory cells decides the value of the cell. The growth in the number of states per cell raises interference between states as the quantized decision levels of the cell start moving close between adjacent states, which results in detection errors. Recent advances in 3D lamination technology make it easier to secure cell-to-cell spacing compared to conventional 2D methods, resulting in a slight increase in P/E cycles compared with those of conventional 2D.

There are two crucial points regarding NAND flash memory. One, if the data are written to a flash memory, write operations should be preceded by erase operations internally. Second, the basic units of erase and write operations are different from each other. To overcome the unit mismatch between write and erase operations and to efficiently use flash memory, special software layers called flash translation layers (FTLs) have been developed [12–17]. Beyond single internal operational issues, other works have focused on the parallelism issue within embedded storage devices to speed them up [18–23].

2.2. Error Correction Codes for Flash Memory

The first-generation NAND flash memory was much simpler than the current generation. However, over time, the storage capacity of NAND flash memory has increased by moving to smaller geometries, and also storing more bits per cell. This required a much better error-correction algorithm to ensure data integrity for NAND flash devices. The P/E cycle value is about 100,000 for SLCs, 10,000 or less for MLCs, and 1000 or less for TLCs. As error rates increase rapidly, the usage of flash devices is emerging as a major issue. The advent of TLC flash and even smaller geometries will further increase the error-correction requirements in NAND flash. Figure 2 shows the raw bit error rate (RBER) for the P/E cycle of the TLC NAND flash memory [24]. As shown in the figure, when there are thousands of P/E cycles, the RBER is so high that it is known to reach the lifecycle limit.

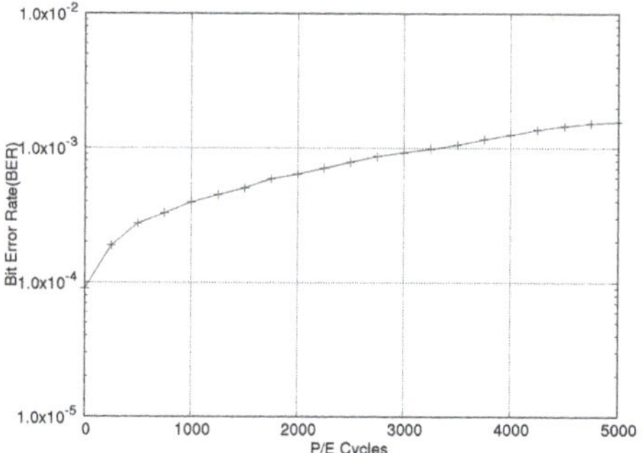

Figure 2. Evaluated bit error rates according to P/E cycles for TLC NAND flash memories [24].

As a method for recovering such errors, ECC methods [3,4,25–28], such as BCH and LDPC, are embedded in the NAND flash memory controller, which can cover page-level errors for programming and retrieving data. The BCH codes form a class of cyclic error-correcting codes that are constructed using polynomials over a finite field. An arbitrary level of error correction is possible, and it includes efficient code for un-correlated error patterns. Low-density parity check (LDPC) codes are considered one of the best choices for current flash memory controllers owing to their excellent error-correction capabilities. LDPC code decoding is an iterative process with the decoding input as certain probability information, which leads to a dependency of LDPC error correction on the accuracy of the input information [4]. ECCs for flash memory storage systems have been widely studied in both BCH and LDPC. Ref. [3] addresses applications using BCH ECC coding driver on a Linux platform, and many other chip designers have presented BCH bit error coding implementations. In contrast, LDPC receives attention from the industry, as there have been many presentations on LDPC in flash summits in the recent past [25,26]. In [27], a LDPC decoding strategy optimized for flash memory was proposed. Tanakamaru et al., have shown that LDPC codes can improve SSD lifetimes by 10 [28]. K. Zhao presents three techniques for mitigating an LDPC-induced response time delay [4].

2.3. Parity with RAID

While ECC can prevent data errors inside a page, many other approaches for preventing data errors outside a page exist. RAID [29] enhances the reliability by using redundant data. RAID creates an additional parity for clustered data to ensure error recovery. Cluster means a group of units associated with parity data. In a general RAID system, this unit is a disk; however, inside an SSD, it can be a block or a chip. The parity is generated by using bitwise exclusive-or operations between blocks that belong to the same cluster, and thus, parity updates usually require two block reads and two block writes in traditional disk-based storage systems. However, in flash-based storage devices, owing to the out-of-place characteristic of flash memory, the approach of RAID techniques is different; they mainly focus on reducing write requests for designing flash-based RAID schemes.

Previous works show that parity achieved by using the RAID5 technique can reduce the page error rate compared with ECC. However, the parity overhead of the RAID architecture has its own limitations, as a lot of reads for existing data must precede new parity calculations. There are several previous works that have discussed ways to reduce the parity overhead. Lee [8] retains the parity blocks in the buffer memory, postponing parity writes until the idle time, which can reduce read operations and write response time; however, system crash at partial stripe write points cannot be recovered as parity updates are postponed until a full stripe is written. Im [30] generates partial parity for partial stripes so that partial write is possible; however, additional non-volatile RAM (NVRAM) hardware is required as partial parity is maintained in NVRAM. Kim [10] employs variable size striping, which constructs a new stripe with data written to portions of a full stripe and writes parity for that partial stripe without any additional hardware support. However, this approach generates too much partial parity, and therefore, degrades write performance and space utilization.

3. Compression-Assisted Adaptive ECC and RAID Scattering

This Section describes, first, the adaptive ECC and RAID-scattering scheme, followed by the metadata structure and operation of FTL for RAID parity management as an implementation issue.

3.1. Compression-Assisted Adaptive ECC

From the perspective of storage efficiency, data compression is an effective way of enabling a high-density storage system, and a significant amount of user's data in storage media could be compressed according to [31,32]. We incorporated this compression mechanism in an attempt to provide the enhancement of error recoverability. In this system, the incoming write request goes through two steps before being programmed into flash cells, that is, RAID scattering and adaptive ECC with compression, as shown in Figure 3. At first, the incoming data are compressed using the compressor module. Then, the compressed data are placed at some position within the page buffer, and the unused region caused by compression is filled with 1 s because the value of 1 s is electrically stable for flash memory.

In general, the read and write units of NAND flash memory are page units; however, a page is divided into sub-pages for ECC encoding/decoding. The size of a page is usually 4 KB, 8 KB, or 16 KB; each page can be composed of four to eight sub-pages. For example, if a page has an 8 KB data area and a spare area for 1 KB of ECC, it consists of eight 1 KB sub-pages and a space area for 128 bytes of ECC. The write operation method of this system is as follows. When write data are sent from the host, the data are compressed by the compression module. If the compressed and reduced size is not larger than the sub-page size, it is not necessary to store data in more than one sub-page. The compressed data are moved to a specific position within the page by the RAID-scattering module. Thereafter, the ECC is generated by the adaptive ECC module.

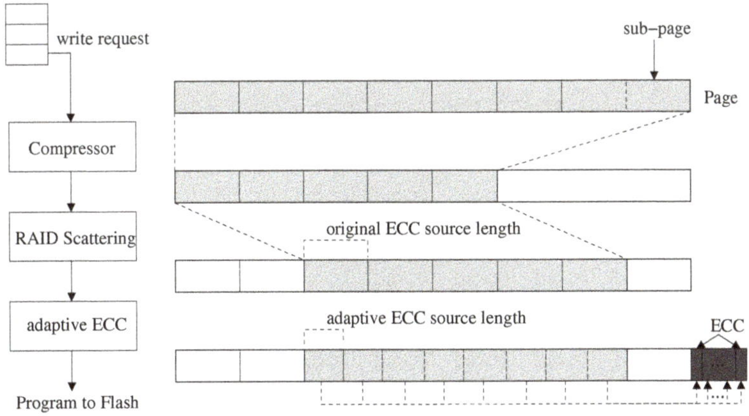

Figure 3. Data write procedure with compression-assisted adaptive ECC and RAID scattering.

Compression modules have already been applied to improve the ECC function [5,6]. However, in existing works, unused areas due to data compression are used to store additional redundancy. In this system, in contrast, we use ECC encoding and decoding only for the compressed region, leaving the unused region intact. This method can be applied with little modification to existing ECC engine modules and encoding schemes. The procedure is as follows. After compression, the size of the compressed data is divided by the number of sub-pages, i.e., N, to obtain the source size of ECC encoding. N is denoted as the number of ECC operations for each page, which is fixed for the implementation of legacy flash controller. This ensures that the source length of the ECC encoding is smaller than the existing sub-page size. With the reduced source by compression, the variable-sized ECC [33] is applied to generate ECC parity, and the generated ECC parity size does not change. The ECC parity size remains the same, and the code rate indicating the ECC parity relative to the source length is lowered, indicating a higher error correction rate.

The instance of the data-writing process and ECC-generation process is described in Figure 3. As shown in the figure, it is assumed that one page, with size 8 KB, is composed of eight sub-pages. The 8 KB of data transmitted from the host were compressed using the compression module, resulting in 5 KB of compressed data. It means that three sub-pages do not need to store data. These data are moved to the third sub-page position by the RAID-scattering scheme. The RAID-scattering scheme is discussed in the next subsection. As the data are reduced from 8 KB to 5 KB by compression, one ECC encoding size of 5 KB data is 640 bytes, which is 5 KB/8. This represents a reduction in code rate, which means that the ECC-correction capability is improved. Because ECC encoding is an error-correction function for data stored in a page, the adaptive ECC scheme improves the in-page error-correction function.

3.2. RAID Scattering

If data are compressed by the compression module, the size of the compressed data would be smaller than the original size, so it occupies only a specific part of the page. In our system, when the compressed data are stored in the flash memory, it is not stored from the beginning of the page but in a specific position in the page. This position is determined according to the stripe number where the page is located in the stripe. Therefore, RAID scattering is defined as positioning compressed data in a specific location on the page according to the position in the corresponding stripe.

For instance, the architecture of RAID scattering with a stripe size four times larger than the page size is described in Figure 4. In the RAID configuration, SG (Stripe Group) is a group of blocks belonging to a stripe, and the notation BI (block index) is a block number in the stripe group; SB (Scatter Base) is a page size divided by the number of blocks in a stripe, and SI (Scatter Index) is the offset for the placement of each compressed data unit in a stripe group, calculated as SB multiplied by BI. For each write request, the compressed data are placed at the offset value SI calculated by SB and BI. If the data do not exceed the page size starting from a specific position, i.e., SI, all of the compressed data can be stored on the page, as appropriate. If the amount of data exceeds the limit of the page, the remaining data are placed from zero offset at the page, in circular manner. For example, as shown in Figure 4, if the number of blocks in a stripe group is four and the page size is 8 KB, data $D5$ are placed at offset zero in the page, data $D6$ are placed at offset 2 KB, data $D7$ is placed from offset 4 KB, and data $D8$ is placed offset by 6 KB in the page, respectively. As a result, all compressed data can be placed at different positions within their stripes by the RAID scattering scheme. Then, RAID parity is created by xor operation of all data in the stripe. In Figure 4, $P2$ is the RAID parity created through the xor operation of $D5$, $D6$, $D7$, and $D8$.

In our system, if data cannot be recovered by in-page ECC, it will be recovered via the outer-page error recovery, i.e., RAID parity. In an existing RAID system, when attempting to recover an error page by using the RAID parity of the stripe group containing the page where the error occurred, all pages in the stripe group must be read to perform the xor operation. However, in the RAID scattering system, it is not necessary to read all pages of the stripe group. Instead of reading all the pages in the stripe group, it is only necessary to read the pages that overlap the actual valid area of the compressed data of the page where the error occurred. For instance, in Figure 4, let us assume errors occur during the read of data $D5$, which means error is not recovered by in-page ECC. In this case, to recover data with RAID parity, the RAID scattering system reads only $D6$ and $D8$ and performs the xor operation to recover $D5$. $D7$ does not overlap with $D5$, and hence reading for the recovery operation is unnecessary.

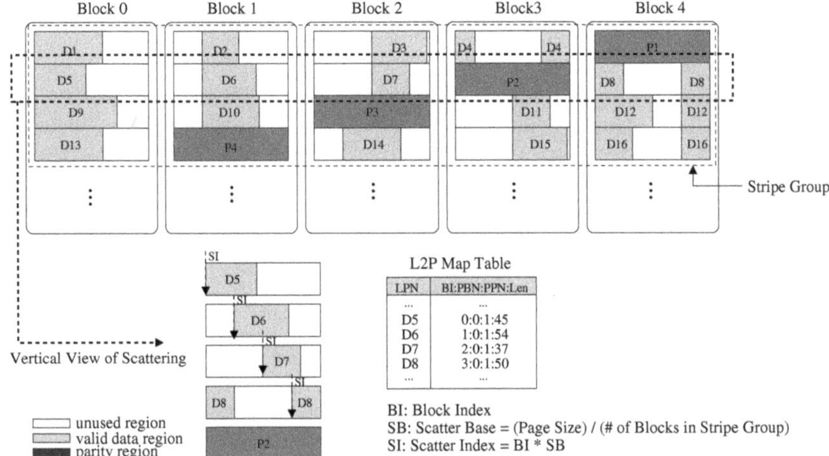

Figure 4. Compression-assisted RAID scattering architecture.

To identify overlap information of pages in a stripe group, the length of the compressed data is added in a mapping-table in the FTL metadata. The compressed length is an important parameter to determine which pages are overlapped for recovery, as well as to decompress data for the read operation. The added length information is depicted in Figure 4. The length of valid data is represented by the actual length of the compressed data. In our implementation, the length field has a compact size for reducing metadata space overhead. For instance, a 12-bit field is required to represent a maximum length of 4 KB page, and a 13-bit field is required to represent a maximum length of 8 KB page. So maximum 3 bytes per entry is required for that.

Using the method of scattering within a stripe group has an effect of reducing the association of pages with each other within a stripe group. In other words, it is possible to increase the size of the stripe group for the same level of association. That is, if a RAID parity in the existing RAID system covered a stripe group with stripe size four, then another RAID parity in the RAID scattering system can cover a stripe group with stripe size eight. This results in reducing the overhead of RAID parity.

3.3. Metadata Structure and Parity Management

The design objective of metadata structure is to reduce flash write operations for metadata and parity data of RAID5 architecture. For this, the metadata changes are updated as a logged fashion with metadata log structure, and parity is always flushed once at all writes of the corresponding stripe is completed. The designed metadata structure for a RAID stripe and parity management is shown in Figure 5, where this metadata are present in the FTL layer. The left part of the figure shows the in-NAND metadata snapshot structure, which is composed of the root block, StripeMap block, PageMap block, and MapLog block. For the root block, either flash memory block 0 or 1 is used as the pivot root block. At first, block 0 is assigned to the root block. The root block contains information of metadata blocks, and it is mainly used to find the metadata blocks at system boot. The last updated page of the root block contains information called RootInfo that contains the location of the blocks where the metadata are stored in Flash memory. Each time the metadata block changes, RootInfo is updated on the one page of the root block as in a log. When the update reaches the last page of the current root block (e.g., block 0), the root block changes to block 1 and the update continues.

The StripeMap block contains the stripe information for each stripe. Those stripe information is managed by three types of data structures: BlkInfo, StripeInfo, and StripeInfoList. The BlkInfo contains the status information for each data block, and StripeInfo has the status information for each stripe group. The size of the stripe group is the number of data blocks included in the stripe.

For example, if the stripe size is S, one stripe group consists of S blocks. StripeInfoList is a hash data structure with a bucket from 0 to $(S \times N)$, where N is the number of valid pages in the stripe group.The stripe information includes stripe number, valid page count for the stripe, page offset, and so on. Each bucket has a list of dual-connected stripe groups with the same valid number of pages. It is used to select the free stripe group for the next write and the victim stripe group to perform when garbage collection (GC) is needed. The PageMap block contains page-mapping table information. Each entry in the page-mapping table is an array of physical page numbers (PPNs). The last part of the metadata is the StripeMapLog block, which is used to log the metadata changes information for data writes. The StripeMapLog block is important for our metadata log architecture. The in-NAND flash metadata blocks represent the latest version of the snapshot, while the main memory contains the latest metadata that has changed since the last snapshot. The metadata changes are written as a logged manner so that the latest metadata can be maintained.

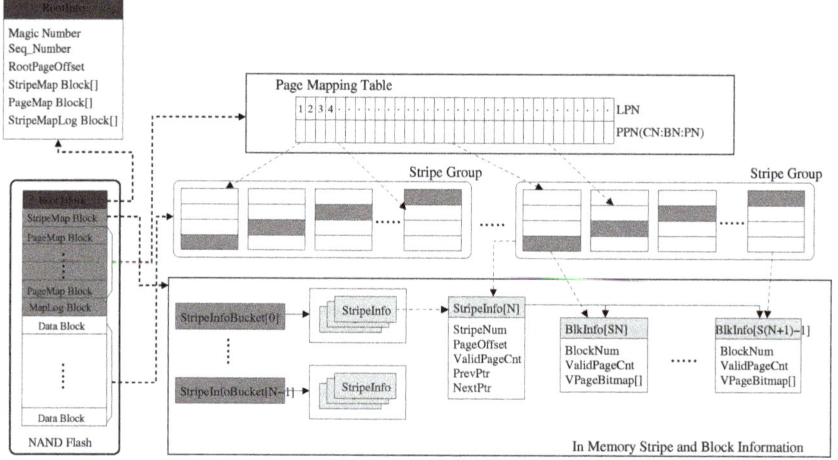

Figure 5. Software metadata structure for RAID and parity management.

A metadata log operation is illustrated in Figure 6. Each metadata log is composed of one StripeMapLogInfo structure, and the StripeMapLogInfo consists of an array of logical addresses of the corresponding stripe group, which represent logical-physical mapping information. A stripe group is a group of blocks grouped together to form a RAID stripe. The stripe group is at the block level; however, the parity for the stripe is at the page level within the group. For example, if a stripe is composed of four blocks, parity is generated based on three pages with the same page offset in the group. The logging strategy is for metadata logging not for data logging. So, whenever data write occurs, the data are just written to flash memory, while its metadata change is not flushed immediately. Each time a new write occurs, new data are allocated to a new page within the stripe group, and the LPN of the corresponding PPN is logged to StripeMapLogInfo in memory. The updated metadata information, that is, the logical addresses for the written data in this stripe group, is just maintained as the StripeMapLogInfo structure information in memory until all pages of the stripe group are used for writes. Once it is filled with a certain amount of metadata updates, the stripeMapLogInfo is flushed to the last available page in the StripeMapLog block. When the StripeMapLog block is full of loaded pages, thus, no longer available for logging, the next available block area is needed. Before allocating a new StripeMapLog block, metadata information in the main memory is synchronized to the flash to be the latest snapshot, and then, a new StripeMapLog block is allocated. After this process, the logging mechanism described above is repeated.

In addition to the metadata logging, full stripe parity update operation is designed and implemented. In our system, parity for the stripe is retained in a buffer memory until full stripe write is done. The full stripe write means that all of the data for the corresponding stripe is written. Since parity is updated in memory buffer with the incoming data of the stripe, it does not need to read old parity and old data to update parity. The parity is just written when a full stripe write is completed. Since partial parity is maintained in memory, it can be lost with power failure. The metadata log information is used during the recovery process of the partial parity from power failure. When power loss occurs during the parity is in partial state retained in buffer, that partial parity can be recovered by scanning the logged information.

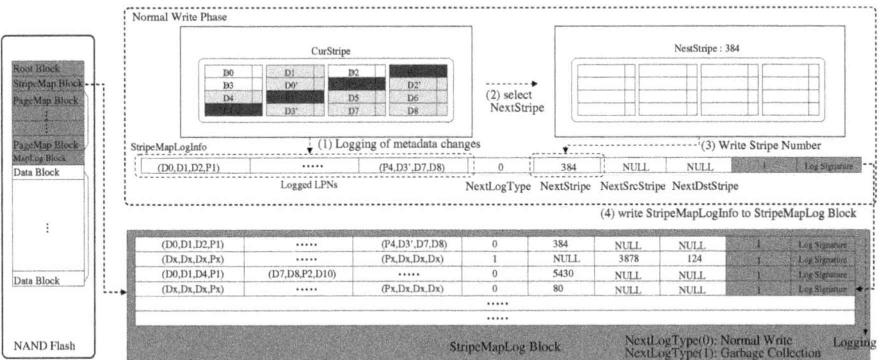

Logged LPNs			NextLogType	NextStripe	NextSrcStripe	NextDstStripe		
(D0,D1,D2,P1)	.....	(P4,D3',D7,D8)	0	384	NULL	NULL	1	Log Signature
(Dx,Dx,Dx,Px)	.....	(Px,Dx,Dx,Dx)	1	NULL	3878	124	1	Log Signature
(D0,D1,D4,P1)	(D7,D8,P2,D10)	.....	0	5430	NULL	NULL	1	Log Signature
(Dx,Dx,Dx,Px)	.....	(Px,Dx,Dx,Dx)	0	80	NULL	NULL	1	Log Signature
.....								
.....								

Figure 6. Stripe map log operations for RAID and parity management.

3.4. Power Loss Recovery Step of Metadata Log and Partial Parity

The metadata log and partial parity retained in memory can be lost from power failure; however, it can be recovered by recovery process. Since the metadata changes after the snapshot are logged to the StripeMapLog block, we can recover metadata changes by scanning the MapLog blocks. For the paritis, almost all parity data are stored in flash memory as a status of parity for full stripe write; the actual portion to be recovered is just the partial parity of write data for the corresponding stripe before power loss.

The metadata log and parity recovery steps are as follows. This step is power loss recovery step, not error recovery step. At the first step, flash storage tries to find the latest RootInfo by scanning *Root* block. In the second step, it recovers the metadata snapshot that consists of StripeInfoList, StripeBlkInfo, BlkInfo, and page mapping table by retrieving StripeMap and PageMap blocks. In the third step, the metadata changes, that occur after the snapshot, were stored in the StripeMapLog blocks, so they are recovered by reading page by page and updating each information. The last step to be recovered is the metadata changes that were losted due to the power loss are recovered. Those were just maintained in-memory StripeMapLogInfo so were not flushed to StripeMapLog block. Metadata changes of the StripeMapLogInfo can be recovered by scanning the corresponding stripe group itself, and we can identify the stripe group number from last valid page in the StripeMapLog Block since we record the number of NextStripe group at last valid StripeMapLog. So, the mapping changes of the NextStripe group can be recovered by scanning the spare region of each page. If the page has valid logical address in the spare region, which means the data are stored safely so that its metadata should be updated. If not, the data are not stored safely and the pages are abandoned. For the scanning of each stripe page, if all the pages of the stripe have valid logical address, which means all the data for the stripe are stored safely, the corresponding parity is also safe. If some pages do not have valid logical addresses, which means not all the data in this stripe are valid, then there is no valid parity for this stripe. For this stripe, each valid page is retrieved and parity is calculated with those. It is the partial parity that is lost by power failure. Therefore, we recovered partial parity for this stripe.

4. Evaluation

In order to prove the validity of the proposed RAID scattering and adaptive ECC, we implemented the proposed schemes inside FlashSim [34,35] that is widely used in the storage system evaluation. This open source simulator includes the simulated NAND flash controller and NAND flash memories that can be set for page size, block size, etc. The simulator also implements well-known FTLs and NAND flash operations, including read, write, and erase, as well as garbage collection. We have added the RAID parity management, compression-based adaptive ECC, and RAID-scattering schemes in this simulator. The simulation configuration set for performance evaluation is as follows. The page-level FTL was used for an FTL mapping management scheme, in which 8 KB was used for page size, and a 1 KB sub-page was used. The compression module uses the zlib [36] compression module. The BCH ECC model is used with 1024BCH16, which is an ECC module that can correct 16 bits per 1024 bytes. That is, 1024BCH16 is applied for the 1 KB sub-page. For the bit error generation and simulation, the BER is sampled and an error is injected according to its P/E cycles. The samples from BER versus P/E cycle data are taken from a previous study on NAND flash devices [24]. The simulated BER sample is depicted in Figure 2.

4.1. Analysis of Adaptive ECC and RAID Scattering

To evaluate the performance of the proposed system in different user environments, various IO traces, such as filesystem workloads, database workloads, web server access data, and media streaming IO, are collected by developers reflecting the real human behavior. The IO traces are collected and categorized as F-1, F-2, F-3, F-4, and F-5 according to the application's characteristics. Among those IO traces, F-1 and F-2 are text-dominant file system and database traces, F-3 and F-4 are media-dominant streaming application traces, and F-5 is a mix of IO traces.

We have measured the impact of compression for various IO traces. We applied page-based data compression using the zlib [36] compression module and measured the compression ratio for each page. The compression ratio distribution for IO traces from F-1 to F-5 is shown in Figure 7. As shown in the figure, compression ratios are distributed differently among all IO traces. For general filesystems and databases, large chunks of data are compressed to less than 50% of the original data, which means much of data for general filesystems and databases are highly compressed. The portion of text-like data in filesystems and databases is large, and these types of data are generally compressed at a high compression rate. On the contrary, for web servers or media streaming applications, not much data are compressed, as depicted in F-3 and F-4. Media-like data, such as videos, images, and audio, are dominant on web servers, which are inherently compressed data, and hence, the compression rate is low, which results in low compression rate.

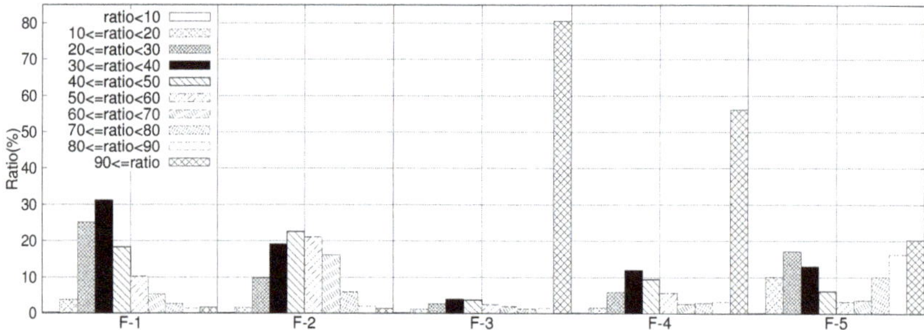

Figure 7. Distribution for compressions for each dataset.

Based on the distribution for the data-compression rate of IO traces, we have evaluated the adaptive ECC scheme and the RAID-scattering scheme. For a fair evaluation, we have simulated IO operations over 10 million times and checked errors for each trace, as bit errors occur rarely. We measured the uncorrectable bit error rate (UBER) [1] for RBER as a performance metric by increasing the P/E cycle from 0 to 5000 for each experiment of each configuration. For UBER a page cannot be recovered by using the error-recovery scheme.

At first, the adaptive ECC scheme is evaluated to show its own performance contribution to error recovery. This is achieved by comparing with the original ECC scheme, that is, fixed-sized ECC encoding. The results of the error-recovery ability for adaptive ECC are plotted in Figure 8. The *x*-axis represents P/E cycles, and the *y*-axis represents UBER on a logarithmic scale. Thus, the result error rate after ECC decoding is presented as UBER. As per the results, the lower the UBER, the better the error-correction capability. As shown in Figure 8, the adaptive ECC scheme results in better UBER over legacy ECC for all IO traces, which means that the adaptive ECC scheme shows a better performance than the fixed ECC scheme in the correction of bit errors. The reason of this increase is the reduction in valid data from compression. From the figure, we can also identify that the performance gap increases for increasing P/E cycles. It means that more errors can be covered by adaptive ECC as more bit errors occur. Among IO traces, F-1 and F-2 result much better UBER than other traces. Because F-1 and F-2 achieve higher compression rates than others, these give smaller source sizes for ECC decoding. On the contrary, F-3 and F-4 show little performance improvement in comparison with that of the original ECC owing to inefficient compression.

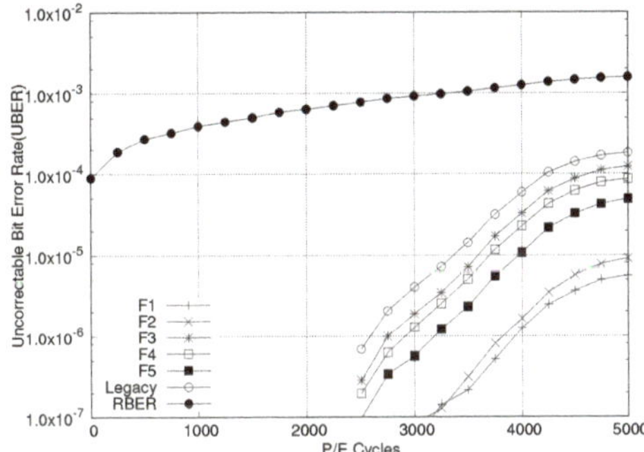

Figure 8. Adaptive ECC scheme for IO traces from F-1 to F-5 in comparison with legacy.

Next, the RAID-scattering scheme is compared with the original RAID scheme having the RAID5 parity model. For this evaluation we have simulated the original RAID and RAID scattering with varying stripe sizes of four, eight, and 16. For all RAID and RAID-scattering configurations, UBER is evaluated as the P/E cycle increases from 0 to 5000. The experimental results are plotted in Figure 9. In all the figures, in accordance with the P/E cycle, RAID-Stripe4, RAID-Stripe8, and RAID-Stripe16 represent UBER for the original RAID configuration with stripe sizes four, eight, and 16, respectively, while Scatter4, Scatter8, and Scatter16 represent UBER for the RAID-scattering scheme with stripe sizes four, eight, and 16, respectively. We have also plotted Legacy UBER to compare results with the RAID-scattering scheme, as well as the RAID system. The Legacy shows the original uncorrected page error rate of the original fixed ECC scheme.

From Figure 9a,b,e, we identify that RAID scattering gives UBER enhancement over the legacy RAID system for F-1, F-2, and F-5. This is due to the high compression ratio of these traces. In the RAID system, the error-containing page can be recovered using the parity page and other data pages within the corresponding stripe. If more than one of those pages also has an uncorrected page error, the error-containing page fails to be recovered. As data are compressed and scattered within a stripe in the RAID-scattering system, there is less probability for overlap between error-containing pages than in the original RAID system. This results in less UBER for RAID-scattering systems for traces F-1, F-2, and F-5. For F-1 and F-2, we identify that RAID scattering with all stripe sizes give less UBER than legacy RAIDs with even smallest stripe size, that is, four. This means that even if the stripe size is large and the parity covers a wider range, the error-correction capability is better than of the original RAID with a smaller stripe size. However, for IO traces with lower compression rates such as F-3 and F-4, as shown in Figure 9c,d, there is no big difference between UBER of the RAID-scattering scheme and UBER of the existing RAID scheme. As compression is less effective for F-3 and F-4, the number of pages to be recovered in the corresponding stripe has an error rate that does not differ much from that of the original RAID scheme. Even so, the figure shows that there is some performance improvement for each P/E cycle.

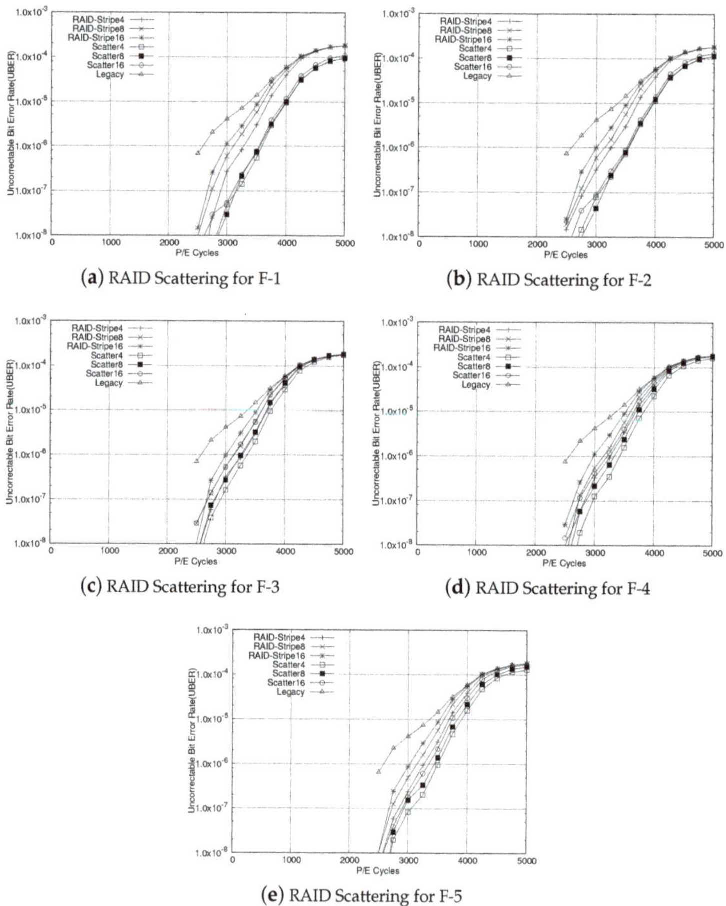

Figure 9. Experimental results for RAID scattering in comparison with original RAID with various stripe sizes. In accordance with P/E cycle, UBERs of RAID Scattering and original RAID are estimated along with stripe sizes four, eight, and 16.(**a**) RAID Scattering for F-1, (**b**) RAID Scattering for F-2, (**c**) RAID Scattering for F-3, (**d**) RAID Scattering for F-4, (**e**) RAID Scattering for F-5.

Finally, the RAID-scattering scheme with the adaptive ECC is compared with the legacy RAID system. In this experiment, each IO trace is modeled as compressed, encoded with adaptive ECC, and stored as scattering within the stripe in the RAID system. Like for the previous evaluation, we have simulated those by varying stripe sizes with four, eight, and 16. For all RAID and RAID scattering with adaptive ECC configurations, UBER is evaluated as the P/E cycle increases from 0 to 5000. The results for traces from F-1 to F-5 are plotted from Figure 10a–e. Similar to the results of the RAID scattering only scheme, all figures show plots for UBER for RAID-Stripe4, RAID-Stripe8, RAID-Stripe16, Adp.ECC-Scatter4, Adp.ECC-Scatter8, and Adp.ECC-Scatter16 configurations in accordance with the P/E cycle. From the figures, we identify that RAID scattering with adaptive ECC shows a lower UBER than that of legacy RAID, as well as RAID-scattering scheme. For all cases, it definitely results in less UBER than legacy RAID systems, regardless of the compression rate of the IO data. Another noticeable point is that the UBER for RAID scattering with adaptive ECC is less than that of the legacy RAID even though the stripe size is larger, which implies that the RAID-scattering scheme can enlarge the stripe size for the RAID system configuration with a higher reliability. In other words, by reducing parity overhead, RAID scattering can reduce write overhead at the same reliability level as it has less parity portion.

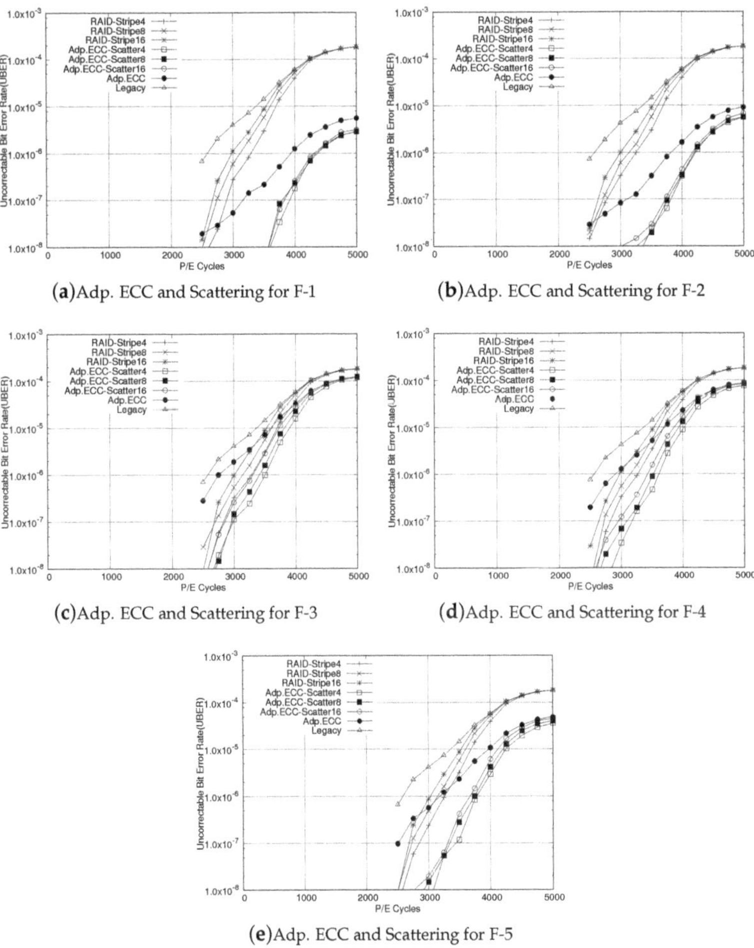

Figure 10. Experimental results for RAID scattering added by adaptive ECC in comparison with the original RAID with various stripe sizes four, eight, and 16. In accordance with P/E cycle, RAID-Stripe4, RAID-Stripe8, RAID-Stripe16, Adp. ECC-Scatter4, Adp. ECC-Scatter8, Ada. ECC-Scatter16 show UBER with stripe sizes four, eight, and 16, respectively. (a) Adp. ECC and Scattering for F-1, (b) Adp. ECC and Scattering for F-2, (c) Adp. ECC and Scattering for F-3, (d) Adp. ECC and Scattering for F-4, (e) Adp. ECC and Scattering for F-5.

To see how much pages can be skipped during the recovery stage in RAID scattering scheme, we have measured the number of pages to be skipped from the recovery operations for the error-occurred page. It was also done with the RAID configuration as stripe size four, eight and 16, respectively, and for each configuration, the five IO traces were performed. Figure 11 shows the ratio of pages skipped during each restoring error-occurred page in the RAID scattering scheme. In the figure, five IO traces are depicted as a group to the same stripe size level, and thus the *x*-axis represents groups of five IO traces according to stripe size level, and the *y*-axis represents the skip ratio. Since the skip ratio represents the ratio of pages that need not be read during recovery; the higher the skip ratio is, the better the recovery rate is as well as the shorter the recovery time is. From the results, it can be inferred that since the compression ratio is high as in the traces F-1 and F-2, the skip ratio is high, which results in reduced UBER. This is an effect of the RAID scattering scheme. On the other hand, the skip ratio of F-3 is low; there is little effect on RAID scattering, which results in little enhancement for UBER.

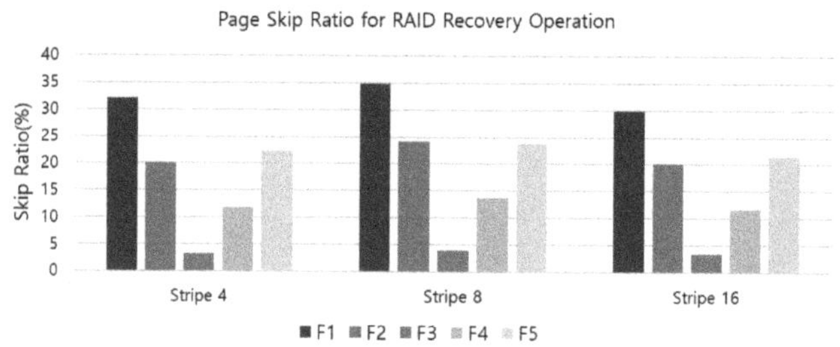

Figure 11. The skip ratio of pages when restoring error-occurred page in the stripe in the RAID scattering scheme.

4.2. Analysis of Metadata and Parity Overhead

The adaptive ECC and RAID-scattering schemes provide an effective error management method that can be achieved by reducing the size of the valid data regions on a page. However, additional overhead is required to support this operation. There is an overhead in storing and managing the size of the compressed data stored in each page. Specifically, a space for recording the compressed data size of the corresponding page is required for each page entry of the FTL mapping table. It takes about 14 bits or 3 bytes per entry. The necessary metadata size can estimated according to page size, block size, and capacity of flash device. For 128 GB flash device, whose page size and block size is 8 KB and 1 MB, respectively, the amount of estimated metadata blocks is about 119–122 blocks which includes two root blocks, four StripeMap blocks, 112 PageMap blocks and 1–4 MapLog blocks. The size is about 120 MB. If the capacity of flash device is double, the amount of estimated metadata blocks is almost double of previous one. For 512 GB flash device, whose page size and block size 8 KB and 3 MB, respectively, the amount of metadata blocks is about 155–158 blocks including two root, two StripeMap, 150 PageMap, and 1–4 MapLog blocks, and whose size is around 237 MB. From the estimation we identify that it is the PageMap block that occupies the largest weight in metadata, and it increases as the number of pages increases. Since we use about 7 bytes per entry of the pagemap to manage the physical address of compressed data, it is considerably larger than the PageMap which does not manage data compression. On the other hand, the additional metadata blocks for stripe management consumes less than 10 blocks for hundreds of GB of flash devices, which is not much overhead.

To see the metadata overhead in the aspect of NAND flash IO, we analyzed the overhead of the metadata logging IO operation. The overhead of the metadata log management is mainly composed of two parts. The first is page write operations that write StripeMapLogInfo, which is a log of metadata changes for the corresponding stripe group. The StripeMaplogInfo is written to a page in the StripeMapLog block whenever the page's current stripe group is used for data write. The other part involves the flushing of whole metadata snapshot, which occurs when all the pages of StripeMapLog block are exhaused. The whole metadata snapshot includes PageMap, StripeMap, and RootInfo. To evaluate the metadata log overhead, we have generated random write operations to the simulator and measured the number of written pages for metadta logs, which include logging of StripeMapLogInfo and the flusing of the entire metadata. Among the metadata log overhead, the duration of writing metadata snapshot is dependent of the number of StripeMapLog blocks. If we have several StripeMapLog blocks, the duration of flushing metadata snapshot is longer since more StripeMapLogInfo can be written to the StripeMapLob blocks. Therefore, the experiment was conducted by changing the number of StripeMapLog blocks from one to four.

Figure 12 shows the results of metadata log overhead for the implemented metadata log system for RAID5 architecture with stripe size four, eight, and 16, respectively. As shown in the figure, metadata log overhead is almost 10 percent if the RAID5 configuration is set with four stripe size and 1 StripeMapLog block. However, the overhead decreases as the stripe size increases and the number of StripeMapLog block increases. As the stripe size increases, the number of pages included in the one stripe group increases, and thus the frequency of flushing the StripeMapLogInfo log increases. In addition, if the number of StripeMapLog blocks is increased, the number of flushing of whole metadata snapshots is reduced. From the results, it can be seen that when the number of StripeMapLog blocks is increased by four or more, the overall metadata log overhead can be reduced to 5% even if the capacity of flash device is hundreds of GB.

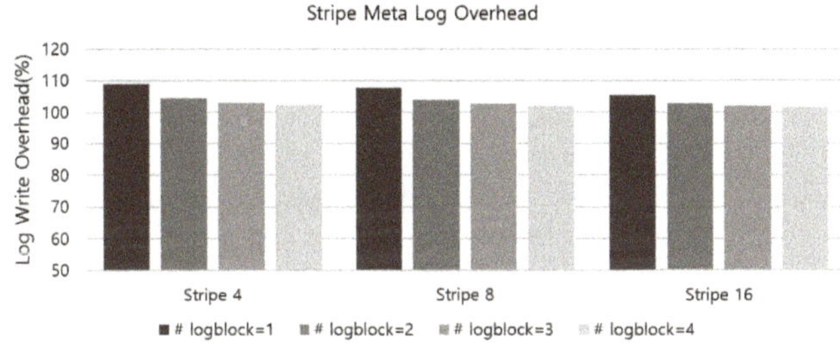

Figure 12. The results of metadata log overhead for the metadata log system for RAID5 architecture with stripe size four, eight, and 16, respectively.

Next, to analyze the parity overhead, we have measured the number of parity writes for random write requests while varying average request size from 4 KB to 128 KB. The experiments are done with two RAID5 configurations, that is, full stripe parity scheme and conventional RAID5 scheme. The full stripe parity scheme implemented in this paper was compared with the conventional RAID5 scheme, in which parities of RAID5 are updated whenever a write request occurs for the corresponding stripe. Those configurations are also set to four, eight, 16 stripe size, respectively.

The result of the parity overhead according to the average random request size is depicted in Figure 13, for each stripe size four, eight, 16, respectively. In the figure, the x-axis represents the average request size and the y-axis represents the number of parity writes per request. The number of parity writes per requests means the number of parities written for each request, and therefore, the lower the value, the smaller the parity overhead. Two RAID configurations, STLog and RAID5, are depicted, where STLog stands for the full stripe parity scheme and RAID5 for the conventional RAID5 scheme. As shown in the Figure 13a, if the stripe size is four, there is little difference between STLog and RAID5. There is lower parity overhead for small requests, and parity overhead becomes similar as the request size increases. However, if the stripe size is bigger, i.e., eight or 16, the parity overhead is largely reduced with the STLog parity management scheme as shown in Figure 13b,c. It is due to the partial parity buffering of the full stripe parity scheme. Particularly, when the request size is smaller than the stripe size, the partial parity buffering applied in STLog is more effective since parity updates are frequently performed in the conventional RAID5 scheme. If the request size exceeds the stripe size, parity for full stripe write is created as it is, so STLog is no different from conventional RAID5.

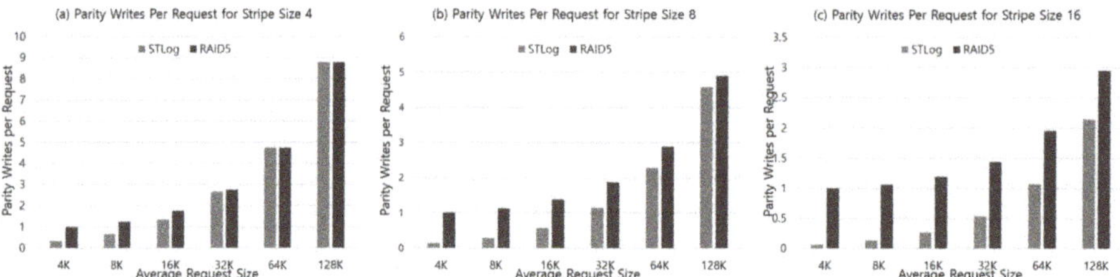

Figure 13. Result of the number of parity writes per request according to the change in the average request size, for each stripe size four, eight, 16, respectively.

4.3. Discussion about Compression Module

The adaptive ECC and RAID scattering is applied based on lossless data compression, and thus the overhead of the data compression module is a performance-deteriorating factor for NAND flash memory devices. Although the software zlib module is used in our system for data compression, the compression module requires a lot of CPU computation, and hence it is not suitable for practical use in software implementations. However, recent compression hardware [37] or accelerator [38] further increases the practical possibilities of lossless data compression. In particular, [37] describes that it supports gzip/zlib/deflate data compression with approximately 100 Gbps bandwidth. If such a compression module is included in the NAND flash controller, it is expected to be able to perform data compression with little effect to the performance degradation of other modules, although it will increase the cost. In this paper, we could not analyze the performance degradation effect by the compression module, but in the future, there is a need to analyze the performance of the overall system including the hardware compression module. This is our further work.

5. Conclusions

The density of flash memory devices has increased by moving to smaller geometries and storing more bits per cell; however, this generates a lot of data errors. The typical approach to check these increasing errors is the use of well-known ECCs; however, the current state of the ECC scheme it not on the same level with the evolution of errors of the current state of flash memory. The RAID technology can be employed for flash storage devices to enhance their reliability. However, RAID has inherent parity update overhead.

In this paper, we propose the enhancement for ECC ability for inside the page and RAID parity management for outside the page, with the help of lossless data compression. The compressed data are encoded with adaptively reduced source length, which increases the recovery ability of the ECC module. The adaptive ECC method can reduce the size of a source length in proportion to the compression ratio. As this reduces the source length to be encoded, the effect of lowering the code rate increases the error-correction rate. For the perspective of outside the page, the compressed data are placed at a specific position in the page. As the unused area in the page generated by compression can be used as a non-overlapping area in RAID scattering, it is effective during the RAID recovery stage. The unused area is spread out within the parity cluster, which reduces the error coverage range due to the possible skipping pages that are not overlapping with the error-occurred page. The experimental results show that adaptive ECC can enhance the ability of error recovery. Furthermore, the RAID-scattering scheme achieves a high reliability at the same RAID stripe level. As the next step of this study, we will apply compression, ECC, and RAID scattering modules as hardware modules to evaluate the actual overhead and performance impact. Therefore, a more rigorous comparison of the performance of the propose scheme versus others could be an important task to improve the completeness of the proposed scheme. Consequently, we set the more rigorous performance evaluations as our further work.

Author Contributions: Conceptualization, S.-H.L. and K.-W.P.; methodology, S.-H.L. and K.-W.P.; software, S.-H.L.; validation, S.-H.L.; formal analysis, S.-H.L. and K.-W.P.; investigation, S.-H.L. and K.-W.P.; resources, S.-H.L. and K.-W.P.; data curation, S.-H.L. and K.-W.P.; writing–original draft preparation, S.-H.L. and K.-W.P.; writing–review and editing, S.-H.L. and K.-W.P.; visualization, S.-H.L. and K.-W.P.; supervision, S.-H.L. and K.-W.P.; project administration, S.-H.L. and K.-W.P. All authors have read and agreed to the published version of the manuscript.

Funding: This work was supported by the National Research Foundation of Korea (NRF) grant funded by the Korean Government (MSIT) (NRF-2019R1F1A1057503). This work was supported by Hankuk University of Foreign Studies Research Fund. Also, this work was supported by the National Research Foundation of Korea (NRF) grant funded by the Korean Government (MSIT) (NRF-2020R1A2C4002737).

Conflicts of Interest: The authors declare no conflict of interest.

Abbreviations

The following abbreviations are used in this manuscript:

ECC	Error Correction Code
RAID	Redundant Array of Independent Disks
SLC	Single-level cell
MLC	Multi-level cell
TLC	Triple-level cell
OOB	Out-of-Band
BCH	Block Hamming Code
LDPC	Low-Density Parity Check
SSD	Solid-State Disks
GC	Garbage Collection
FTL	Flash Translation Layer
LPN	Logical Page Number
PPN	Physical Page Number
SG	Stripe Group
BER	Bit Error Rate
RBER	Raw Bit Error Rate
UBER	Uncorrectable Bit Error Rate
NVRAM	Non-Volatile Random Access Memory
GC	Garbase Collection

References

1. Mielke, N.; Marquart, T.; Wu, N.; Kessenich, J.; Belgal, H.; Schares, E.; Trivedi, F.; Goodness, E.; Nevill, L.R. Bit error rate in NAND Flash memories. In Proceedings of the 2008 IEEE International Reliability Physics Symposium, Phoenix, AZ, USA, 27 April–1 May 2008; doi:10.1109/RELPHY.2008.4558857. [CrossRef]
2. Cai, Y.; Mutlu, O.; Haratsch, E.F.; Mai, K. Program interference in MLC NAND flash memory: Characterization, modeling, and mitigation. In Proceedings of the 2013 IEEE 31st International Conference on Computer Design (ICCD), Asheville, NC, USA, 6–9 October 2013; pp. 123–130.
3. Micron. *Enabling Software BCH ECC on a Linux Platform*; Technical Note, TN-29-71; Micron: Boise, ID, USA, 2012.
4. Zhao, K.; Zhao, W.; Sun, H.; Zhang, X.; Zheng, N.; Zhang, T. LDPC-in-SSD: Making advanced Error Correction Codes Work Effectively in Solid State Drives. In Proceedings of the 11th USENIX Conference on File and Storage Technologies, San Jose, CA, USA, 12–15 February 2013; pp. 243–256.
5. Xie, N.; Dong, G.; Zhang, T. Using lossless data compression in data storage systems: Not for saving space. *IEEE Trans. Comput.* **2011**, *60*, 335. [CrossRef]
6. Ahrens, T.; Rajab, M.; Freudenberger, J. Compression of short data blocks to improve the reliability of non-volatile flash memories. In Proceedings of the 2016 International Conference on Information and Digital Technologies (IDT), Rzeszow, Poland, 5–7 July 2016; pp. 1–4.

7. Chen, P.M.; Lee, E.K. Striping in a RAID level 5 disk array? In Proceedings of the 1995 ACM SIGMETRICS Joint International Conference on Measurement and Modeling of Computer Systems, Ottawa, ON, Canada, 15–19 May 1995; p. 136.

8. Lee, Y.; Jung, S.; Song, Y.H. FRA: A flash-aware redundancy array of flash storage devices. In Proceedings of the 7th IEEE/ACM International Conference on Hardware/Software Codesign and System Synthesis, Grenoble, France, 11–16 October 2009; p. 163.

9. Qin, Y.; Feng, D.; Liu, J.; Tong, W.; Hu, Y.; Zhu, Z. A Parity Scheme to Enhance Reliability for SSDs. In Proceedings of the 7th International Conference on Networking, Architecture, and Storage, Xiamen, China, 28–30 June 2012; pp. 293–297.

10. Kim, J.; Lee, E.; Choi, J.; Lee, D.; Noh, S.H. Chip-Level RAID with Flexible Stripe Size and Parity Placement for Enhanced SSD Reliability. *IEEE Trans. Comput.* **2016**, *65*, 1116. [CrossRef]

11. Kim, K.; Lim, S.H. Compression and Variable-Sized ECC Scheme for the Reliable Flash Memory System. In *Advances in Computer Science and Ubiquitous Computing*; Springer: Singapore, 2017.

12. Intel Corporation. *Understanding the Flash Translation Layer (FTL) Specification*; Application Note; Intel Corporation: Santa Clara, CA, USA, 1998.

13. Kim, J.; Kim, J.M.; Noh, S.H.; Min, S.L.; Cho, Y. A space-efficient flash translation layer for CompactFlash systems. *IEEE Trans. Consum. Electron.* **2002**, *48*, 366–375.

14. Kang, J.U.; Jo, H.; Kim, J.S.; Lee, J. A superblock-based flash translation layer for NAND Flash memory. In Proceedings of the 6th ACM & IEEE International Conference on Embedded Software, Seoul, Korea, 22–25 October 2006; pp. 161–170.

15. Lee, S.W.; Choi, W.K.; Park, D.J. FAST: An efficient flash translation layer for flash memory. In *International Conference on Embedded and Ubiquitous Computing*; Springer: Berlin/Heidelberg, Germany, 2006; pp. 879–887.

16. Gupta, A.; Kim, Y.; Urgaonkar, B. DFTL: A Flash Translation Layer Employing Demand-based Selective Caching of Page-level Address Mappings. In Proceedings of the 14th International Conference on Architectural Support for Programming Languages and Operating Systems, Washington, DC, USA, 7–11 March 2009; pp. 229–240.

17. Ma, D.; Feng, J.; Li, G. LazyFTL: A Page-level Flash Translation Layer Optimized for NAND Flash Memory. In Proceesings of the ACM SIGMOD, Athens, Greece, 12–16 June 2011; pp. 1–12.

18. Chen, F.; Lee, R.; Zhang, X. Essential roles of exploiting internal parallelism of flash memory based solid state drives in high-speed data processing. In Proceedings of the International Symposium on High Performance Computer Architecture (HPCA), San Antonio, TX, USA, 12–16 February 2011; pp. 266–277.

19. Agrawal, N.; Prabhakaran, V.; Wobber, T.; Davis, J.D.; Manasse, M.; Panigrahy, R. Design tradeoffs for SSD performance. In Proceedings of the USENIX Annual Technical Conference, Boston, MA, USA, 22–27 June 2008; pp. 57–70.

20. Seong, Y.J.; Nam, E.H.; Yoon, J.H.; Kim, H.; Choi, J.Y.; Lee, S.; Bae, Y.H.; Lee, J.; Cho, Y.; Min, S.L. Hydra: A Block-Mapped Parallel Flash Memory Solid-State Disk Architecture. *IEEE Trans. Comput.* **2010**, *59*, 905–921. [CrossRef]

21. Hu, Y.; Jiang, H.; Feng, D.; Tian, L.; Luo, H.; Zhang, S. Performance impact and interplay of ssd parallelism through advanced commands, allocation strategy and data granularity. In Proceedings of the International Conference on Supercomputing, Tucson, AZ, USA, 31 May–4 June 2011; pp. 96–107.

22. Jung, M.; Kandemir, M. An evaluation of different page allocation strategies on high-speed SSDs. In Proceedings of HotStorage, Boston, MA, USA, 13–14 June 2012.

23. Bjorling, M.; Axboe, J.; Nellans, D.; Bonnet, P. Linux Block IO: Introducing Multi-queue SSD Access on Multi-core Systems. In Proceedings of the 6th Intetnational Systems and Storage Conference, Haifa, Israel, 30 June–2 July 2013; pp. 1–10.

24. Yaakobi, E.; Grupp, L.; Siegel, P.H.; Swanson, S.; Wolf, J.K. Characterization and error-correcting codes for TLC flash memories. In Proceedings of the International Conference on Computing, Networking and Communications (ICNC), Maui, HI, USA, 30 January–2 February 2012; pp. 486–491.

25. Yang, J. Novel ECC architecture enhances storage system reliability. In Proceedings of the Flash Memory Summit, Santa Clara, CA, USA, 22–24 August 2012.

26. Yeo, E. An LDPC-enabled flash controller in 40 nm CMOS. In Proceedings of the Flash Memory Summit, Santa Clara, CA, USA, 22–24 August 2012.

27. Motwani, R.; Ong, C. Robust decoder architecture for multi-level flash memory storage channels. In Proceedings of the International Conference on Computing, Networking and Communications (ICNC), Maui, HI, USA, 30 January–2 February 2012; pp. 492–496.

28. Tanakamaru, S.; Yanagihara, Y.; Takeuchi, K. Over-10x-extended-lifetime 76%-reduced-error solid-state drives (SSDs) with error-prediction LDPC architecture and error-recovery scheme. In Proceedings of the IEEE International Solid-State Circuits Conference (ISSCC), San Francisco, CA, USA, 19–23 February 2012; pp. 424–426.

29. Chen, P.M.; Lee, E.K.; Gibson, G.A.; Katz, R.H.; Patterson, D.A. Patterson, RAID: High-Performance, Reliable Secondary Storage. *ACM Comput. Surv.* **1994**, *26*, 145–185. [CrossRef]

30. Im, S.; Shin, D. Flash-Aware RAID Techniques for Dependable and High-Performance Flash Memory SSD. *IEEE Trans. Comput.* **2011**, *60*, 80–92. [CrossRef]

31. Kulkarni, P.; Douglis, F.; LaVoie, J.D.; Tracey, J.M. Redundancy elimination within large collections of files. In Proceedings of the USENIX 2004 Annual Technical Conference, Boston, MA, USA, 27 June–2 July 2004; pp. 59–72.

32. Meyer, D.T.; Bolosky, W.J. A study of practical deduplication. *ACM Trans. Storage* **2012**, *7*, 14. [CrossRef]

33. Park, J.; Park, J.; Bhunia, S. VL-ECC: Variable Data-Length Error Correction Code for Embedded Memory in DSP Applications. *IEEE Trans. Circ. Syst. II* **2014**, *6*, 120–124. [CrossRef]

34. Kim, Y.; Tauras, B.; Gupta, A.; Urgaonkar, B. FlashSim: A Simulator for NAND Flash-based Solid-State Drives. In Proceedings of the First International Conference on Advances in System Simulation, Porto, Portugal, 20–25 September 2009; pp. 125–131.

35. Matias Bjorling. Extended FlashSim. GitHub. Available online: https://github.com/MatiasBjorling/flashsim (accessed on 1 August 2018).

36. Gailly, J.; Adler, M. zlib: A Massively Spiffy Yet Delicately Unobtrusive Compression Library. Available online: https://zlib.net (accessed on 1 December 2018).

37. Xilinx Corporation. GZIP/ZLIB/Deflate Data Compression Core. Available online: https://www.xilinx.com/products/intellectual-property/1-7aisy9.html#overview (accessed on 20 March 2020).

38. Intel Corporation. Intelligent Storage Acceleration Library. Available online: https://software.intel.com/en-us/isa-l (accessed on 20 March 2020).

© 2020 by the authors. Licensee MDPI, Basel, Switzerland. This article is an open access article distributed under the terms and conditions of the Creative Commons Attribution (CC BY) license (http://creativecommons.org/licenses/by/4.0/).

Article

An Efficient Data-Hiding Scheme Based on Multidimensional Mini-SuDoKu

Ji-Hwei Horng [1], Shuying Xu [2,*], Ching-Chun Chang [3] and Chin-Chen Chang [2,4]

1   Department of Electronic Engineering, National Quemoy University, Kinmen 89250, Taiwan; horng@email.nqu.edu.tw
2   Department of Information Engineering and Computer Science, Feng Chia University, Taichung 40724, Taiwan; alan3c@gmail.com
3   Department of Electronic Engineering, Tsinghua University, Beijing 100084, China; c.c.chang.phd@gmail.com
4   School of Computer Science and Technology, Hangzhou Dianzi University, Hangzhou 310018, China
*   Correspondence: sy15160082345@gmail.com

Received: 16 April 2020; Accepted: 9 May 2020; Published: 11 May 2020

Abstract: The massive Internet of Things (IoT) connecting various types of intelligent sensors for goods tracking in logistics, environmental monitoring and smart grid management is a crucial future ICT. High-end security and low power consumption are major requirements in scaling up the IoT. In this research, we propose an efficient data-hiding scheme to deal with the security problems and power saving issues of multimedia communication among IoT devises. Data hiding is the practice of hiding secret data into cover images in order to conceal and prevent secret data from being intercepted by malicious attackers. One of the established research streams of data-hiding methods is based on reference matrices (RM). In this study, we propose an efficient data-hiding scheme based on multidimensional mini-SuDoKu RM. The proposed RM possesses high complexity and can effectively improve the security of data hiding. In addition, this study also defines a range locator function which can significantly improve the embedding efficiency of multidimensional RM. Experimental results show that our data-hiding scheme can not only obtain better image quality, but also achieve higher embedding capacity than other related schemes.

Keywords: data hiding; multidimensional; embedding efficiency; mini-SuDoKu; security

---

1. Introduction

We live in the information age in which no such immense amounts of digital information have ever before been consistently transmitted over open communication channels. As a consequence, information security has become a research hotspot. Today, there are two types of strategies for securing information against unauthorized access during transmission. The first type of methods encrypts data by cryptographic algorithms such as RSA [1], DES [2], elliptic-curve signcryption [3] and the blockchain-based solution [4]. However, the use of encryption would easily attract the attention of malicious attackers, causing them to intercept the encrypted data and subsequently use computers of sufficient power to break the encryption [5]. By contrast, the second type of methods hides secret data into cover images by steganographic algorithms and therefore conceals the existence of secret data [6]. After the recipient obtains the stego images, the secret data can be decoded through the corresponding algorithm. Steganographic methods can effectively prevent the interception of the secret data because they conceal the very fact that secret data exists. Indeed, this type of methods has attracted an increasing amount of research attention.

Most data-hiding schemes are performed in the following three domains: frequency domain [7,8], compression domain [9–12] and spatial domain [13–16]. Most developers are devoted to devise data-hiding schemes in the spatial domain due to its explicitness and convenience for implementations [17–19].

For spatial domain-based data-hiding schemes, reference matrices (RM), as a means of modifying pixels, can achieve low distortion and high embedding capacity. The concept of RM originated from the exploiting modification-direction (EMD) scheme proposed by Zhang and Wang [16] in 2006. Kim et al. proposed an improved version called EMD-2, which modifies the value of up to two pixels in a unit [14]. Compared to the original EMD scheme, this scheme improves the embedding capacity while ensuring the image quality. As another follow-up work, Chang et al. utilized SuDoKu tables as the RM [20]. In this scheme, each pixel pair in the cover image can hide a 9-ary binary secret data, which greatly increases the hiding capacity. Hong et al. [21] proposed a scheme that calculates the distance between pixels by the nearest Euclidean distance, which obtained even better image quality. Turtle shell-based RM is characterized by hexagon shaped shells and is able to hide 3 bits of secret information per pixel [22]. Liu et al. classified the locations on the turtle shell matrix into 16 situations to further improve the embedding capacity with the aid of a location table [23]. Jin et al. combined the data-hiding schemes of the turtle shell and swarm optimization algorithm to improve the visual quality of the image [24]. Our method is inspired by He et al.'s mini-SuDoKu matrix (MSM) [25]. More details of related works will be presented in the following section.

In this study, we propose a 3D RM based on the MSM. Using the 3D-MSM for data hiding can hide more secret data while ensuring image quality. The main contributions of this study are as follows: First, this study proposes a novel 3D reference matrix based on the MSM matrix. Second, it achieves good image quality and embedding capacity. Third, an efficient algorithm is devised to embed secret data. Finally, the proposed algorithm can be generalized to RM of arbitrary N dimensions.

The rest of this study is organized as follows: Section 2 briefly introduces two data-hiding algorithms based on reference matrices. Section 3 introduces the 3D MSM proposed in this study, presents the embedding process, and analyzes the time efficiency of the proposed algorithm. Section 4 compares the proposed scheme with other RM-based data-hiding schemes.

2. Related Work

Recent RM-based data-hiding schemes can be characterized into turtle shell-based schemes and SuDoKu-based schemes. In order to pave the way for our idea of multidimensional mini-SuDoKu RM, these two types of matrices are briefly reviewed.

2.1. Turtle Shell Matrix Data Hiding

In the turtle shell-based scheme proposed by Liu et al. [23], the turtle shell matrix $M = \left[m(i,j)_{i,j \in \{0,1,\cdots,255\}} \right]$ is consisted of a number of hexagons, called turtle shells, with a size of 256×256, as shown in Figure 1. The RM is filled with 8-ary digits, the incremental value of each row is always 1, while the incremental value of each column change of 2 and 3 in turn. Hence, each turtle shell structure in Figure 1 contains distinct values from 0 to 7. In order to further improve the hiding capacity, a location table is constructed as shown in Figure 2. The location table T contains all 16 possible situations of turtle shell in the RM. The 16 situations in the location table can be grouped into four categories, as shown in Figure 3. According to the characteristics of the RM, the set of values of elements matching location 1 and location 4 is always {1,3,5,7} and matching location 2 and location 3 is always {0,2,4,6}. Each location in the location table T can be represented by $T(s_i, s_{i+1})$, where s_i indicates the i-th row, and s_{i+1} indicates the (i+1)-th column, and s_i and s_{i+1} belong to {00,01,10,11}.

Next, the process of data hiding is described below. First, the original image is cut into pixel pairs (P_i, P_{i+1}), and the binary secret data stream is cut into two sub-streams s_j and s_{j+1}, each of which contains a 2-bit binary number. Second, a pair of cover pixels (P_i, P_{i+1}) is applied to locate an element $m(P_i, P_{i+1})$ in the RM. After that, employ (s_j, s_{j+1}) as coordinates to find the corresponding $T(s_j, s_{j+1})$ in the location table. Then, find the closest element, making $m(P_i', P_{i+1}') = T(s_j, s_{j+1})$. Finally, modify the values of the pixel pair to embed it.

By taking relative location of a number in the turtle shell into account, this data-hiding scheme improves the embedding capacity of the original turtle shell scheme from 1.5 to 2. However, the image quality is degraded due to increase of embedding area.

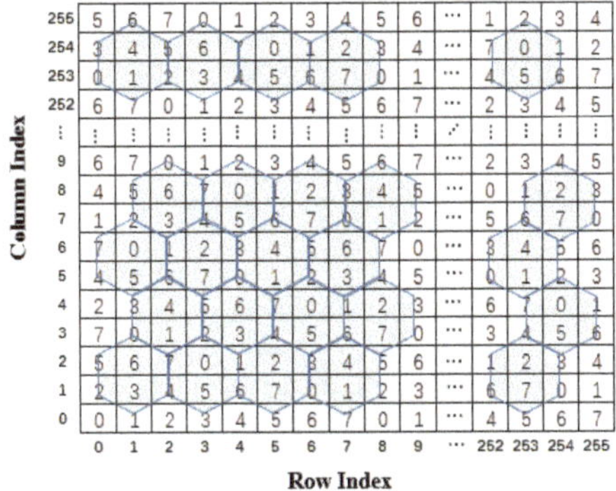

Figure 1. Turtle shell reference matrix.

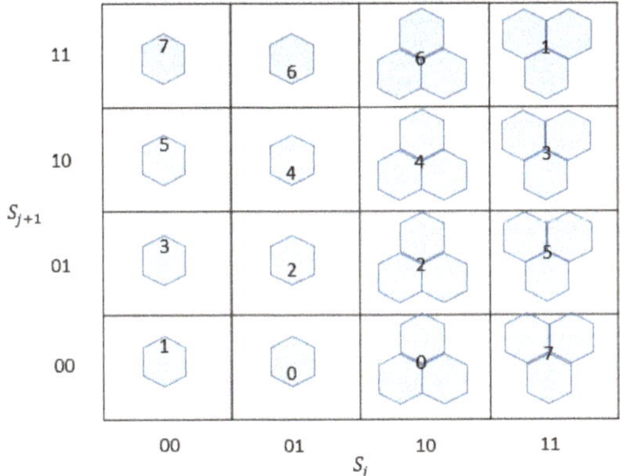

Figure 2. Location table T.

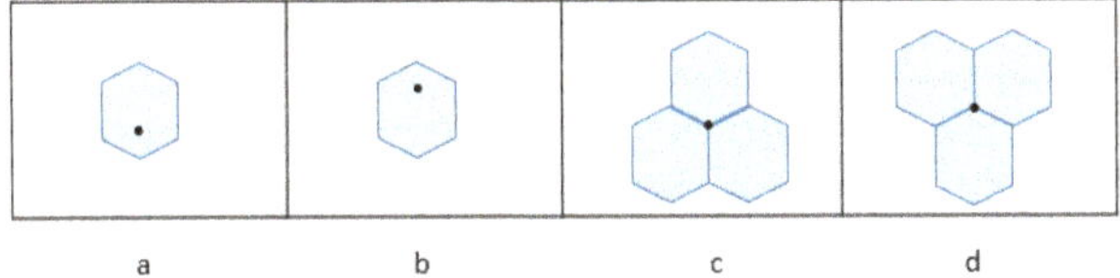

Figure 3. Four locations of elements in M.

2.2. Mini SuDoKu Matrix-Based Data Hiding

The SuDoKu is a matrix which contains nine 3 × 3 sub-matrices with numbers from 1 to 9. In addition, each number is used only once in each raw and each column. Inspired by the conventional SuDoKu, the mini-SuDoKu matrix (MSM) [25] was proposed.

As shown in Figure 4, the MSM is a matrix that contains 4096 4 × 4 submatrices, and each submatrix contains 4 basic structures. Each basic structure is filled with numbers from 0 to 3. In addition, the digits from 0 to 3 must occur just once in each row and each column of the submatrix.

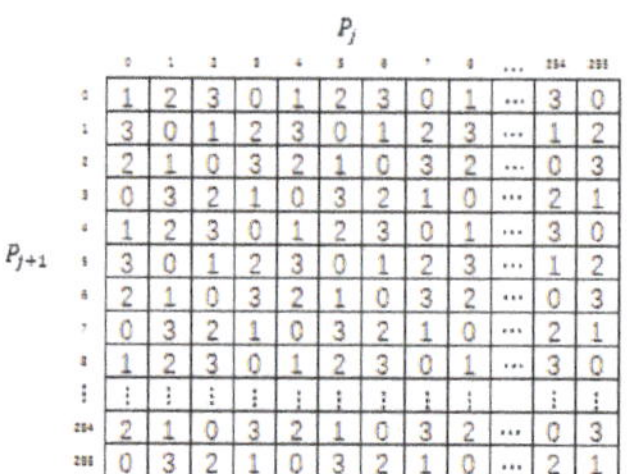

Figure 4. Mini-SuDoKu reference matrix.

The data-hiding scheme using MSM is briefly described as follows: First, the original image is cut into pixel pairs (p_k, p_{k+1}). The secret data stream is cut into four-bit substreams $sg_d = \{s_{4d-3}, s_{4d-2}, s_{4d-1}, s_{4d}\}$, and then we divide sg_d into three groups, i.e., $B_1 = s_{4d-3}s_{4d-2}$, $B_2 = s_{4d-1}$, $B_3 = s_{4d}$. Thus, B_1 is a quaternary digit from 0 to 3 and B_2/B_3 is a bit of 0 or 1. Second, take the pixel pair as coordinates and locate $MSM(p_k, p_{k+1})$ in the MSM. Then, a 4×4 candidate block G is determined by (1). Find the element $MSM(p_k', p_{k+1}')$ within G satisfying $m(p_k', p_{k+1}') = B_1$, $mod(p_k', 2) = B_2$ and $mod(p_{k+1}', 2) = B_3$. Finally, modify the pair (p_k, p_{k+1}) to (p_k', p_{k+1}'). We can embed the secret data and get the stego image by repeating these steps.

$$G = \begin{cases} MSM(0:3, 0:3), & \text{if } p_k \le 1 \text{ and } p_{k+1} \le 1; \\ MSM(0:3, p_{k+1}-2:p_{k+1}+1), & \text{if } p_k \le 1 \text{ and } 1 < p_{k+1} < 255; \\ MSM(0:3, 252:255), & \text{if } p_k \le 1 \text{ and } p_{k+1} = 255; \\ MSM(p_k-2:p_k+1, 0:3), & \text{if } 1 < p_k < 255 \text{ and } p_{k+1} \le 1; \\ MSM(p_k-2:p_k+1, p_{k+1}-2:p_{k+1}+1), & \text{if } 1 < p_k < 255 \text{ and } 1 < p_{k+1} < 255; \\ MSM(p_k-2:p_k+1, 252:255), & \text{if } 1 < p_k < 255 \text{ and } p_{k+1} = 255; \\ MSM(252:255, 0:3), & \text{if } p_k = 255 \text{ and } p_{k+1} \le 1; \\ MSM(252:255, p_{k+1}-2:p_{k+1}+1), & \text{if } p_k = 255 \text{ and } 1 < p_{k+1} < 255; \\ MSM(252:255, 252:255), & \text{otherwise} \end{cases} \quad (1)$$

The mini-SuDoKu matrix suffers from the problem of low security level. There are too many constraints on the construction rules. Each row, each column and each basic structure of size 2×2 must contain distinct values of 0 to 3. To obtain a better PSNR (peak signal to noise ratio) performance, the candidate block for a regular element is defined as $MSM(p_k-2:p_k+1, p_{k+1}-2:p_{k+1}+1)$. The two axial ranges of the candidate block do not always coincide with an original 4×4 submatrix. To satisfy the translational invariant requirement for embedding, the whole MSM should repeat the same 4×4 submatrix. In addition, the candidates of embedding should satisfy $mod(p_k', 2) = B_2$ and $mod(p_{k+1}', 2) = B_3$. An example grouping of embedding candidates for different combinations of B_2 and B_3 is shown in Figure 5. The elements in the same group also should contain all values of 0 to 3. These requirements severely restrict the variety of the MSM and thus threaten the security of data hiding.

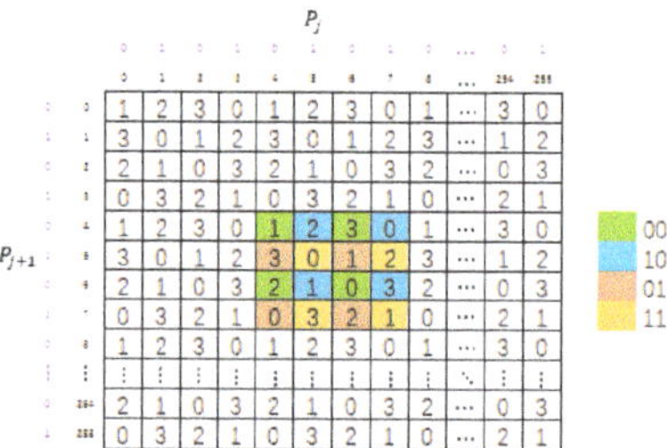

Figure 5. Candidates of embedding for different combinations of B_2 and B_3.

3. The Proposed Scheme

In this section, we will introduce the proposed cubic mini-SuDoKu matrix and a two-layered data-hiding scheme based on the proposed matrix. Then, the matrix and its corresponding data-embedding and extraction algorithm will be generalized to n-dimensional version. Some mechanisms for improving the time efficiency will also be presented.

3.1. Cubic Mini-SuDoKu Matrix (CMSM)

In this study, we propose a two-layered hiding scheme based on a cubic mini-SuDoKu matrix. By leveraging the proposed cubic mini-SuDoKu matrix, the proposed data-hiding scheme can embed secret data with an efficient way and produce stego images of good visual quality.

3.1.1. Construction of the Cubic Mini-SuDoKu Matrix

The cubic mini-SuDoKu matrix is a $256 \times 256 \times 256$ matrix that contains $64 \times 64 \times 64$ sub-cubes of size $4 \times 4 \times 4$. Each sub-cube contains eight basic structures of size $2 \times 2 \times 2$. The basic structures are labeled with bold Arabic numerals as shown in Figure 6. Elements of each basic structure are randomly assigned with distinct values of 0 to 7. The resulting RM is denoted as $M(x, y, z)$, $x, y, z = 0, 1, \ldots, 255$.

Before embedding, secret data and cover image should be prepared. Binary secret stream is divided into segments of 6 digits each, while the pixels of the cover image are grouped into triplets. A set of three pixels in a triplet $\left(p_{xi}, p_{yi}, p_{zi}\right)$ is used to embed a secret segment of 6 digits $s_j = \left(d_5^j d_4^j d_3^j d_2^j d_1^j d_0^j\right)$. First, the values of the three cover pixels are applied as the coordinates to locate a reference element in the 3D RM. Then, a two-layered embedding scheme is executed. The outer layer is to obtain a matched basic structure by using the three most significant bits (3 MSBs) of secret segment $s_j^M = \left(d_5^j d_4^j d_3^j\right)$. Subsequently, the inner layer of embedding is to find an element within the obtained basic structure with a value matching the three least significant bits (3 LSBs) $s_j^L = \left(d_2^j d_1^j d_0^j\right)$. Finally, the pixel values are modified to the indices of the matched element.

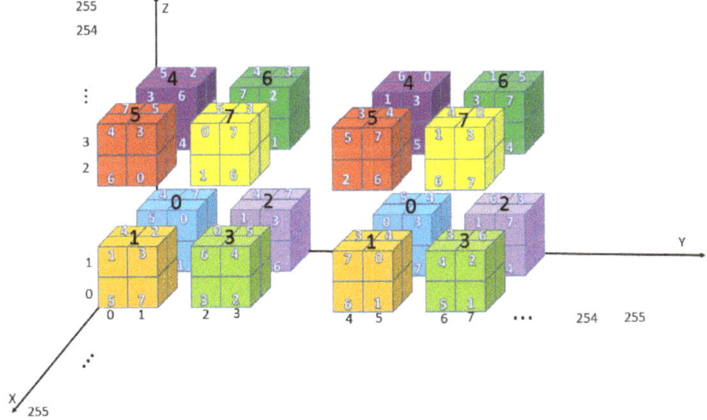

Figure 6. Architecture of cubic matrix.

To improve the efficiency of outer layer embedding, we define a range locator function G to identify the precise searching range for each direction of axis. By combining ranges of all axes, the basic structure matching the 3 MSBs can be determined.

Let p be the pixel value and $w = mod(p, 4)$ as shown in Figure 7. We always apply the segment [0:1] of w to embed secret digit $d = 0$, while apply the segment [2:3] of w to embed $d = 1$. To meet this constraint and minimize the modification distortion, we choose the nearest formal segment to embed. For convenience, we define a range front matrix Δ to record the offset values from the current pixel to the range front of embedding. In the case of $d = 0$, the offset is -2 for $mod(p, 4) = 2$; $+1$ for $mod(p, 4) = 3$; 0 and -1 for $mod(p, 4) = 0$ and 1, respectively. In the case of $d = 1$, the offset is -2 for $mod(p, 4) = 0$; $+1$ for $mod(p, 4) = 1$; 0 and -1 for $mod(p, 4) = 2$ and 3, respectively. The resulting offset matrix Δ and range locator function G for the entire axis are given in Equations (2) and (3), respectively.

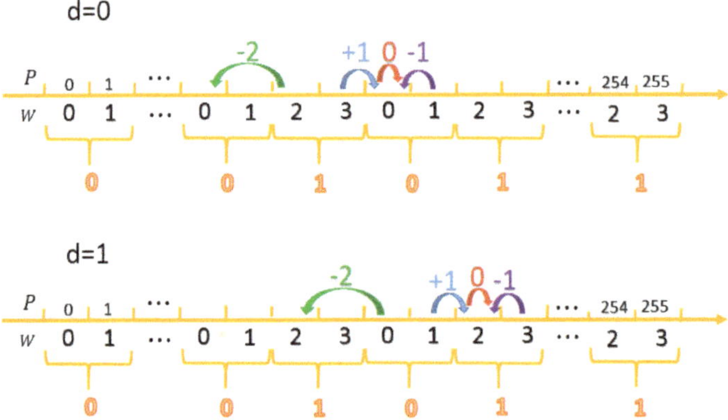

Figure 7. Illustration of incremental value Δ for different situations.

$$\Delta = \left[\begin{array}{cccc} 0 & -1 & -2 & +1 \\ -2 & +1 & 0 & -1 \end{array} \right] \tag{2}$$

$$G(p, d) = \begin{cases} [0 : 1], & \text{if } p = 0 \text{ and } d = 0; \\ [2 : 3], & \text{if } p = 0 \text{ and } d = 1; \\ [252 : 253], & \text{if } p = 255 \text{ and } d = 0; \\ [254 : 255], & \text{if } p = 255 \text{ and } d = 1; \\ [p' : p' + 1] \text{ with } p' = p + \Delta(d, mod(p, 4)), & \text{if } 1 \leq p \leq 254. \end{cases} \tag{3}$$

Figure 8 illustrates an example of combining ranges of three axes. Assuming the cover triplet is $(p_{xi}, p_{yi}, p_{zi}) = (2, 3, 2)$ and the 3 MSBs of the secret segment to be embedded is $s_j^M = (d_5^j d_4^j d_3^j) = (110)_2$. By applying the range locater, the embedding range of each axis can be determined independently as $G(p_{xi}, d_3^j) = G(2, 0) = [2 + \Delta(0, 2) : 2 + \Delta(0, 2) + 1] = [0 : 1]$, $G(p_{yi}, d_4^j) = G(3, 1) = [3 + \Delta(1, 3) : 3 + \Delta(1, 3) + 1] = [2 : 3]$ and $G(p_{zi}, d_5^j) = G(2, 1) = [2 + \Delta(1, 2) : 2 + \Delta(1, 2) + 1] = [2 : 3]$. As shown in the figure, the basic structure obtained by combining the located ranges is $M(0 : 1, 2 : 3, 2 : 3)$. Comparing with Figure 6, the basic structure obtained by applying $s_j^M = (110)_2$ to range locater coincides with the structure labeled $6 = (110)_2$. This result demonstrates that the range locater can be treated as an efficient tool for the outer layer of embedding scheme.

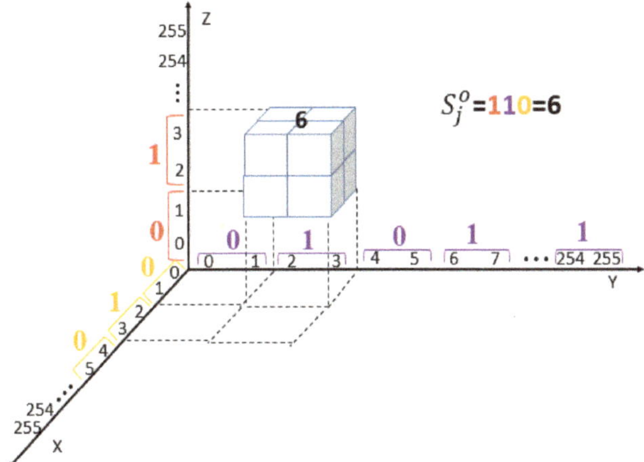

Figure 8. Locating the basic structure for embedding by combining ranges of three axes.

3.1.2. Secret Data Embedding

As mentioned in the previous subsection, the proposed data-embedding scheme is composed of two hiding layers. The outer layer uses the 3 MSBs to locate the nearest formal basic structure for embedding. The inner layer seeks to find the element with a value matching the 3 LSBs and embeds the 6 digits in total by modifying the pixel values. The details of secret data embedding are as follows:

The secret data-embedding algorithm based on the cubic mini-SuDoKu matrix (CMSM)

Input: cover image P, secret stream S, secret key K

Output: stego image P' Step 1: Construct the CMSM M using the secret key K (details are given in Appendix A)

(a)   Apply the secret key K to initialize the random number generator;

(b)   Allocate an empty array of size $256 \times 256 \times 256$ and divide it into blocks of size $2 \times 2 \times 2$;

(c)   Fill in each block with random ordered 0 to 7, consecutively;

Step 2: Group the cover pixels into triplets $P = \{(p_{xi}, p_{yi}, p_{zi}) | i = 1, 2, \ldots, (H \times W)/3\}$;

Step 3: Segment secret digits $S = \{s_j = (d_5^j d_4^j d_3^j d_2^j d_1^j d_0^j) | j = 1, 2, \ldots, L/6\}$;

Step 4: Locate $M(G(p_{xi}, d_3^j), G(p_{yi}, d_4^j), G(p_{zi}, d_5^j))$ by applying Equations (2) and (3);

Step 5: Search the matching element in the located basic structure;

$$M(p'_{xi}, p'_{yi}, p'_{zi}) = (2^2 \times d_2^j + 2^1 \times d_1^j + 2^0 \times d_0^j);$$

Step 6: Record $(p'_{xi}, p'_{yi}, p'_{zi})$ to stego image P';

Step 7: Repeat Steps 4–6, until all secret digits are embedded.

In the embedding algorithm, we use a secret key K to initialize the random number generator. Each basic structure is stored with a random permutation of 0 to 7. The number of different CMSM is $(8!)^{128 \times 128 \times 128}$. To reduce the computational load of the matrix, we can produce a randomly generated matrix of size, for example, $16 \times 16 \times 16$ and repeat it to obtain a $256 \times 256 \times 256$ CMSM. The number of different permutations is $(8!)^{16 \times 16 \times 16}$, which is still much secure than the 2D mini-SuDoKu version [23]. By sharing the initialization key for the random number generator, the receiver can reconstruct the CMSM using the same rule.

In the following, three examples are used to further explain the secret data-embedding process as shown in Figure 9. Assume the three triplets are $\{(1,1,2), (1,3,2), (1,5,2)\}$, which will be used to hide the secret segments $\{(001010)_2, (101010)_2, (000000)_2\}$.

For the triplet (1,1,2) as shown by the purple submatrix in Figure 9a, the secret data $s_j = (d_5^j d_4^j d_3^j d_2^j d_1^j d_0^j) = (001010)_2$ is to be hidden. First, use triplet (1,1,2) and 3 MSBs $s_j^M = (001)_2$ to locate the basic structure $G\left(p_{xi}, d_3^j\right) = G(1,1) = [1 + \Delta(1,1) : 1 + \Delta(1,1) + 1] = [2:3]$, $G\left(p_{yi}, d_4^j\right) = G(1,0) = [1 + \Delta(0,1) : 1 + \Delta(0,1) + 1] = [0:1]$, $G\left(p_{zi}, d_5^j\right) = G(2,0) = [2 + \Delta(0,2) : 2 + \Delta(0,2) + 1] = [0:1]$, and $M\left(G\left(p_{xi}, d_3^j\right), G\left(p_{yi}, d_4^j\right), G\left(p_{zi}, d_5^j\right)\right) = M(2:3, 0:1, 0:1)$. Then, $M(2,1,1) = 2^2 \times 0 + 2^1 \times 1 + 2^0 \times 0 = 2$ can be found in the located basic structure and record $(2,1,1)$ to stego image P'.

For the second example of (1,3,2) as shown by the red submatrix in Figure 9b, the secret data $s_j = \left(d_5^j d_4^j d_3^j d_2^j d_1^j d_0^j\right) = (101010)_2$ is to be hidden. First, use triplet (1,3,2) and 3 MSBs $s_j^M = (101)_2$ to locate the basic structure $G\left(p_{xi}, d_3^j\right) = G(1,1) = [1 + \Delta(1,1) : 1 + \Delta(1,1) + 1] = [2:3]$, $G\left(p_{yi}, d_4^j\right) = G(3,0) = [3 + \Delta(0,3) : 3 + \Delta(0,3) + 1] = [4:5]$, $G\left(p_{zi}, d_5^j\right) = G(2,1) = [2 + \Delta(1,2) : 2 + \Delta(1,2) + 1] = [2:3]$, and $M\left(G\left(p_{xi}, d_3^j\right), G\left(p_{yi}, d_4^j\right), G\left(p_{zi}, d_5^j\right)\right) = M(2:3, 4:5, 2:3)$. Then, $M(3,4,2) = 2^2 \times 0 + 2^1 \times 1 + 2^0 \times 0 = 2$ can be found in the located basic structure and record $(3,4,2)$ to stego image P'.

For the third triplet (1,5,2) as shown by the green submatrix in Figure 9c, the secret data $s_j = \left(d_5^j d_4^j d_3^j d_2^j d_1^j d_0^j\right) = (000000)_2$ is to be hidden. First, use triplet (1,5,2) and 3 MSBs $s_j^M = (000)_2$ to locate the basic structure $G\left(p_{xi}, d_3^j\right) = G(1,0) = [1 + \Delta(0,1) : 1 + \Delta(0,1) + 1] = [0:1]$, $G\left(p_{yi}, d_4^j\right) = G(5,0) = [5 + \Delta(0,5) : 5 + \Delta(0,5) + 1] = [4:5]$, $G\left(p_{zi}, d_5^j\right) = G(2,0) = [2 + \Delta(0,2) : 2 + \Delta(0,2) + 1] = [0:1]$, and $M\left(G\left(p_{xi}, d_3^j\right), G\left(p_{yi}, d_4^j\right), G\left(p_{zi}, d_5^j\right)\right) = M(0:1, 4:5, 0:1)$. Then, $M(1,4,1) = 2^2 \times 0 + 2^1 \times 0 + 2^0 \times 0 = 0$ can be found in the located basic structure and record $(1,4,1)$ to stego image P'.

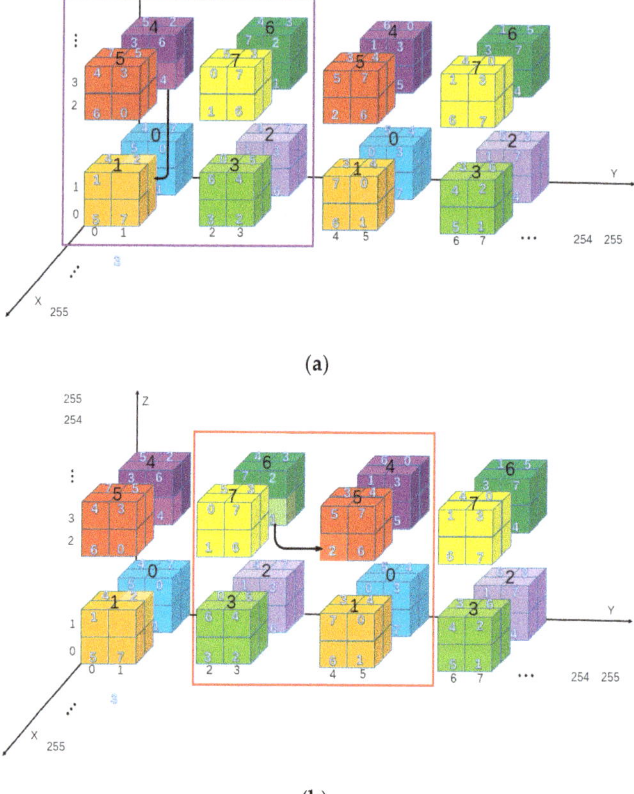

(a)

(b)

Figure 9. *Cont.*

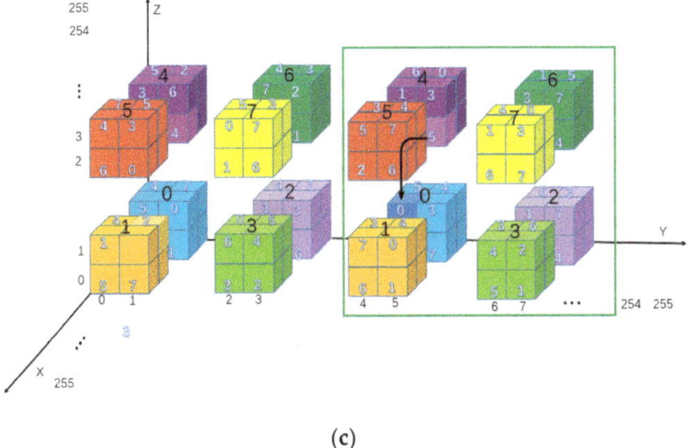

(c)

Figure 9. Examples of data hiding with cubic mini-SuDoKu matrix (CMSM). (**a**) Hide $(001010)_2$ into cover triplet $(1, 1, 2)$; (**b**) Hide $(101010)_2$ into cover triplet $(1, 3, 2)$; (**c**) Hide $(000000)_2$ into cover triplet $(1, 5, 2)$.

3.1.3. Secret Data Extraction

After receiving the stego image P', the recipient first groups the stego pixels into triplets. Then, the secret segments can be obtained by mapping the triplets into the CMSM. For a located element, its corresponding secret segment includes the 3 MSBs determined by the label of basic structure it belongs to and the 3 LSBs determined by its value. The details of the extraction process are provided as follows:

The secret data extraction algorithm based on the cubic mini-SuDoKu matrix (CMSM)

Input: stego image P', secret key K

Output: secret stream S Step 1: Construct the CMSM M using the secret key K (details are given in Appendix A)

(a) Apply the secret key K to initialize the random number generator.

(b) Allocate an empty array of size $256 \times 256 \times 256$ and divide it into blocks of size $2 \times 2 \times 2$.

(c) Fill in each block with random ordered 0 to 7, consecutively.

Step 2: Group the stego pixels into triplets $P = \left\{ \left(p'_{xi}, p'_{yi}, p'_{zi} \right) \middle| i = 1,\ 2,\ \ldots,\ (H \times W)/3 \right\}$.

Step 3: Extract the 3 LSBs by $s_j^L = \left[M\left(p'_{xi}, p'_{yi}, p'_{zi} \right) \right]_2$.

Step 4: Extract the 3 MSBs by

$$s_j^M = \left(d_5^j, d_4^j, d_3^j \right) = \left(mod(p'_{zi}, 4)/2, mod(p'_{yi}, 4)/2, mod(p'_{xi}, 4)/2 \right).$$

Step 5: Concatenate $s_j = s_j^M s_j^L$.

Step 6: Repeat Steps 3–5, until all secret digits are extracted.

3.2. N-Dimensional MSM (NMSM)

In this section, we introduce the construction of NMSM and the secret data-embedding and extraction algorithm, based on the NMSM. In addition, a fast algorithm for the inner embedding layer is proposed to improve the time efficiency.

3.2.1. The Data-Embedding and Extraction Algorithm

To boost the efficiency of data embedding and extraction, the proposed secret data-embedding scheme can be generalized to an n-dimensional version. In the NMSM, a basic structure consists of 2^n elements and 2^n basic structures constitute a submatrix. An n-tuple pixel group can uniquely map to an element in the NMSM. Therefore, by applying the same embedding rule, we can hide n MSBs with the label of basic structure and n LSBs with the element value. The pseudo codes for the NMSM-based embedding and extraction schemes are given as follows:

The secret data-embedding algorithm based on the n-dimensional mini-SuDoKu matrix (NMSM)

Input: cover image P, secret stream S, secret key K

Output: stego image P' Step 1: Construct the NMSM M using the secret key K (details are given in Appendix B);

(a) Make an n-dimensional array of size 16^n, which consists of 8^n submatrix;

(b) Repeat the array to obtain NMSM;

Step 2: Group the cover pixels into $P = \left\{ \left(p_{X(0)i}, p_{X(1)i}, \ldots, p_{X(n-1)i} \right) \middle| i = 1,\ 2,\ \ldots,\ (H \times W)/n \right\}$;

Step 3: Segment secret digits $S = \left\{ s_j = \left(d^j_{2n-1} d^j_{2n-2} \ldots d^j_0 \right) \middle| j = 1,\ 2,\ \ldots, L/2n \right\}$;

Step 4: Locate the basic structure $M\!\left(G\!\left(p_{X(0)i}, d^j_n \right), G\!\left(p_{X(1)i}, d^j_{n+1} \right), \ldots, G\!\left(p_{X(n-1)i}, d^j_{2n-1} \right) \right)$
　　　　by applying Equations (2) and (3);

Step 5: Search the matching element in the located basic structure

$$M\!\left(p'_{X(0)i}, p'_{X(1)i}, \ldots, p'_{X(n-1)i} \right) = \left(2^{n-1} \times d^j_{n-1} + 2^{n-2} \times d^j_{n-2} + \ldots + 2^0 \times d^j_0 \right);$$

Step 6: Record $\left(p'_{X(0)i}, p'_{X(1)i}, \ldots, p'_{X(n-1)i} \right)$ to stego image P';

Step 7: Repeat Steps 4–6, until all secret digits are embedded.

The secret data extraction algorithm based on the n-dimensional mini-SuDoKu matrix (NMSM)

Input: stego image P', secret key K

Output: secret stream S

Step 1: Construct the NMSM by the same process as Step 1 of embedding;

Step 2: Group the stego pixels into $P' = \left\{ \left(p'_{X(0)i}, p'_{X(1)i}, \ldots, p'_{X(n-1)i} \right) \middle| i = 1,\ 2,\ \ldots,\ (H \times W)/n \right\}$;

Step 3: Extract the n LSBs by $s^L_j = \left[M\!\left(p'_{X(0)i}, p'_{X(1)i}, \ldots, p'_{X(n-1)i} \right) \right]_2$;

Step 4: Extract the n MSBs by

$$s^M_j = \left(mod(p'_{X(n-1)i}, 4)/2, mod(p'_{X(n-2)i}, 4)/2, \ldots, mod(p'_{X(0)i}, 4)/2 \right);$$

Step 5: Concatenate $s_j = s^M_j s^L_j$;

Step 6: Repeat Steps 3–5, until all secret digits are extracted.

3.2.2. Fast Algorithm for the Inner Layer of Embedding

As the MSM generalized to n-dimensions, the basic structure for embedding can be efficiently determined by the range locator. However, the searching process in the Step 5 of embedding algorithm becomes burdensome. To improve time efficiency of the inner embedding layer, we devise a fast algorithm to overcome this burden. Its key idea is to leverage the matrix operation supported by the MATLAB language. The pseudo code of the fast algorithm is given as follows: More precise pseudo code expressed in MATLAB instructions is given in Appendix C.

Fast Algorithm for the Inner Embedding Layer (details are given in Appendix C)

Input: n LSBs of secret segment s^L_j, basic structure for embedding

$$A = M\!\left(G\!\left(p_{X(0)i}, d^j_n \right), G\!\left(p_{X(1)i}, d^j_{n+1} \right), \ldots, G\!\left(p_{X(n-1)i}, d^j_{2n-1} \right) \right)$$

Output: stego pixel values $\left(p'_{X(0)i}, p'_{X(1)i}, \ldots, p'_{X(n-1)i} \right)$

(a)　Construct an n-dimensional basic structure B with all elements valued with s^L_j;

(b)　Using matrix operation to find the only matched element in both A and B;

(c)　Project the element to all axes and obtain the coordinates for embedding.

4. Experimental Results

In the following, the experimental results of the proposed scheme will be presented and compared with the related works. As shown in Figure 10, this study uses eight standard 512×512 grayscale images and four standard 512×512 true color images as the cover image P for secret data embedding. All experiments are implemented with MATLAB R2017b. Figure 11 shows the twelve stego images obtained by applying the two-layered hiding scheme based on the CMSM.

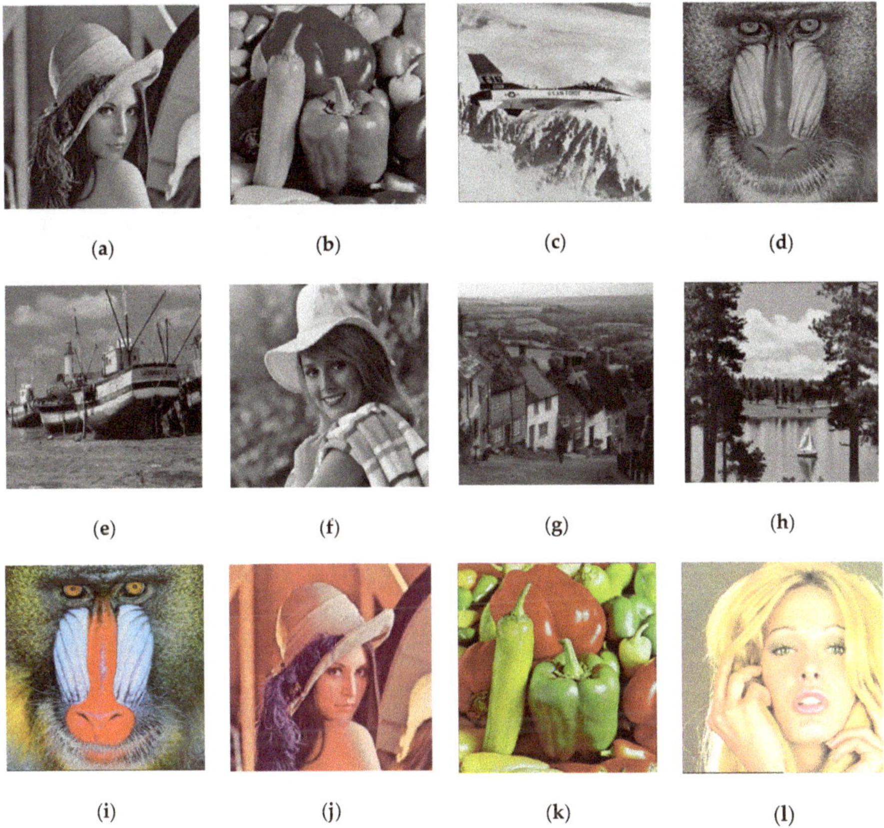

Figure 10. Twelve cover images with size 512×512: (**a**) Lena; (**b**) peppers; (**c**) airplane; (**d**) baboon; (**e**) boat; (**f**) Elaine; (**g**) Gledhill; (**h**) sailboat; (**i**) baboon (RGB); (**j**) Lena (RGB); (**k**) peppers (RGB); and (**l**) Tiffany (RGB).

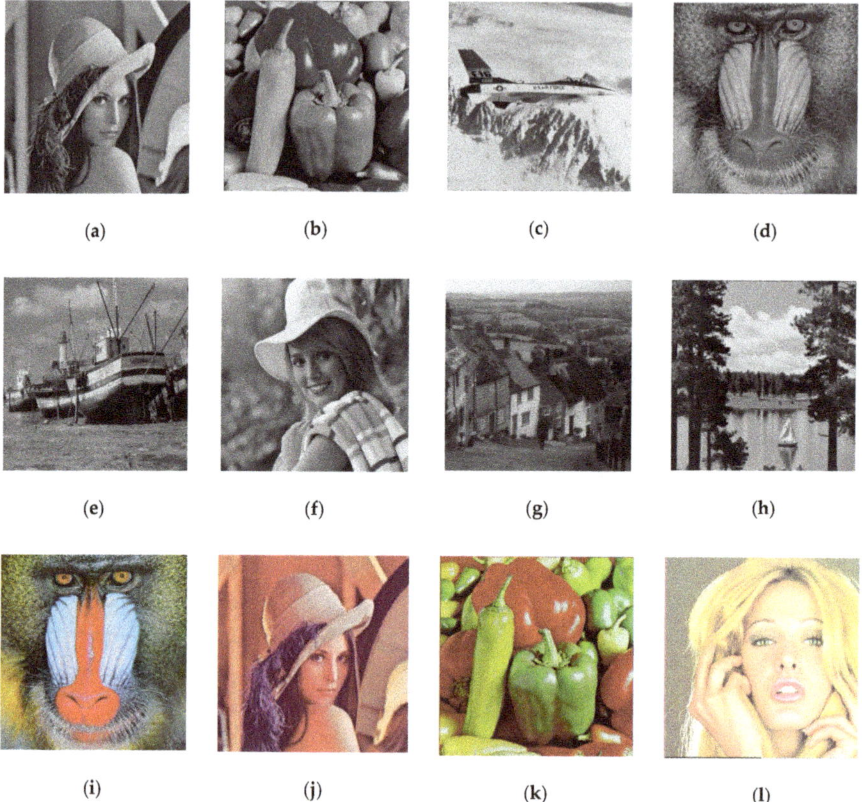

Figure 11. Twelve stego images with size 512×512: (**a**) Lena; (**b**) peppers; (**c**) airplane; (**d**) baboon; (**e**) boat; (**f**) Elaine; (**g**) Gledhill; (**h**) sailboat; (**i**) baboon (RGB); (**j**) Lena (RGB); (**k**) peppers (RGB); and (**l**) Tiffany (RGB).

The stego images look like the cover images and cannot be distinguished by human eyes. The peak signal-to-noise ratio (PSNR) is used to measure the quality of a stego image. Its calculation is given by Equation (4). In the equation, P and P' represent the cover image and the stego image, respectively, H and W represent their height and width and (m, n) represents the coordinate of the pixel.

$$PSNR = 10 \times \log_{10} \frac{255^2 \times H \times W}{\sum_{m=1}^{H} \sum_{n=1}^{W} [P(m,n) - P'(m,n)]^2}. \tag{4}$$

In addition, we also use structural similarity index (SSIM) to measure the similarity between a cover image and its corresponding stego image. Let P and P' represent the cover image and the stego image, respectively, the SSIM of the two images can be obtained according to Equation (5), where μ_P is the average of P, $\mu_{P'}$ is the average of P', σ_P^2 is the variance of P, $\sigma_{P'}^2$ is the variance of P' and $\sigma_{PP'}$ is the covariance of P and P'. In addition, c_1 and c_2 are constants used to maintain stability and can be obtained by Equations (6) and (7), respectively, where L is the dynamic range of pixel values, $k_1 = 0.01$, $k_2 = 0.03$.

$$SSIM(P, P') = \frac{(2\mu_P\mu_{P'} + c_1)(2\sigma_{PP'} + c_2)}{(\mu_P^2 + \mu_{P'}^2 + c_1)(\sigma_P^2 + \sigma_{P'}^2 + c_2)}. \tag{5}$$

$$c_1 = (k_1 L)^2. \tag{6}$$

$$c_2 = (k_2 L)^2. \tag{7}$$

Embedding capacity (EC) is another important issue in the image steganography. It is used to measure the maximum amount of secret data that can be embedded in an image by a data-hiding scheme. Since EC is dependent on the image size, we further define the embedding rate (ER) to express the average number of secret bits that each pixel can embed. ER is defined as (8), where $\|S\|$ represents the total amount of secret data embedded in the entire stego image.

$$ER = \frac{\|S\|}{M \times N}. \tag{8}$$

Table 1 shows the experimental results of the grayscale images. In our hiding scheme, three cover pixels are applied to embed a secret segment of six digits. Therefore, the ER measure is 6 bits/3 pixels = 2 bits/pixel, the embedding capacity for an applied grayscale cover image is therefore $512 \times 512 \times 2 = 524{,}288$ bits as shown in the Under full embedding, the proposed scheme achieves a high image quality of PSNR = 46.37 dB and SSIM = 0.9923 in average. The quality of stego image is irrelevant to features of the cover image. For the true color images, each pixel consists of three channels, including red, green and blue. Each of the three channels is represented by one byte. By mapping the three bytes (r, g, b) of a pixel into the CMSM, we can hide a secret segment of six bits by applying the proposed embedding algorithm. Therefore, the embedding capacity of an applied true color image is $512 \times 512 \times 6 = 1{,}572{,}864$ bits. The PSNR and SSIM of the true color stego images are very close to the experimental values of grayscale images as shown in Table 2.

Table 1. Experimental results for grayscale images.

Image	EC (bits)	PSNR (dB)	SSIM
Lena	524,288	46.38	0.9918
Peppers	524,288	46.37	0.9906
Airplane	524,288	46.37	0.9883
Baboon	524,288	46.36	0.9958
Boat	524,288	46.37	0.9938
Elaine	524,288	46.38	0.9919
Gledhill	524,288	46.36	0.9936
Sailboat	524,288	46.37	0.9929
Average	524.288	46.37	0.9923

Table 2. Experimental results for color images.

Image	EC (bits)	PSNR (dB)	SSIM
Baboon (RGB)	1,572,864	46.36	0.9930
Lena (RGB)	1,572,864	46.38	0.9926
Peppers (RGB)	1,572,864	46.37	0.9908
Elaine (RGB)	1,572,864	46.37	0.9916
Average (RGB)	1,572,864	46.37	0.9920

4.1. EC and PSNR Comparison

Table 3 compares the PSNR of different reference matrices under the same EC. As shown in Table 3, Xie et al.'s scheme has the lowest average PSNR at 41.87 dB. The average PSNR of the proposed scheme is 46.37 dB, which is nearly 4 dB higher than the average PSNR of the Xie et al.'s scheme. Moreover, compared with the other two schemes, our scheme also achieves the highest PSNR with the same EC. It can be seen that the proposed scheme outperforms the related works.

Table 3. Comparison with related works.

Image	Jin et al. [24]		Liu et al. [23]		Xie et al. [17]		Proposed	
	EC	PSNR	EC	PSNR	EC	PSNR	EC	PSNR
Lena	524,288	45.57	524,288	45.55	524,288	41.87	524,288	46.38
Peppers	524,288	45.56	524,288	45.54	524,288	41.86	524,288	46.37
Airplane	524,288	45.56	524,288	45.58	524,288	41.87	524,288	46.37
Baboon	524,288	45.57	524,288	45.55	524,288	41.86	524,288	46.36
Boat	524,288	45.58	524,288	45.54	524,288	41.87	524,288	46.37
Elaine	524,288	45.56	524,288	45.49	524,288	41.87	524,288	46.38
Gledhill	524,288	45.49	524,288	45.49	524,288	41.87	524,288	46.36
Sailboat	524,288	45.58	524,288	45.55	524,288	41.86	524,288	46.37
Average	524,288	45.56	524,288	45.54	524,288	41.87	524,288	46.37

In order to further understand the performance of our scheme, it is compared with three other schemes [19,22,23] based on the SuDoKu reference matrix. As shown in Table 4, the EC of the Chang et al.'s scheme is 393,216 bits, and its average PSNR is 44.83 dB. Regardless of the EC or image quality, the proposed scheme outperforms the Chang et al.'s scheme. Comparing with the other two schemes, although their image quality are better than the proposed scheme, their embedding capacity are far lower than our scheme. Therefore, it can be concluded that the overall performance of the proposed scheme is better than the SuDoKu-based data-hiding schemes.

Table 4. Comparison among SuDoKu-based data-hiding schemes.

Image	Chang et al. [20]		Hong et al. [21]		Lin et al. [19]		Proposed	
	EC	PSNR	EC	PSNR	EC	PSNR	EC	PSNR
Lena	393,216	44.96	393,216	48.68	393,216	49.90	524,288	46.38
Peppers	393,216	44.67	393,216	48.67	393,216	49.91	524,288	46.37
Airplane	393,216	44.99	393,216	48.68	393,216	49.92	524,288	46.37
Baboon	393,216	44.68	393,216	48.66	393,216	.49.89	524,288	46.36
Boat	393,216	44.90	393,216	48.67	393,216	49.91	524,288	46.37
Elaine	393,216	44.92	393,216	48.68	393,216	49.91	524,288	46.38
Gledhill	393,216	44.85	393,216	48.67	393,216	49.90	524,288	46.36
Sailboat	393,216	44.67	393,216	48.67	393,216	49.90	524,288	46.37
Average	393,216	44.83	393,216	48.67	393,216	49.90	524,288	46.37

In addition, we also compare the proposed CMSM-based scheme with the 3D SuDoKu-based scheme. As shown in Table 5, the average PSNR of the scheme proposed by Xia et al. is 41.31 dB, while the average PSNR of our scheme is 5 dB higher than their scheme. Based on the same frame structure of using 3D reference matrix, our scheme has a relatively small modification of pixel values under the same EC.

Table 5. Comparison with the 3D SuDoKu matrix-based scheme.

Image	Xia et al. [18]		Proposed	
	EC	PSNR	EC	PSNR
Lena	524,288	41.31	524,288	46.38
Peppers	524,288	41.30	524,288	46.37
Airplane	524,288	41.28	524,288	46.37
Baboon	524,288	41.25	524,288	46.36
Boat	524,288	41.23	524,288	46.37
Elaine	524,288	41.26	524,288	46.38
Gledhill	524,288	41.29	524,288	46.36
Sailboat	524,288	41.27	524,288	46.37
Average	524,288	41.27	524,288	46.37

Our scheme is inspired by the He et al.'s scheme. Except for expanding the 2D mini-SuDoKu to a 3D CMSM, our scheme effectively improved the complexity of reference matrix and efficiently reduced the computation time. As shown in Table 6, although our scheme has no advantage over the He et al.'s scheme in terms of EC and PSNR. However, the improvement in time consumption (TC) is obvious. As discussed in Section 2.2, to achieve a translation invariant property for minimizing the distortion of pixel value modification, the mini-SuDoKu has to repeat a same basic 4×4 submatrix. This severely damages the complexity of a reference matrix. In our scheme, in spite of CMSM or NMSM, each basic structure is a completely random permutation of 2^n distinct numbers. The possible combinations of a CMSM is an enormous figure. In the construction of NMSM, referring to Section 3.2.1, we even truncate the size of a randomly generated array and make the whole NMSM by repetition to reduce computational load. The key difference with the He et al.'s scheme is the design of range locator. It releases the constraint on the diversity of reference matrix. An additional benefit of the range locator is that the basic structure for embedding can be efficiently located without applying a time-consuming searching process. The cooperation of range locater in the outer embedding and the matrix operation in the inner embedding frees the hiding scheme from intensive loops of searching. According to the experimental data in Table 6, the required embedding time is less than half of the compared scheme.

Table 6. Comparison with the mini-SuDoKu matrix-based scheme.

Image	He et al. [25]			Proposed		
	EC	PSNR	TC	EC	PSNR	TC
Lena	524,288	46.37	2.71 s	524,288	46.38	1.07 s
Peppers	524,288	46.37	2.72 s	524,288	46.37	1.12 s
Airplane	524,288	46.37	2.70 s	524,288	46.37	1.10 s
Baboon	524,288	46.36	2.73 s	524,288	46.36	1.09 s
Boat	524,288	46.36	2.71 s	524,288	46.37	1.11 s
Elaine	524,288	46.38	2.71 s	524,288	46.38	1.12 s
Gledhill	524,288	46.37	2.72 s	524,288	46.36	1.09 s
Sailboat	524,288	46.35	2.71 s	524,288	46.37	1.12 s
Average	524,288	46.37	2.71 s	524,288	46.37	1.10 s

4.2. Time Efficiency Comparison

To investigate the time efficiency of the proposed algorithm, we try to compare the time consumption of the proposed algorithm with the traditional approach. Although we present the three-dimensional (3D) CMSM and the n-dimensional (n−D) NMSM, our approach can also be de-generalized back to two-dimensional (2D) and one-dimensional (1D) version.

This experiment uses 8 typical grayscale images for testing and compares the proposed algorithm with a traditional one. A computer with a Dual i7-920 CPU and 8 GB memory is adopted for the experiment. The tic and toc commands in MATLAB are used to record the time cost in seconds. As shown in Table 7, due to the large number of loops used in the search step of the traditional algorithm, it takes significantly longer time to embed secret data. As the dimension rises, the time consumption increases rapidly. On the other hand, our algorithm has consistent performance as the dimension rises. Note that, as the dimension rises, the complexity of the reference matrix increases and thus the security level raises.

Table 7. Comparison of time efficiency between the proposed scheme and traditional algorithm.

Image	Trad-2D	Trad-3D	Proposed-2D	Proposed-3D	Proposed-4D
Lena	2.80 s	4.48 s	1.09 s	1.07 s	1.11 s
Peppers	2.81 s	4.51 s	1.11 s	1.12 s	1.13 s
Airplane	2.83 s	4.52 s	1.10 s	1.10 s	1.12 s
Baboon	2.82 s	4.49 s	1.11 s	1.09 s	1.11 s
Boat	2.75 s	4.47 s	1.10 s	1.11 s	1.12 s
Elaine	2.94 s	4.54 s	1.13 s	1.12 s	1.14 s
Gledhill	2.86 s	4.54 s	1.08 s	1.09 s	1.11 s
Sailboat	2.83 s	4.49 s	1.11 s	1.13 s	1.13 s

5. Conclusions

This study introduces an efficient multidimensional secret data-embedding scheme based on the mini SuDoKu matrix. In the proposed scheme, a CMSM RM with high complexity is first constructed to guarantee the security, and then a range locator function and the matrix operation are adopted to enhance the embedding efficiency. The reference matrix is further expanded to multidimension in order to obtain even higher embedding capacity and, meanwhile, still preserve good security and efficiency. The proposed scheme is compared with state-of-the-art RM-based data-hiding schemes and the experimental results show that the proposed scheme achieved higher than 46 dB in terms of the image quality and two bits per pixel in terms of the embedding capacity. In addition, the time consumption of the proposed algorithm is less than half of the traditional approach and keeps consistency as the dimension and security level raises. It is shown that the proposed scheme is advantageous in both embedding efficiency and security compared to the original mini-SuDoKu matrix.

We also provide a set of true color test images to demonstrate that the proposed scheme performs equal well to multi-channel images. By leveraging CMSM, each pixel of three-color channels—i.e., R, G, and B—can exactly match with the requirement of embedding a secret segment of data. The flexibility in dimension of RM meets the diverse data structure of cover media in the future word of massive IoT.

Author Contributions: Conceptualization, J.-H.H. and S.X.; Data curation, S.X.; Formal analysis, S.X.; Funding acquisition, J.-H.H.; Investigation, S.X.; Methodology, J.-H.H. and S.X.; Project administration, C.-C.C. (Chin-Chen Chang); Resources, S.X.; Software, S.X.; Supervision, C.-C.C. (Chin-Chen Chang); Validation, S.X.; Visualization, J.-H.H., C.-C.C. (Ching-Chun Chang) and C.-C.C. (Chin-Chen Chang); Writing: original draft, S.X.; Writing: review & editing, J.-H.H., C.-C.C. (Ching-Chun Chang) and C.-C.C. (Chin-Chen Chang). All authors have read and agreed to the published version of the manuscript.

Funding: This research received no external funding.

Conflicts of Interest: The authors declare no competing interests.

Appendix A

The pseudo code for constructing the CMSM is provided as follows:

(a) Apply the secret key to initialize the random number generator.

$\quad$ rng(K);

(b) Allocate an empty array of size $256 \times 256 \times 256$ and divide it into blocks of size $2 \times 2 \times 2$.

$\quad M = $ zero$(256, 256, 256)$;

(c) Fill in each block with random ordered 0 to 7, consecutively;

$\quad$ for $x = 0 : 2 : 255$,

$\quad$ for $y = 0 : 2 : 255$,

$\quad$ for $z = 0 : 2 : 255$,

$$V = \text{randperm}(8) - 1;$$

$\quad A = [V(0) \ V(1); V(2) \ V(3)]; B = [V(4) \ V(5); V(6) \ V(7)];$

$$M(x : x + 1, y : y + 1, z : z + 1) = A;$$

$\quad$ end

```
        end
end
```

The MATLAB functions rng() and randperm(8) are used to initialize random number generator and produce random permutations of 1 to 8, respectively. Although we apply the coding format and the functions of MATLAB programming language, the index of an array in the pseudo code follows the convention of zero leading value.

Appendix B

The pseudo code for constructing the NMSM is provided as follows:
(a) Make an n-dimensional array of size 16^n using the secret key K

```
rng(K);
for X(0) = 0 : 2 : 15,
for X(1) = 0 : 2 : 15,
...
  for X(n − 1) = 0 : 2 : 15,
      V = randperm(2^n) − 1;
      k = 0;
      for X'(0) = 0 : 1,
      for X'(1) = 0 : 1,
      ...
        for X'(n − 1) = 0 : 1,
            M(X'(0), X'(1), ..., X'(n − 1)) = V(k);
            Mk = k + 1;
        end
      ...
        end
        end
    end
  ...
    end
    end
```

(b) Repeat the array to obtain NMSM

```
    for X(1) = 0 : 16 : 255,
    for X(2) = 0 : 16 : 255,
    ...
      for X(n) = 0 : 16 : 255,
          M(X(0) : X(0) + 15, X(1) : X(1) + 15, ..., X(n − 1) : X(n − 1) + 15)
                        = M(0 : 15, 0 : 15, ..., 0 : 15);
      end
    ...
      end
    end
```

Appendix C

The pseudo code for fast algorithm of inner embedding is given as follows:
Input: n LSBs of secret segment s_j^L, basic structure for embedding

$$A = M\left(G\left(p_{X(0)i}, d_n^j\right), G\left(p_{X(1)i}, d_{n+1}^j\right), \ldots, G\left(p_{X(n-1)i}, d_{2n-1}^j\right)\right)$$

Output: stego pixel values $\left(p'_{X(0)i}, p'_{X(1)i}, \ldots, p'_{X(n-1)i}\right)$
(a) Construct an n-dimensional basic structure B with all elements valued with s_j^L;

$$B = s^L_j \times ones\big(2, 2, \ldots, 2_{(n)}\big);$$

(b) Using matrix operation to find the only matched element in both A and B;

$$C = ismember(A, B);$$

(c) Project the element to all axes and obtain the coordinates for embedding;

for $k = 0 : n - 1$,

$$\big[v, p'_{X(k)i}\big] = \max \Sigma_{q \neq k} C\big(:, :, \ldots, :_{(q)}, \ldots, :\big);$$

end

The algorithm utilizes an array of unique value s^L_j to compare with the basic structure for embedding A using the function $ismember()$. As a result, the unique '1' in the array C indicates the location of s^L_j in A. By projecting the summations to each axis can obtain the index of the corresponding axis. Thus, the values of stego pixels are determined.

References

1. Matsui, M. Linear cryptanalysis scheme for DES cipher. In *Workshop on the Theory and Application of Cryptographic Techniques*; Springer: Berlin/Heidelberg, Germany, 1993; pp. 386–397.
2. Rivest, R.; Shamir, A.; Adleman, L. A scheme for obtaining digital signatures and public-key cryptosystems. *Commun. ACM* **1978**, *21*, 120–126. [CrossRef]
3. Ullah, S.; Marcenaro, L.; Rinner, B. Secure smart cameras by aggregate-signcryption with decryption fairness for multi-receiver IoT applications. *Sensors* **2019**, *19*, 327. [CrossRef] [PubMed]
4. Li, Y.; Tu, Y.; Lu, J.; Wang, Y. A security transmission and storage solution about sensing image for blockchain in the Internet of Things. *Sensors* **2020**, *20*, 916. [CrossRef] [PubMed]
5. Ker, A. Improved detection of LSB steganography in grayscale images. In *International Workshop on Information Hiding*; Springer: Berlin/Heidelberg, Germany, 2004; Volume 3200, pp. 97–115.
6. Qin, C.; Chang, C.; Huang, Y. An inpainting-assisted reversible steganographic scheme using a histogram shifting mechanism. *IEEE Trans. Circuits Syst. Video Technol.* **2012**, *23*, 1109–1118. [CrossRef]
7. Chang, C.; Lin, C.; Tseng, C.; Tai, W. Reversible hiding in DCT-based compressed images. *Inf. Sci.* **2007**, *177*, 2768–2786. [CrossRef]
8. Huang, F.; Qu, X.; Kim, H.; Huang, J. Reversible data hiding in JPEG images. *IEEE Trans. Circuits Syst. Video Technol.* **2016**, *26*, 1610–1621. [CrossRef]
9. Chang, C.; Kieu, T.; Wu, W. A loss-less data embedding technique by joint neighboring coding. *Pattern Recog.* **2009**, *42*, 1597–1603. [CrossRef]
10. Hu, Y. High capacity image hiding scheme based on vector quantization. *Pattern Recogn.* **2006**, *39*, 1715–1724. [CrossRef]
11. Lin, Y.; Hsia, C.; Chen, B.; Chen, Y. Visual IoT security: Data hiding in AMBTC images using block-wise embedding strategy. *Sensors* **2019**, *19*, 1974. [CrossRef] [PubMed]
12. Chang, C.; Wang, X.; Horng, J. A hybrid data hiding method for strict AMBTC format images with high-fidelity. *Symmetry* **2019**, *11*, 1314. [CrossRef]
13. Bender, W.; Gruhl, D.; Morimoto, N.; Lu, A. Techniques for data hiding. *IBM Syst. J.* **1996**, *35*, 313–336. [CrossRef]
14. Kim, H.; Kim, C.; Choi, Y.; Wang, S.; Zhang, X. Improved modification direction schemes. *Comput. Math. Appl.* **2010**, *60*, 319–325. [CrossRef]
15. Mielikainen, J. LSB matching revisited. *IEEE Signal Proc. Lett.* **2006**, *13*, 285–287. [CrossRef]
16. Zhang, X.; Wang, S. Efficient steganographic embedding by exploiting modification direction. *IEEE Commun. Lett.* **2006**, *10*, 781–783. [CrossRef]
17. Xie, X.; Liu, Y.; Chang, C. Extended squared magic matrix for embedding secret information with large payload. *Multimed. Tools Appl.* **2019**, *78*, 19045–19059. [CrossRef]
18. Xia, B.; Wang, H.; Chang, C.; Liu, L. An image steganography scheme using 3D-Sudoku. *J. Inf. Hiding Multimed. Signal Process.* **2016**, *7*, 836–845.
19. Lin, C.; Chang, C.; Lee, W.; Lin, J. A Novel Secure Data Hiding Scheme Using a Secret Reference Matrix. In Proceedings of the Fifth International Conference on Intelligent Information Hiding and Multimedia Signal Processing, Kyoto, Japan, 12–14 September 2009; pp. 369–373.

20. Chang, C.; Chou, Y.; Kieu, T. An Information Hiding Scheme Using Sudoku. In Proceedings of the 3rd International Conference on Innovative Computing Information and Control (ICICIC), Dalian, China, 18–20 June 2008; pp. 17–22.
21. Hong, W.; Chen, T.; Shiu, C. A Minimal Euclidean Distance Searching Technique for Sudoku Steganography. In Proceedings of the International Symposium on Information Science and Engineering (ISISE2008), Shanghai, China, 20–22 December 2008; pp. 515–518.
22. Chang, C.; Liu, Y.; Nguyen, T. A Novel Turtle Shell Based Scheme for Data Hiding. In Proceedings of the Tenth International Conference on Intelligent Information Hiding and Multimedia Signal Processing (IIHMSP 2014), Kitakyushu, Japan, 27–29 August 2014; pp. 89–93.
23. Liu, Y.; Chang, C.; Nguyen, T. High capacity turtle shell-based data hiding. *IET Image Process.* **2016**, *10*, 130–137. [CrossRef]
24. Jin, Q.; Li, Z.; Chang, C.; Wang, A.; Liu, L. Minimizing turtle shell matrix based stego image distortion using particle swarm optimization. *Netw. Secur.* **2017**, *19*, 154–162.
25. He, M.; Liu, Y.; Chang, C. A mini-Sudoku matrix-based data embedding scheme with high payload. *IEEE Access* **2019**, *7*, 141414–141425. [CrossRef]

© 2020 by the authors. Licensee MDPI, Basel, Switzerland. This article is an open access article distributed under the terms and conditions of the Creative Commons Attribution (CC BY) license (http://creativecommons.org/licenses/by/4.0/).

Article

Improving Census Transform by High-Pass with Haar Wavelet Transform and Edge Detection

Jiun-Jian Liaw [1], Chuan-Pin Lu [2,*], Yung-Fa Huang [1], Yu-Hsien Liao [1] and Shih-Cian Huang [1]

[1]   Department of Information and Communication Engineering, Chaoyang University of Technology, Taichung 413310, Taiwan; jjliaw@cyut.edu.tw (J.-J.L.); yfahuang@cyut.edu.tw (Y.-F.H.); s10630608@gm.cyut.edu.tw (Y.-H.L.); s10230612@gm.cyut.edu.tw (S.-C.H.)

[2]   Department of Information Technology, Meiho University, Pingtung 912009, Taiwan

*   Correspondence: chuan.pin.lu@gmail.com; Tel.: +886-8-779-9821

Received: 29 March 2020; Accepted: 27 April 2020; Published: 29 April 2020

Abstract: One of the common methods for measuring distance is to use a camera and image processing algorithm, such as an eye and brain. Mechanical stereo vision uses two cameras to shoot the same object and analyzes the disparity of the stereo vision. One of the most robust methods to calculate disparity is the well-known census transform, which has the problem of conversion window selection. In this paper, three methods are proposed to improve the performance of the census transform. The first one uses a low-pass band of the wavelet to reduce the computation loading and a high-pass band of the wavelet to modify the disparity. The main idea of the second method is the adaptive size selection of the conversion window by edge information. The third proposed method is to apply the adaptive window size to the previous sparse census transform. In the experiments, two indexes, percentage of bad matching pixels (PoBMP) and root mean squared (RMS), are used to evaluate the performance with the known ground truth data. According to the results, the computation required can be reduced by the multiresolution feature of the wavelet transform. The accuracy is also improved with the modified disparity processing. Compared with previous methods, the number of operation points is reduced by the proposed adaptive window size method.

Keywords: census transform; sparse census transform; disparity; stereo vision

---

1. Introduction

The development of automatic equipment has always been one of the focuses in the field of computer science. The features of automatic equipment are its self-moving and exploring characteristics that can be used to reduce the risk of participation of personnel. In many automated devices, the distance of the device from the object in the surrounding environment is an important parameter, such as using vision to avoid obstacles. In modern automated or intelligent devices, the distance from the object is an important indicator. The corresponding action can be performed by determining the distance, whether this is using a robot arm to grab the object [1], automatic car driving to determine the road condition [2,3], or using a robot that can self-plan a path to shuttle through the environment [4]. All of these studies show that measuring distance is an essential part of achieving automation.

The methods of measuring distance are roughly divided into two types. The first one uses ray or waves as the direct measurement method [5]. These methods use laser, infrared, or ultrasonic means to shoot toward the object and simultaneously record the time of transmission and the time of receiving the reflection. The distance of the object can be calculated by the time difference between the transmission and reception. The second method uses a camera and image processing or machine vision technologies [6]. These methods use a camera to capture an image of an object, and then analyze the pixels in the image and measure the distance of the object. This type of method is mainly divided into two types: monocular vision with additional reference objects and two-eye stereoscopic vision.

The method of using ray or sound waves is more accurate and faster than the method of using a camera with an algorithm, but the equipment is expensive and the application is not easy to popularize. In recent years, in the case of the popularity of digital cameras, the technology for obtaining distance by image processing schemes has gradually received attention. Furthermore, as the advancement of the computer has improved the speed of this in recent years, the performance of two-eye stereoscopic processing has also improved. Even when using an accessible style camera, the stereo vision system can also be applied to measure the target distance [7].

In the steps of using image processing or machine vision technology to measure distance, the most-time consuming loading step is matching. The matching includes object recognition or feature extraction [8]. The problems of object recognition or feature extraction can be divided into hardware and algorithm domains. In the hardware methods, a field-programmable gate array can be used to implement the spike-based method [9]. The circuit can be also designed to achieve the efficiency of low power consumption by combining an active continuous-time front-end logarithmic photoreceptor [10]. In the algorithm methods, the visual information (such as an image) can be calculated by the address-event-representation [11] or by constructing stereo vision with cameras [12]. However, the devices for matching are more expensive than consumer cameras. It is easier and more effective when we use the algorithms with general cameras to solve the matching problem.

Most common cameras record two-dimensional images, with only the horizontal axis and the vertical axis. Since the camera takes images only in two dimensions, the distance measurement function that can be achieved is limited. To solve this problem, scholars have proposed mechanical stereo vision. This uses two cameras to shoot the same object from different positions and analyze the distance between the camera and the object by the algorithms of image processing or machine vision. The axis perpendicular to the image plane and the two axes of the image plane constitute the three-dimensional relationship between the camera and the object [12].

When we use two cameras to observe the same object, the positions of the object on the two images will be slightly different. This difference is called disparity. A simple two-eye stereo vision model is shown in Figure 1, where P is the target, CAM_L and CAM_R are the two cameras, b is the distance between two cameras, f is the focal length, I_L and I_R are the imaging planes, d_L and d_R are the distances between targets on the image planes and the centers of the images, O_L and O_R represent the center lines of the lens, and Z is the distance we are looking for. We can see that Z can be calculated by the relationship of similar triangles [13]:

$$\frac{b}{Z} = \frac{b - (d_L + d_R)}{Z - f} = \frac{d_L + d_R}{f} \tag{1}$$

The disparity is denoted as $d_L + d_R$, and it is the amount of horizontal displacement that is produced by the same object that is imaged by two cameras. Since both f and b are known, it can be seen that it is quite important to obtain the disparity in the stereo vision system. The key to obtaining the disparity is matching the same object in the two images [14,15].

As shown in the above description, the method for obtaining three-dimensional information by the mechanical stereo vision system is to analyze and obtain the disparity between the two images. Census transform (CT) is one of the most robust algorithms for calculating the disparity of two images [16]. When we use CT, the size of the conversion window directly affects the computational load and accuracy. In a previous study, it was confirmed that a larger conversion window makes the result of object matching more accurate [17,18]. When the conversion window is larger than a certain size, the matching performance is not as significant as the window becomes larger. However, an oversized window not only consumes computational resources but also makes too many errors in matching. Therefore, the size of the window in CT is one of the important keys to determining the performance [19].

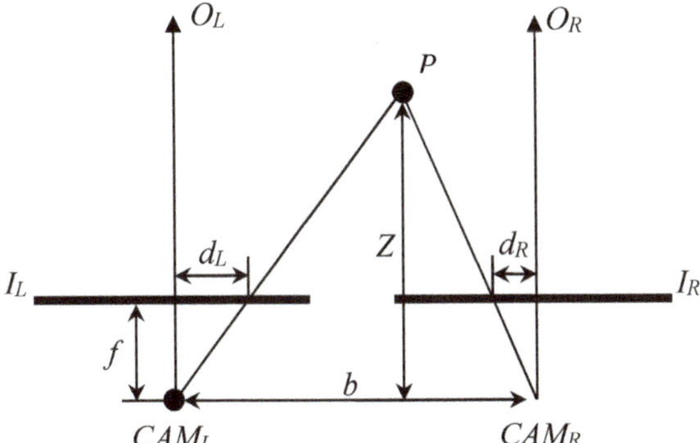

Figure 1. Schematic diagram of a three-dimensional model with disparity.

The calculation of each pixel by CT requires a large computational load and memory requirements. This makes it difficult for CT to be applied in real-time systems. Since the object in the image is bound by its edge, which is a sudden change in intensity, the edge (the high-frequency information in the image) is an important image feature [20]. This edge detection and the high-frequency information method are very important parts of image processing. These have also been applied in many real applications, such as oil spill detection for offshore drilling platforms [21], vehicle license plate location identification [22], pedestrian detection [23], and gesture recognition [24]. In the stereo vision system, a boundary can be used to identify whether the region is flat or has texture. The boundaries in the image can be obtained by gradients. Changing the length and width of the window according to the vertical and horizontal gradients can be used to reduce the bad matching of CT [25]. However, the quality of the disparity and operation loading are not discussed. The disparity quality can be improved by matching with a variable window and p post-processing with sub-pixel interpolation after CT [26]. This method does not adjust the window size when performing CT. However, using the sub-band of the high-frequency (such as edge) to improve the performance of CT is one of the feasible methods. Wavelet transform is a multi-resolution analysis method. The image data can be transformed into different sub-bands according to the defined wavelet [17]. The Haar wavelet is a well-known method to analyze the frequency information from sub-bands [27].

In this paper, two methods are proposed to improve the performance of CT using edge information. The first method is named census transform with Haar wavelet (CTHW) and uses edge information that is extracted by a wavelet. Since the edge information provides more accurate object information, the high-passed data is used to modify the disparity. The second method is called an adaptive window census transform (AWCT). The AWCT can determine whether the boundary of the window is increased or not when the window is enlarged. The increased rate of the boundary pixels in the window is used to determine the suitable window size. Moreover, since the sparse census transforms can be used to enhance the CT's performance by the designed mapping positions [28]; we also applied sparse census transforms to AWCT. AWCT and adaptive window sparse census transform (AWSCT) are applied to avoid using the oversized window and improving the performance.

2. Related Methods

2.1. The CT Algorithm

CT converts the grayscale intensity represented by each pixel in the grayscale image into the grayscale intensity relationship of each pixel to the neighbor pixels. The relationship can be treated as a feature of the pixel and used to find the two most similar features in the left and right images by the Hamming distance. The positions of the most similar points can be used to compute the disparity by Equation (1).

CT is defined as an order in a local neighborhood (conversion window) by comparing the relationship between the center pixel and the neighborhood pixels. The relationship between the center point (denoted as p) and the neighbor point (denoted as p') can be described by the conversion function [16,29]:

$$\xi(p,p\prime) = \begin{cases} 1, & if\ I(p) > I(p\prime) \\ 0, & otherwise \end{cases} \tag{2}$$

where $I()$ is the intensity of the pixel. A conversion window is defined to select neighbor pixels. In Equation (2), p is located at the center of the window, and the other pixels located in this window are selected to be p' in turn. Usually, the shape of the converted window is square, and the size is user-defined. The CT at the pixel p can be written as

$$C(p_{xy}) = \underset{p_{ij} \in w}{\otimes} \xi(I(p_{xy}), I(p_{ij})) \tag{3}$$

where $\otimes$ is the concatenation operator and w is the conversion window. In the stereo vision with CT, two images (such as I_L and I_R in Figure 1) are transformed by CT and hamming distance is applied to obtain the disparity between two transformed images. The disparity can be determined by using winner-takes-all to find the minimum value among all possible disparity values [30].

2.2. The Haar Wavelet

The wavelet transform converts signals into small waves and performs signal processing and signal analysis with multi-resolution. It is widely used in compression, transmission and image analysis [31]. The Haar wavelet was proposed by Alfréd Haar in 1909 [27]. It is the earliest proposed and simplest type of wavelet transforms [32]. With two pixels (p_1 and p_2) in the image, the Haar wavelet can be implemented to the low-band and high-band by

$$\text{Low band} = (p_1 + p_2)/\sqrt{2} \tag{4}$$

and

$$\text{High band} = (p_1 - p_2)/\sqrt{2} \tag{5}$$

respectively. In practice, the Haar wavelet can be described as a transformation matrix [32]:

$$Haar = \frac{\sqrt{2}}{2}\begin{bmatrix} 1 & 1 \\ 1 & -1 \end{bmatrix} \tag{6}$$

We can see that the image is high-passed and low-passed in the x-direction with down-sampling. The result obtained is also high-passed and low-passed in the y-direction with down-sampling. Finally, we obtain four sub-bands of LL (horizontal low-band and vertical low-band), LH (horizontal low-band and vertical high-band), HL (horizontal high-band and vertical low-band), and HH (horizontal high-band and vertical high-band).

2.3. Edge Detection

The boundary information can be extracted by edge detection and regarded as a result of high-pass filtering. The result of edge detection is mainly used to highlight whether the image is an area where the pixel changes significantly. In this study, the boundary information is used to classify and determine the complexity in the vicinity of the pixel. If the gray scale intensity changes significantly, a smaller conversion window can be used; otherwise, if the gray scale intensity of the area near the pixel does not change significantly, a larger conversion window must be used. In this paper, the Canny edge detection method is used for boundary detection [33,34]. First, the noise must be filtered by a two-dimensional Gaussian filter. The Gaussian function can be described as

$$G(x,y) = e^{-\frac{x^2+y^2}{2\sigma^2}}$$

(7)

where σ is the variance of the Gaussian function and it can be regarded as a smoothing factor in the filtering. The Gaussian function and the image can be computed by convolution to obtain the amount of change in the x and y directions, which are denoted as g_x and g_y. The gradient based on the pixel value in the image can be expressed as the gradient magnitude and the gradient direction, by

$$I_{gm}(x,y) = \sqrt{g_x^2 + g_y^2}$$

(8)

and

$$I_{g\theta}(x,y) = \tan^{-1}\left(\frac{g_y}{g_x}\right)$$

(9)

respectively. Since the edge can be described as the gradient, the boundary points can be detected as the larger gradient magnitude on the gradient direction. We compare the pixels on the gradient direction with pixels not on the gradient direction. If the value of a pixel on the direction is larger than the value of pixel not on the direction, the pixel is regarded as a boundary point; otherwise, it is a non-boundary point.

3. Proposed Methods

3.1. Census Transform with Haar Wavelet (CTHW)

In this section, the multi-resolution of the image with a Haar wavelet is used to reduce the computational time of the census transform (CT) and Hamming distance operations. The flow of this method is shown in Figure 2. First, the left and right view images are input. For both images, Haar wavelets are performed to obtain the frequency domain data of LL, HL, LH, and HH-bands. The outputs of LL-bands are converted by CT and the disparity calculated using the Hamming distance. The output of the HH-band from the left image is performed by path searching after binarization. The stop point of the path searching is determined by the high values in the HH-band. The disparity is modified to find the largest disparity that appears in the search paths.

Since the size of the LL-band is smaller than the original image, one main idea of this method is to use the LL-bands of original images to reduce the computational load of CT. Moreover, since the down-sampled image may result in errors, another main idea is to use the HH-band of one original image to modify the disparity. The path searches of the pixel in the HH-band are in four directions: up, down, right, and left. An example of an HH-band image is shown in Figure 3 and a point is colored in red. Since the mechanical stereo vision is applied by two horizontal cameras, the horizontal direction is the main searching path. In order to reduce the effect of large areas without borders, the vertical direction is added to the searching paths. The four searching paths are from the red point to four directions, which are denoted as A, B, C, and D with green arrows. Each searching path is stopped when the path arrives at the edge (white point). The corresponding disparity values (the output of CT with Hamming distance) on the paths are recorded and counted. The modified disparity is set as the disparity value with the largest number.

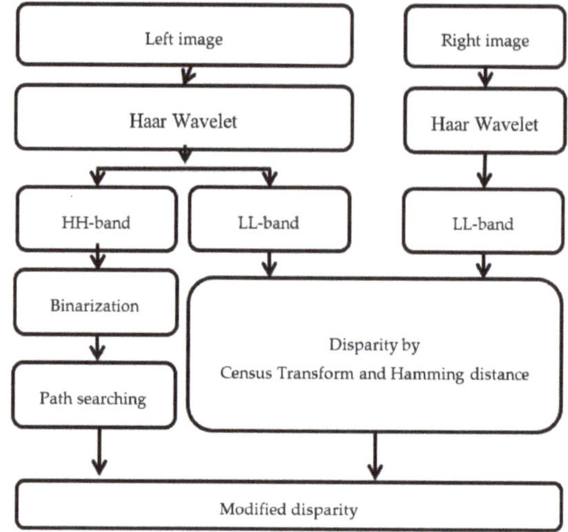

Figure 2. The flowchart of the census transform with Haar Wavelet.

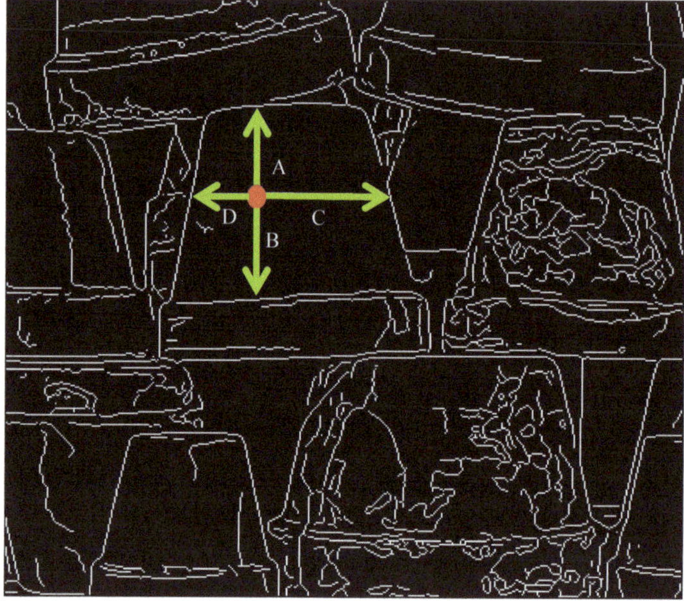

Figure 3. Example of the path searching of a (red) point.

3.2. Adaptive Window Census Transform (AWCT)

The size of the conversion window will affect the computational load and accuracy when we use CT. The larger the conversion window, the more accurate the results, but this also consumes the computational resources. In this section, the proposed method, AWCT, changes the selection of the conversion window size. The conversion window size changes from a fixed size of all pixels to an adaptive size by the boundary around the pixel. The edge information is used to select the window size.

The flow chart of AWCT is shown in Figure 4. First, the edge information can be obtained from the left view image by edge detection. The conversion window size of each point can be determined by the edge information. The selected window sizes are applied for CT, and the disparity can be computed through a Hamming distance computation.

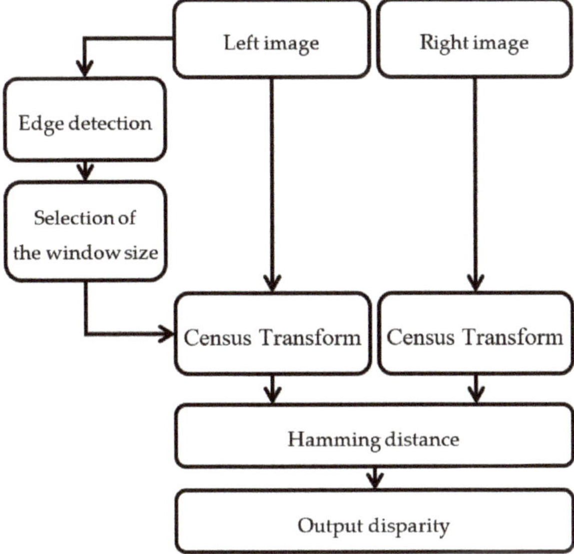

Figure 4. The flowchart of the adaptive window census transform (AWCT).

The conversion window sizes can be $3 \times 3, 5 \times 5, 7 \times 7, \ldots, 21 \times 21$. After the edge detection, each point will be the center of the windows in turn, and we count the number of edge points in each window. We also count the proportion of the number of edge points in the window. The window size ranges from small to large, and the size is selected when the window contains edge points. We divide the window size selection into two types. The first type is no edge, in which there are no edge points in the window. The image texture representation in this area is unclear, even without texture. The largest window size is used in this type. The other type is one in which the edge points are in the window. In this type, we record the proportion of edge points in the window when the window changes from small to large. We record it as one-time negative growth when the window size increases by one level and the ratio decreases. The window size will be selected when we have negative growth N consecutive times. In this study, the value of N is set to 5 through experience. This value can be set by the user for different cases.

Since the conversion window size is adaptive, the used window size of each pixel may be different. The smaller windows size is selected to compute the hamming distance when the sizes are different. In order to make the comparison of hamming distance reasonable, the calculation order of pixels is counter-clockwise from the center outward. An example of the pixels' order is shown in Figure 5. Even if the two windows are different in size, this order will make the relative positions of the pixels the same. The comparison of Hamming distance is up to the length of the small window. This allows the hamming distance to be used in the same window size.

42	41	40	39	38	37	36
43	20	19	18	17	16	35
44	21	6	5	4	15	34
45	22	7		3	14	33
46	23	8	1	2	13	32
47	24	9	10	11	12	31
48	25	26	27	28	29	30

Figure 5. The order of pixel calculation (with a 7×7 window size).

3.3. Adaptive Window Sparse Census Transform (AWSCT)

According to Equation (3), we can see that the number of points to be computed will increase as the window size increases. Since some points may be ignored to reduce the operation times of the computer, the sub-set of points of the conversion window is applied to determine which points are calculated for CT. This modified method is called a sparse census transform, and it is defined as [28]

$$C(p_{xy}) = \bigotimes_{p_{ij} \in w_s} \xi(I(p_{xy}), I(p_{ij}))$$ (10)

where w_s is the sub-set of points of the conversion window. We can see that the sparse census transform takes a part of the points to convert instead of all the points in the window. According to the results of the sparse census transform [28], the neighbor points are selected to be symmetric. The 16-points are selected with a 7×7 window as shown in Figure 6c, which maximizes the performance. In this paper, the same pattern of Figure 6c is used and expanded for the selected points to the different windows. The selected points with different windows are shown in Figure 6. In this section, the sparse census transform is combined with AWCT. The flowchart of AWSCT is the same as AWCT (Figure 4), but the CT is changed to SCT (sparse census transform).

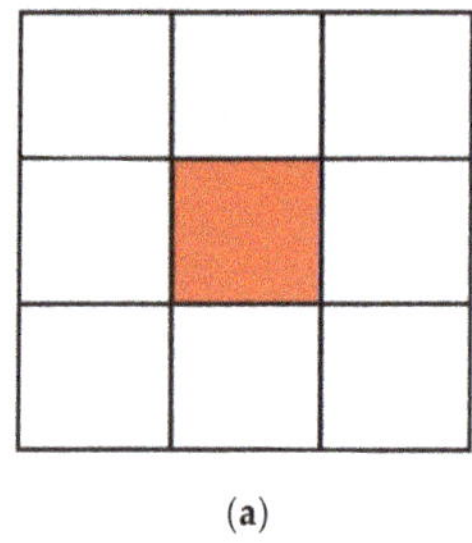

(a)

(b)

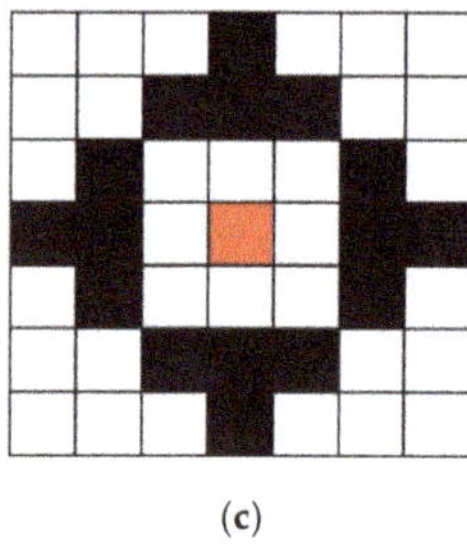

(c)

Figure 6. *Cont.*

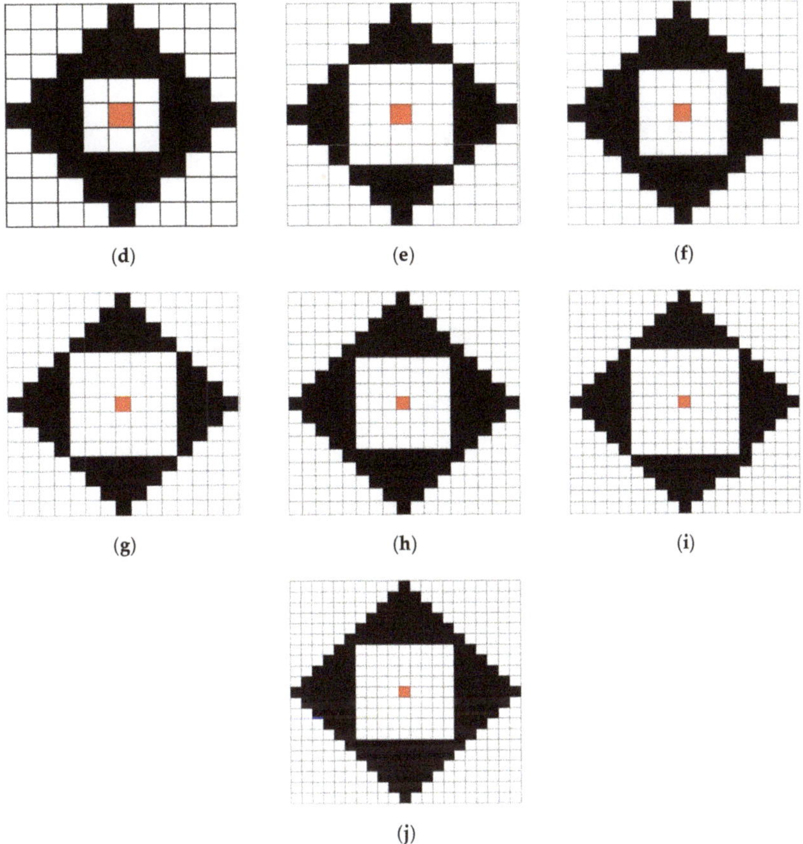

Figure 6. The selected points in the windows with a sparse census transform. (**a**) AWSCT 3×3, (**b**) AWSCT 5×5, (**c**) AWSCT 7×7, (**d**) AWSCT 9×9, (**e**) AWSCT 11×11, (**f**) AWSCT 13×13, (**g**) AWSCT 15×15, (**h**) AWSCT 17×17, (**i**) AWSCT 19×19, and (**j**) AWSCT 21×21.

4. Experiments and Results

The results were compared with the ground truth data by PoBMP (percentage of bad matching pixels) and RMS (root mean squared) [35]. The PoBMP is defined by

$$\text{PoBMP} = \frac{1}{N} \sum_{(x,\, y)} \left(|d_C(x,\, y) - d_T(x,\, y)| \right) > \delta_d \tag{11}$$

where d_C is the disparity with the proposed method, d_T is the disparity with the ground truth and δ_d is the allowable error which is set as 3 in this paper. The RMS can be obtained by

$$\text{RMS} = \left(\frac{1}{N} \sum_{(x,y)} |d_C(x,y) - d_T(x,y)|^2 \right)^{\frac{1}{2}} \tag{12}$$

Six images (Moebius, Flowerpots, Reindeer, Cloth2, Midd1, and Baby1), which were provided by Middlebury Stereo Datasets [36], were used to show the performances of the proposed methods. These six images and their ground truth are shown in Figure 7.

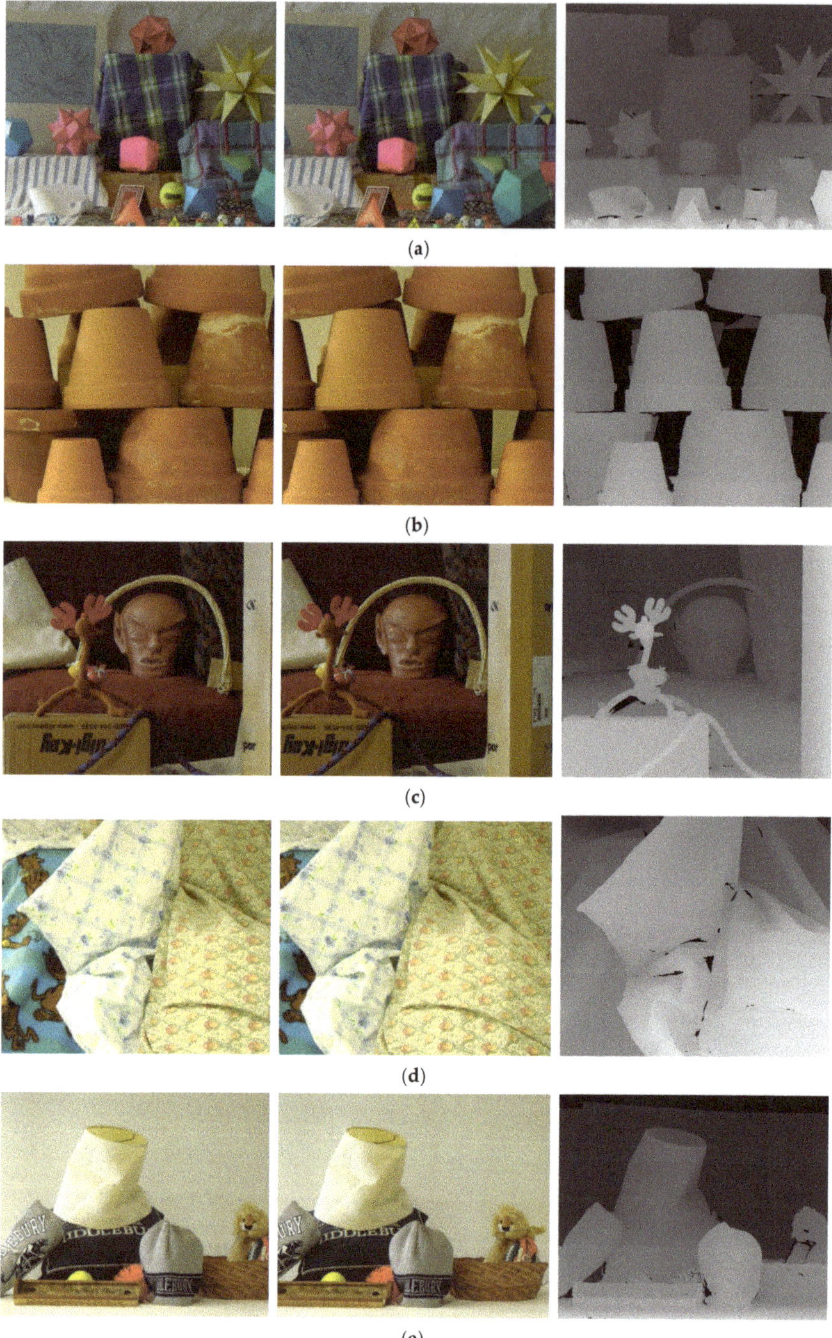

(a)

(b)

(c)

(d)

(e)

Figure 7. *Cont.*

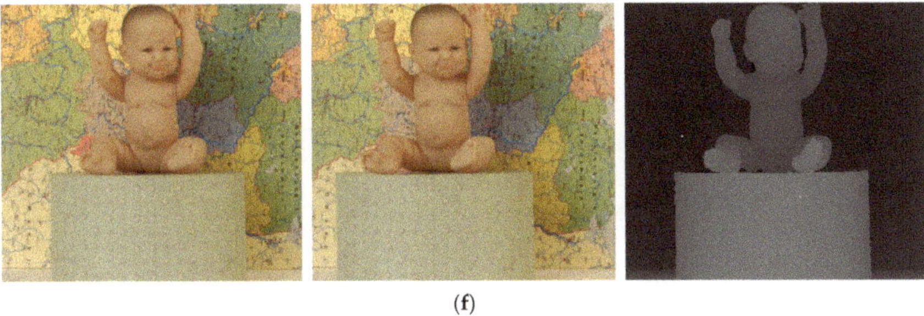

(**f**)

Figure 7. The experimental images: sequentially, the right image, left image and ground truth. (**a**) Moebius, (**b**) Flowerpots, (**c**) Reindeer, (**d**) Cloth2, (**e**) Midd1, and (**f**) Baby1.

4.1. Results of CTHW

The disparity results of six images by CTHW and CT with a 21 × 21 window size are shown in Figure 8, and the PoBMP results are shown in Table 1. According to the results, we can see that we obtained a better PoBMP with the proposed CTHW. Especially in the case of small conversion windows, such as 3 × 3 and 5 × 5, the PoBMP was less than 10% lower than the CT. In some exceptional cases—for example, Reindeer and Cloth2 with 13 × 13 window sizes—although the PoBMP of CTHW was higher than the PoBMP of CT, the accuracy of CTHW was still higher with a small window size. In the disparity results, the black points represent unknown disparity. We can see that the disparity images of CTHW was better than that of CT because there were significantly fewer black points. The results show that CTHW obtained better disparity results.

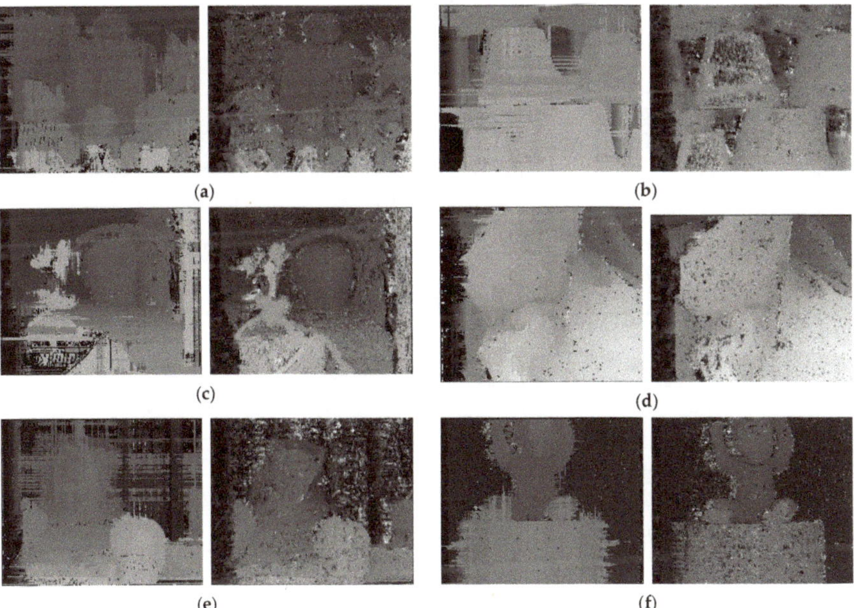

Figure 8. The results, sequentially, of CTHW and CT. (**a**) Moebius, (**b**) Flowerpots, (**c**) Reindeer, (**d**) Cloth2, (**e**) Midd1, and (**f**) Baby1.

Table 1. The results of CTHW and CT with percentages of bad matching pixels (PoBMP) (%).

Window Size	3×3		5×5		7×7		9×9		11×11		13×13	
Image Name	CT	CTHW	CT	CTHW	CT	CTHW	CT	CTHW	CT	CTHW	CT	CTHW
Moebius	78.75	41.95	53.97	24.17	39.35	21.33	31.72	19.85	27.34	19.36	24.66	19.40
Flowerpots	79.11	51.92	67.51	39.55	59.14	36.23	53.59	35.39	49.90	34.78	47.18	34.28
Reindeer	82.84	52.41	59.88	38.22	45.58	34.17	38.55	32.50	34.28	32.10	32.23	32.70
Cloth2	71.59	46.78	43.44	29.84	31.84	26.69	26.95	25.87	24.62	25.52	23.27	25.28
Midd1	84.97	70.29	68.38	54.44	58.33	46.47	53.21	41.88	50.12	38.82	48.17	37.10
Baby1	72.06	37.45	52.70	21.54	40.11	20.81	31.69	20.80	26.63	20.68	23.37	20.59

4.2. Results of AWCT

The results of Moebius by AWCT are shown in Figure 9. The edge detection result is shown in Figure 9a. The main idea of the AWCT is that the windows can be adapted. The worst option is to select the largest window size. In order to show the worst case for the size of each window, the pixels which are adapted to the largest windows size in 13×13, 15×15, 17×17, 19×19, and 21×21 are set as white and shown in Figure 9b–f, respectively. We can observe that when using a 13×13 and 15×15 window size, not all edge areas are applied to the largest window. At a 17×17 and 19×19 window size, the largest window was used for almost all edge points. Similar experimental results of Flowerpots, Reindeer, Cloth2, Midd1 and Baby1 are also shown in Figures 10–14, respectively. The results of six images with RMS are shown in Figures 15–20, respectively. These results show that AWCT's RMS is equivalent to the results of the largest windows (7×17, 19×19 or 21×21) with CT. The detailed results of AWCT and the results of CT with the largest window are listed in Table 2. According to the results, we can see that when we use the windows sizes with 13×13, 15×15, 17×17, or 19×19, the number used of the largest window is significantly less. This means that the AWCT can effectively adjust the window size and reduce the number of the largest windows. The results also show that the accuracy (PoBMP and RMS) of AWCT is similar to that of CT, but the reduction ratio of the operation number of calculation points (total pixels are calculated in Equation (3)) is about 4%–7%.

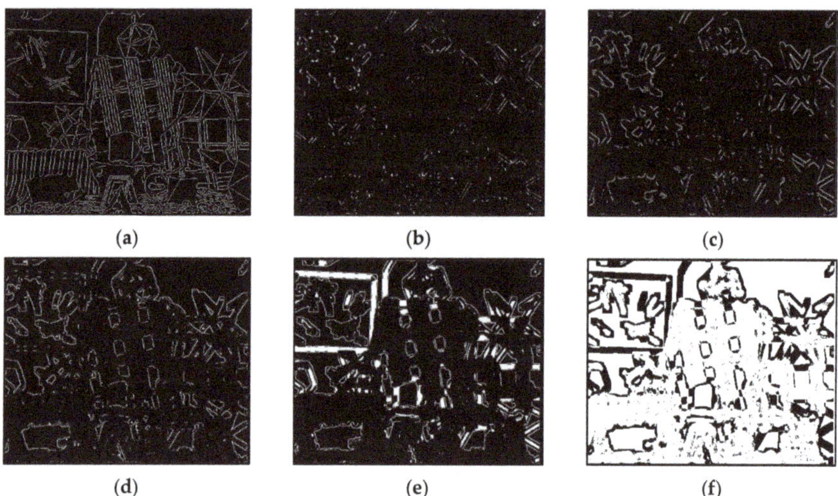

(a)　　　　　　(b)　　　　　　(c)

(d)　　　　　　(e)　　　　　　(f)

Figure 9. The results of Moebius by AWCT. (**a**) Edge information and schematic diagrams of the biggest window size in (**b**) 13×13, (**c**) 15×15, (**d**) 17×17, (**e**) 19×19, and (**f**) 21×21.

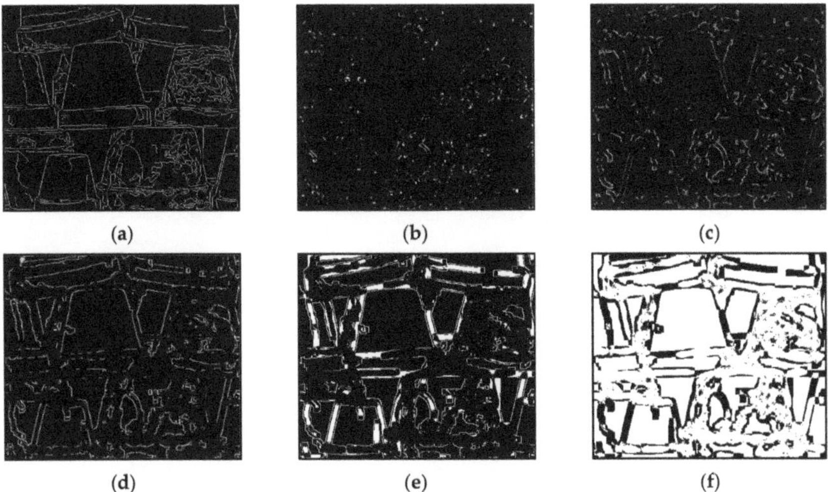

Figure 10. The results of Flowerpots by AWCT. (**a**) Edge information and schematic diagrams of the biggest window size in (**b**) 13 × 13, (**c**) 15 × 15, (**d**) 17 × 17, (**e**) 19 × 19, and (**f**) 21 × 21.

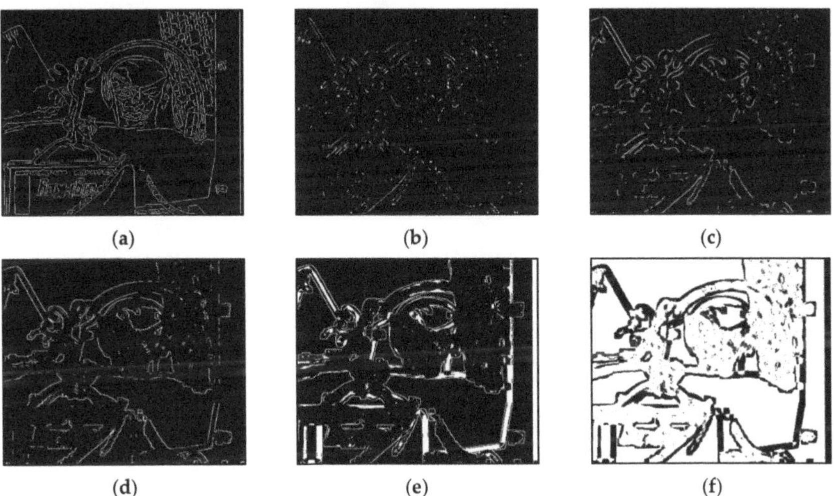

Figure 11. The results of Reindeer by AWCT. (**a**) Edge information and schematic diagrams of the biggest window size in (**b**) 13 × 13, (**c**) 15 × 15, (**d**) 17 × 17, (**e**) 19 × 19, and (**f**) 21 × 21.

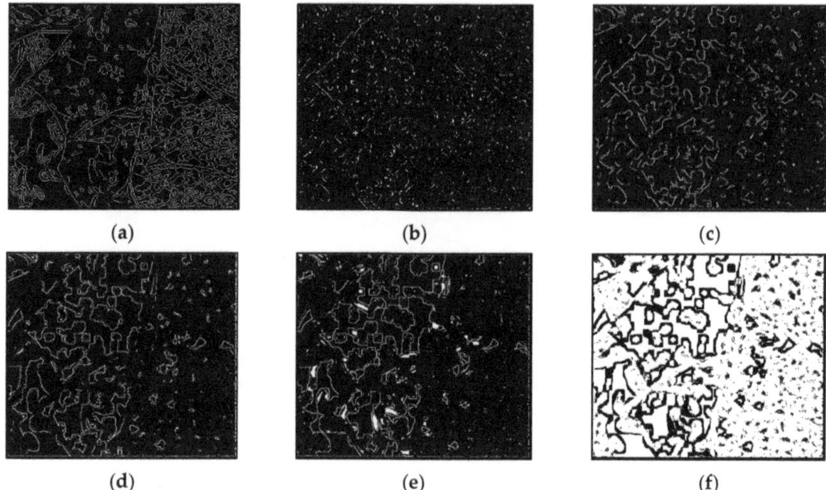

Figure 12. The results of Cloth2 by AWCT. (**a**) Edge information and schematic diagrams of the biggest window size in (**b**) 13 × 13, (**c**) 15 × 15, (**d**) 17 × 17, (**e**) 19 × 19, and (**f**) 21 × 21.

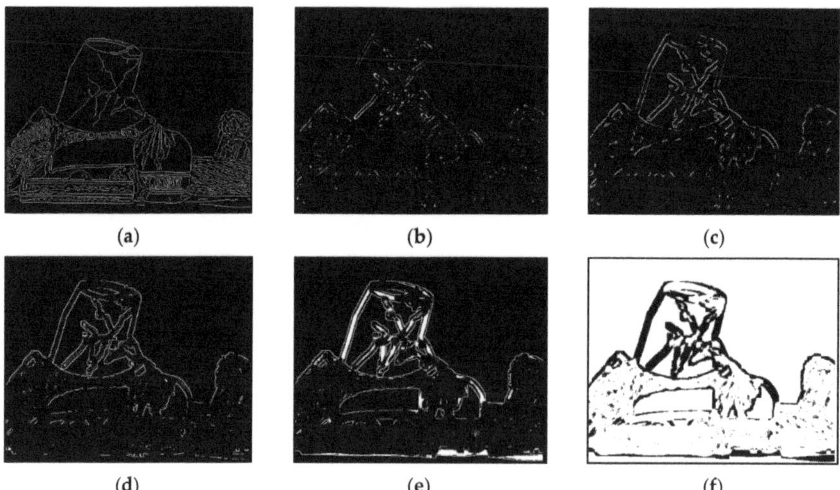

Figure 13. The results of Midd1 by AWCT. (**a**) Edge information and schematic diagrams of the biggest window size in (**b**) 13 × 13, (**c**) 15 × 15, (**d**) 17 × 17, (**e**) 19 × 19, and (**f**) 21 × 21.

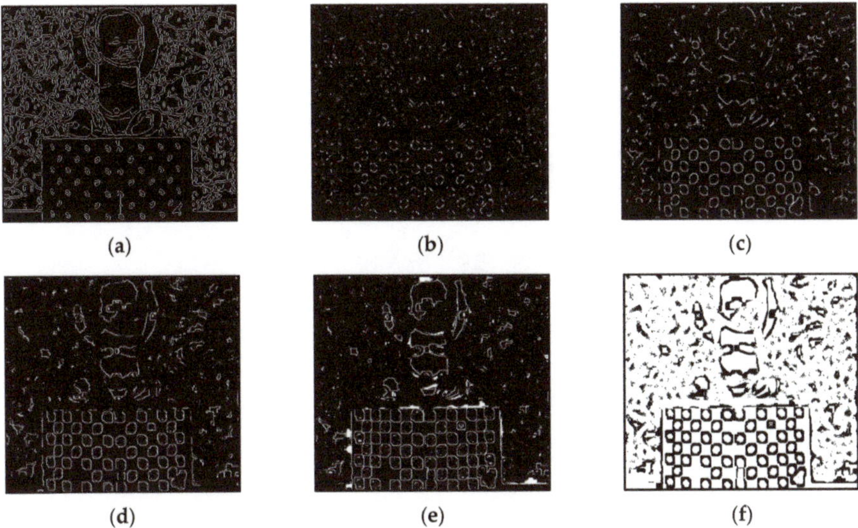

(a)   (b)   (c)

(d)   (e)   (f)

Figure 14. The results of Baby1 by AWCT. (**a**) Edge information and schematic diagrams of the biggest window size in (**b**) 13×13, (**c**) 15×15, (**d**) 17×17, (**e**) 19×19, and (**f**) 21×21.

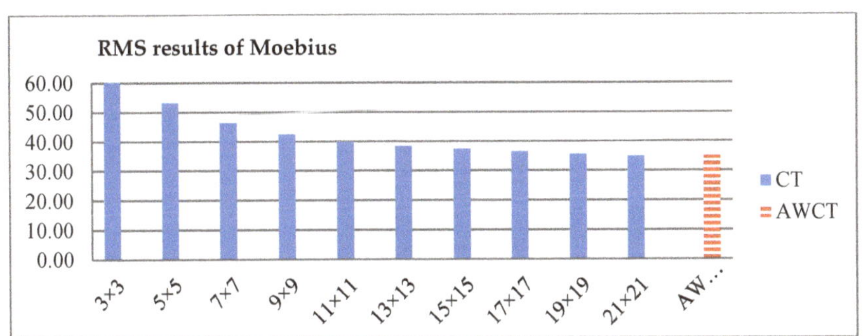

Figure 15. RMS comparison results of Moebius between AWCT and CT.

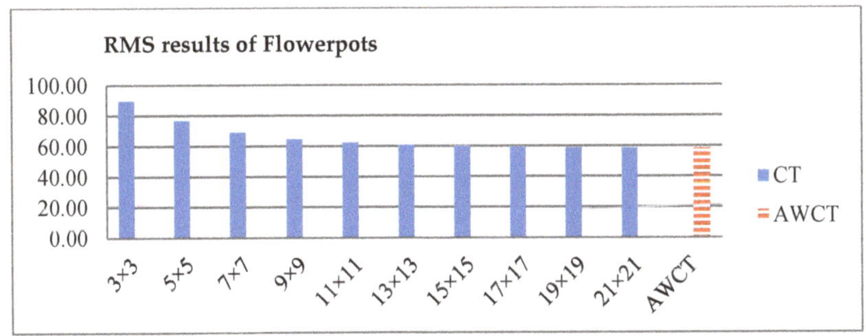

Figure 16. RMS comparison results of Flowerpots between AWCT and CT.

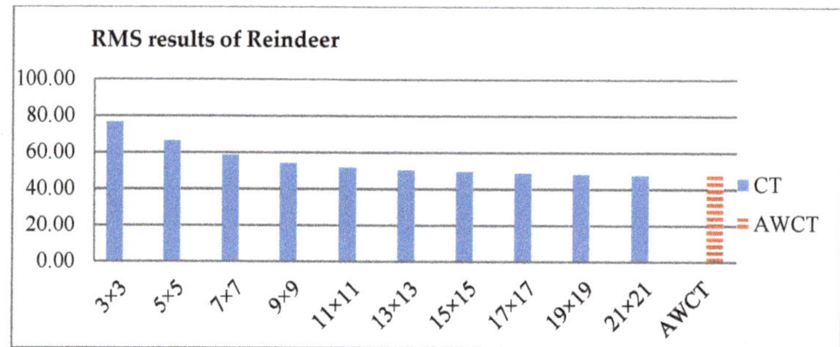

Figure 17. RMS comparison results of Reindeer between AWCT and CT.

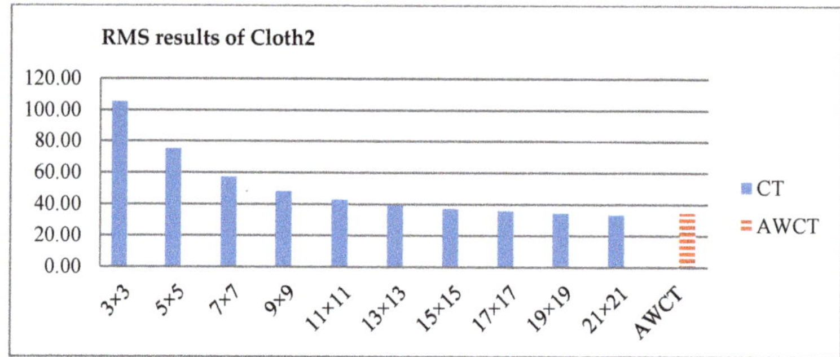

Figure 18. RMS comparison results of Cloth2 between AWCT and CT.

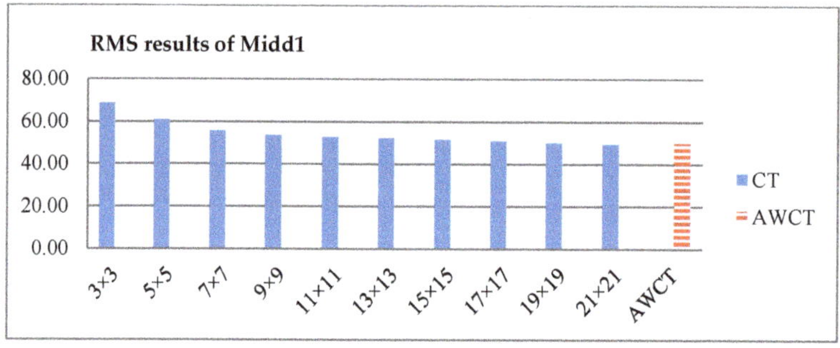

Figure 19. RMS comparison results of Midd1 between AWCT and CT.

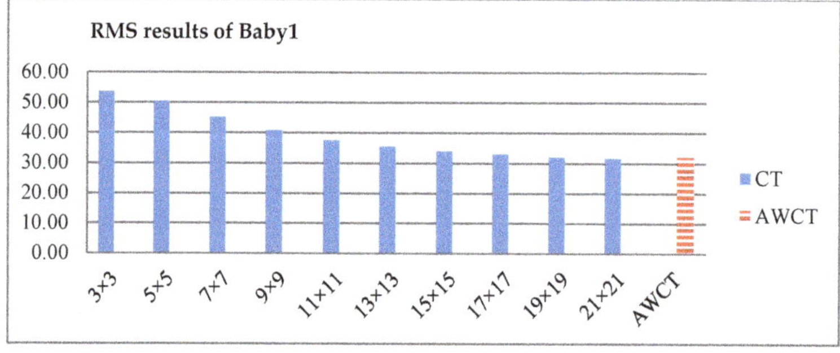

Figure 20. RMS comparison results of Baby1 between AWCT and CT.

Table 2. Comparison of AWCT and CT.

Image Name	PoBMP of CT (21 × 21)	PoBMP of AWCT	RMS of CT (21 × 21)	RMS of AWCT	Reduction Ratio of Operation
Moebius	20.12	20.23	34.93	34.90	6.98%
Flowerpots	32.25	32.52	58.55	58.49	6.97%
Reindeer	29.55	29.42	47.48	47.49	6.11%
Cloth2	16.82	17.11	33.18	34.12	8.72%
Midd1	43.71	43.49	49.37	49.86	3.94%
Baby1	18.64	19.17	31.57	32.02	7.72%

4.3. Results of AWSCT

The results of AWSCT and the results of SCT with the largest window are listed in Table 3. The experimental results show that the accuracies (PoBMP and RMS) of the two methods are similar, but the proposed AWSCT is better within the terms of operational requirements. This means that AWSCT can use fewer computing resources to achieve the same accuracy.

Table 3. Comparison of AWSCT and SCT.

Image Name	PoBMP of SCT (21 × 21)	PoBMP of AWSCT	RMS of SCT (21 × 21)	RMS of AWSCT	Reduction Ratio of Operation
Moebius	25.00	25.82	38.81	39.48	6.16%
Flowerpots	39.06	39.80	61.80	61.98	8.08%
Reindeer	34.24	34.69	51.13	51.54	7.23%
Cloth2	22.36	23.18	43.86	45.09	9.42%
Midd1	48.33	48.89	52.99	53.24	5.03%
Baby1	24.89	26.24	36.49	37.33	9%

4.4. Discussion of Results

The results of CTHW show that using wavelet's high-frequency band with path searching to modify disparity can effectively reduce PoBMP. This is because most bad matching is replaced by other disparities, but the modified disparity may not be accurate. Since the main problem of using CT is conversion window selection, it is easy to understand that CTHW (without adjusting the window) is not better in RMS and operation reduction. Based on high-frequency technology, we propose an AWCT method that uses edges to adjust the window size. The results show that AWCT's quality (PoBMP and RMS) is acceptable with a reduction of 4%–7% operation. Applying the sparse concept to AWCT can also reduce the operation by 5%–9% compared to SCT.

5. Conclusions

One of the well-known methods for obtaining disparity is called CT. We discussed the key problem of CT, which is the size of the conversion window. The larger the conversion window, the more accurate the process; however, an oversized window may not only consume computational resources but also make too many errors in matching. In this paper, we proposed one method, CTHW, to increase the accuracy with a wavelet transform and another one, AWCT, to enable the conversion window size to be adjusted for every point. In the results of CTHW, only the bad matching is improved, which does not reduce the RMS and operation loading. We can see that the proposed CTHW can provide a better result with a small window size and be suitably applied to a system with low computational resources. AWCT further finds the number of edge points to select the suitable window size for each point. According to the results, AWCT achieves a better performance in reducing the operation times with acceptable quality. Compared with CT, its average reduction ratio of operation was found to be about 6.6%. When we applied the sparse census transform to AWCT, as AWSCT, and compared this with SCT, the average reduction ratio of operation was about 7.5%. In the future, it is worth studying the use of high-frequency information to improve the quality and reduce the operation, and further enhance the performance, of CT.

Sensors **2020**, *20*, 2537

Author Contributions: Conceptualization, J.-J.L. and C.-P.L.; Investigation, J.-J.L., Y.-F.H. and Y.-H.L.; Methodology, J.-J.L. and S.-C.H.; Supervision, J.-J.L.; Visualization, C.-P.L. and Y.-H.L.; Writing—original draft, J.-J.L., Y.-H.L. and S.-C.H.; Writing—review and editing, C.-P.L. and Y.-F.H. All authors have read and agreed to the published version of the manuscript.

Funding: This research was supported in part by the Ministry of Science and Technology, Taiwan, under grant MOST 105-2221-E-324-008-MY2.

Acknowledgments: This research is partially sponsored by Chaoyang University of Technology (CYUT) and Higher Education Sprout Project, Ministry of Education, Taiwan, under the project name: "The R&D and the cultivation of talent for Health-Enhancement Products."

Conflicts of Interest: The authors declare no conflict of interest.

References

1. Li, L.; Zhang, M.; Guo, L.; Zhao, Y. Stereo Vision Based Obstacle Avoidance Path-Planning for Cross-Country Intelligent Vehicle. In Proceedings of the Sixth International Conference on Fuzzy Systems and Knowledge Discovery, Tianjin, China, 14–16 August 2009.
2. Hou, A.-L.; Cui, X.; Geng, Y.; Yuan, W.-J.; Hou, J. Measurement of Safe Driving Distance Based on Stereo Vision. In Proceedings of the Sixth International Conference on Image and Graphics, Hefei, China, 12–15 August 2011.
3. Kheng, E.S.; Hassan, A.H.A.; Ranjbaran, A. Stereo Vision with 3D Coordinates for Robot Arm Application Guide. In Proceedings of the IEEE Conference on Sustainable Utilization and Development in Engineering and Technology, Petaling Jaya, Malaysia, 20–21 November 2010.
4. Kanellakis, C.; Nikolakopoulos, G. Guidance for autonomous aerial manipulator using stereo vision. *J. Intell. Robot. Syst.* **2019**. [CrossRef]
5. Fu, Y.; Wang, C.; Liu, R.; Liang, G.; Zhang, H.; Rehman, S.U. Moving object localization based on UHF RFID phase and laser clustering. *Sensors* **2018**, *18*, 825. [CrossRef] [PubMed]
6. Khaleghi, B.; Ahuja, S.; Wu, Q.M.J. An Improved Real-Time Miniaturized Embedded Stereo Vision System (MESVS-II). In Proceedings of the IEEE Computer Society Conference on Computer Vision and Pattern Recognition Workshops, Anchorage, AK, USA, 21–28 June 2008.
7. Holzmann, C.; Hochgatterer, M. Measuring Distance with Mobile Phones Using Single-Camera Stereovision. In Proceedings of the 32nd International Conference on Distributed Computing Systems Workshops, Macau, China, 18–21 June 2012.
8. Domínguez-Morales, M.; Cerezuela-Escudero, E.; Jiménez-Fernández, A.; Paz-Vicente, R.; Font-Calvo, J.L.; Iñigo-Blasco, P.; Linares-Barranco, A.; Jiménez-Moreno, G. Image Matching Algorithms in Stereo Vision Using Address-Event-Representation: A Theoretical Study and Evaluation of the Different Algorithms. In Proceedings of the International Conference on Signal Processing and Multimedia Applications, Seville, Spain, 18–21 July 2011.
9. Domínguez-Morales, M.; Domínguez-MoralesOrcID, J.P.; Jiménez-FernándezOrcID, A.; Linares-Barranco, A.; Jiménez-Moreno, G. Stereo Matching in Address-Event-Representation (AER) Bio-Inspired Binocular Systems in a Field-Programmable Gate Array (FPGA). *Electronics* **2019**, *8*, 410. [CrossRef]
10. Lichtsteiner, P.; Posch, C.; Delbruck, T. A 128 × 128 120 dB 15 µs Latency Asynchronous Temporal Contrast Vision Sensor. *IEEE J. Solid-State Circuits* **2008**, *43*, 566–576. [CrossRef]
11. Domínguez-Morales, M.; Jiménez-Fernández, A.; Paz-Vicente, R.; Jiménez, G.; Linares-Barran, A. Live Demonstration: On the Distance Estimation of Moving Targets with a Stereo-Vision AER System. In Proceedings of the 2012 IEEE International Symposium on Circuits and Systems, Seoul, Korea, 20–23 May 2012.
12. Burschka, D.; Brown, M.Z.; Hager, G.D. Advances in Computational Stereo. *IEEE Trans. Pattern Anal. Mach. Intell.* **2003**, *25*, 993–1008.
13. Liaw, J.-J.; Chen, K.-L.; Huang, M.-S. Using Different Baselines to Improve Image Depth Measurement Accuracy. In Proceedings of the International Conference on Engineering & Technology, Computer, Basic & Applied Sciences, Osaka, Japan, 24–25 January 2017.
14. Ogale, A.S.; Aloimonos, Y. Shape and the stereo correspondence problem. *Int. J. Comput. Vis.* **2005**, *65*, 147–162. [CrossRef]
15. Juang, C.-F.; Chen, G.-C.; Liang, C.-W.; Lee, D. Stereo-camera-based object detection using fuzzy color histograms and a fuzzy classifier with depth and shape estimations. *Appl. Soft Comput.* **2016**, *46*, 753–766. [CrossRef]

16. Zabih, R.; Woodfill, J. Non-parametric local transforms for computing visual correspondence. *Comput. Vis.-Eccv'94* **2005**, *801*, 151–158. [CrossRef]

17. Huang, S.-C.; Liaw, J.-J.; Chu, H.-C. Modified Census Transform Using Haar Wavelet Transform. In Proceedings of the International Conference on Applied System Innovation, Osaka, Japan, 22–27 May 2015.

18. Huang, S.-C. Improved Census Transform's Disparity Map by Using Edge Information. Master's Thesis, Chaoyang University of Technology, Taichung, Taiwan, 2016.

19. Rahnama, O.; Frost, D.; Miksik, O.; Torr, P.H.S. Real-time dense stereo matching with ELAS on FPGA-accelerated embedded devices. *IEEE Robot. Autom. Lett.* **2018**, *3*, 2008–2015. [CrossRef]

20. Mahmood, N.; Shah, A.; Waqas, A.; Abubakar, A.; Kamran, S.; Zaidi, S.B. Image segmentation methods and edge detection: An application to knee joint articular cartilage edge detection. *J. Theor. Appl. Inf. Technol.* **2015**, *71*, 87–96.

21. Jing, Y.; An, J.; Li, L. The Edge Detection of Oil Spills Image Using Self-Adaptive Dynamic Block Threshold Algorithm Based on Non-Maximal Suppression. In Proceedings of the 2nd International Congress on Image and Signal Processing, Tianjin, China, 17–19 October 2009.

22. Rashedi, E.; Nezamabadi-pour, H. A hierarchical algorithm for vehicle license plate localization. *Multimed. Tools Appl.* **2018**, *77*, 2771–2790. [CrossRef]

23. Zhang, C.; Tan, N.; Lin, Y. Remote Pedestrian Detection Algorithm Based on Edge Information Input CNN. In Proceedings of the 3rd High Performance Computing and Cluster Technologies Conference, Guangzhou, China, 22–24 June 2019.

24. Pansare, J.R.; Ingle, M. Vision-Based Approach for American Sign Language Recognition Using Edge Orientation Histogram. In Proceedings of the International Conference on Image, Vision and Computing, Portsmouth, UK, 3–5 August 2016.

25. Ko, J.; Ho, Y.-S. Stereo Matching Using Census Transform of Adaptive Window Sizes with Gradient Images. In Proceedings of the Asia-Pacific Signal and Information Processing Association Annual Summit and Conference (APSIPA), Jeju Island, Korea, 13–16 December 2016.

26. Dinh, V.Q.; Nguyen, D.D.; Nguyen, V.D.; Jeon, J.W. Local Stereo Matching Using an Variable Window, Census Transform and an Edge-Preserving Filter. In Proceedings of the 12th International Conference on Control, Automation and Systems, Jeju Island, Korea, 17–21 October 2012.

27. Haar, A. Zur Theorie der orthogonalen Funktionensysteme. *Math. Ann.* **1910**, *69*, 331–371. [CrossRef]

28. Fife, W.S.; Archibald, J.K. Improved Census Transforms for Resource-Optimized Stereo Vision. *IEEE Trans. Circuits Syst. Video Technol.* **2013**, *23*, 60–73. [CrossRef]

29. Froba, B.; Emst, A. Face Detection with the Modified Census Transform. In Proceedings of the Sixth IEEE International Conference on Automatic Face and Gesture Recognition, Seoul, Korea, 19 May 2004.

30. Jo, H.-W.; Moon, B. A Modified Census Transform Using the Representative Intensity Values. In Proceedings of the International SoC Design Conference, Gyungju, Korea, 2–5 November 2015.

31. Daubechies, I. The wavelet transform, time-frequency localization and signal analysis. *IEEE Trans. Inf. Theory* **1990**, *36*, 961–1005. [CrossRef]

32. Alemohammad, M.; Stround, J.R.; Bosworth, B.T.; Foster, M.A. High-speed all-optical Haar wavelet transform for real-time image compression. *Opt. Express* **2017**, *25*, 9802–9811. [CrossRef] [PubMed]

33. Canny, J. A computational approach to edge detection. *IEEE Trans. Pattern Anal. Mach. Intell.* **1986**, *PAMI-8*, 679–698. [CrossRef]

34. Fujimoto, T.R.; Kawasaki, T.; Kitamura, K. Canny-edge-detection/rankine-hugoniot-conditions unified shock sensor for inviscid and viscous flows. *J. Comput. Phys.* **2019**, *396*, 264–279. [CrossRef]

35. Scharstein, D.; Szeliski, R. A taxonomy and evaluation of dense two-frame stereo correspondence algorithms. *Int. J. Comput. Vis.* **2002**, *47*, 7–42. [CrossRef]

36. Scharstein, D.; Hirschmuller, H.; Kiajima, Y.; Krathwohi, G.; Nesic, N.; Wang, X.; Westling, P. High-Resolution Stereo Datasets with Subpixel-Accurate Ground Truth. In Proceedings of the German Conference on Pattern Recognition, Münster, Germany, 3–5 September 2014.

© 2020 by the authors. Licensee MDPI, Basel, Switzerland. This article is an open access article distributed under the terms and conditions of the Creative Commons Attribution (CC BY) license (http://creativecommons.org/licenses/by/4.0/).

Article

PM$_{2.5}$ Concentration Estimation Based on Image Processing Schemes and Simple Linear Regression

Jiun-Jian Liaw [1], Yung-Fa Huang [1], Cheng-Hsiung Hsieh [2,*], Dung-Ching Lin [1] and Chin-Hsiang Luo [3]

[1]   Department of Information and Communication Engineering, Chaoyang University of Technology, 168, Jifeng E. Rd., Wufeng District, Taichung 413310, Taiwan; jjliaw@cyut.edu.tw (J.-J.L.); yfahuang@cyut.edu.tw (Y.-F.H.); boom88754032@gmail.com (D.-C.L.)

[2]   Department of Computer Science and Information Engineering, Chaoyang University of Technology, 168, Jifeng E. Rd., Wufeng District, Taichung 413310, Taiwan

[3]   Department of Safety, Health and Environmental Engineering, Hungkuang University, 1018, Sec. 6, Taiwan Blvd., Shalu District, Taichung 433304, Taiwan; andyluo@sunrise.hk.edu.tw

*   Correspondence: chhsieh@cyut.edu.tw

Received: 28 February 2020; Accepted: 23 April 2020; Published: 24 April 2020

Abstract: Fine aerosols with a diameter of less than 2.5 microns (PM$_{2.5}$) have a significant negative impact on human health. However, their measurement devices or instruments are usually expensive and complicated operations are required, so a simple and effective way for measuring the PM$_{2.5}$ concentration is needed. To relieve this problem, this paper attempts to provide an easy alternative approach to PM$_{2.5}$ concentration estimation. The proposed approach is based on image processing schemes and a simple linear regression model. It uses images with a high and low PM$_{2.5}$ concentration to obtain the difference between these images. The difference is applied to find the region with the greatest impact. The approach is described in two stages. First, a series of image processing schemes are employed to automatically select the region of interest (RoI) for PM$_{2.5}$ concentration estimation. Through the selected RoI, a single feature is obtained. Second, by employing the single feature, a simple linear regression model is used and applied to PM$_{2.5}$ concentration estimation. The proposed approach is verified by the real-world open data released by Taiwan's government. The proposed scheme is not expected to replace component analysis using physical or chemical techniques. We have tried to provide a cheaper and easier way to conduct PM$_{2.5}$ estimation with an acceptable performance more efficiently. To achieve this, further work will be conducted and is summarized at the end of this paper.

Keywords: PM$_{2.5}$ concentration estimation; digital image processing; automatic region of interest selection; data exclusion; linear regression

---

1. Introduction

Air pollution has been reported to significantly affect human health [1], causing issues such as premature death, bronchitis, asthma, cardiovascular disease, and lung cancer [2]. Pollutants in the air include CO, NO$_2$, and particulate matter. Among them, particulate matter with a diameter of less than 2.5 microns (PM$_{2.5}$) is a key component which severely affects human health in many ways. For example, PM$_{2.5}$ aerosols are able to directly enter the lungs through the respiratory tract and affect a person's health [3]. According to the World Health Organization report, more than 90% of the world's population inhales large amounts of pollutants every day, which results in approximately seven million deaths each year [4]. Consequently, PM$_{2.5}$ concentration estimation is required and has become an important concern for human health [5,6].

Many techniques have been developed to measure the $PM_{2.5}$ concentration, such as the filter-based gravimetric method [7], tapered element oscillating microbalance method [8], beta attenuation monitoring method [9], optical analysis method [10,11], and black smoke measurement [12]. These methods require expensive instruments and professional operations. Some more comprehensive methods analyze the relationship between human activities and $PM_{2.5}$ by satellite and big data [13,14]. However, satellite and big data are not available to the common user. Therefore, a simple and effective method should be sought for $PM_{2.5}$ concentration estimation.

In urban environments, researchers have developed low-cost sensors. These sensors are widely deployed throughout the city to monitor the $PM_{2.5}$ concentration [15]. Although one sensor is low in cost, it is not effective when widely deployed in a city requiring many sensors. The portable $PM_{2.5}$ sensor can be used to monitor the $PM_{2.5}$ concentration at different locations [16]. The portable device reduces the cost of employing a large number of sensors, but requires more manpower to move the sensors. Optical sensors, such as TEOM 1400a analyzer, SDS011 (Nova Fitness, Jinan, China), ZH03A (Winsen, Zhengzhou, China), PMS7003 (Plantower, Beijing China), and OPC-N2 (Alphasense, Braintree, UK), have been introduced to monitor $PM_{2.5}$ [17]. However, these optical sensors are more expensive than ordinary cameras. Since a camera is installed on the top floor of environmental monitoring stations in Taiwan, using the camera to estimate $PM_{2.5}$ is a simpler and more effective approach than employing extra devices.

It should be noted that air pollution is usually characterized by a poor visibility due to light scattering, such as Rayleigh scattering and Mie scattering, caused by the interaction between light and airborne particles [18]. In other words, the visibility is reduced, as a large amount of aerosol pollution scatter the atmospheric light [19], and vice versa. In previous decades, some researchers proposed methods to estimate the visibility through image processing schemes [20,21]. Recently, an expensive digital camera was used to take high-quality photos for visibility estimation [22]. These studies have shown that image processing schemes can be applied to visibility estimation. Furthermore, it was reported that the $PM_{2.5}$ concentration is related to visibility reduction [23]. However, these studies did not develop image processing technologies to estimate the $PM_{2.5}$ concentration. Therefore, it gives us hope that the $PM_{2.5}$ concentration may be estimated through image processing schemes.

The rapid development of computers, algorithms, and artificial intelligence has meant that image processing methods using machine learning have been widely applied. The main advantage of using machine learning is that it requires training and does not require defining too many features. Two types of the training-based algorithms are neural network methods [24] and linear regression schemes [25]. The neural network methods require a very fast and expensive graphics processing unit [26]. By contrast, compared to neural network methods, the estimation of spatial variations by linear regression could be performed by a consumer computer, as economical and predictive performance were both acceptable [27]. Nowadays, high-quality images can be taken by a commercial digital camera. This facilitates $PM_{2.5}$ concentration estimation by image processing schemes.

In order to understand which features can affect $PM_{2.5}$ concentrations when using image processing methods, previous research has pointed out that the $PM_{2.5}$ concentrations may affect image characteristics, including the distance, hazy model, entropy, contrast, sky color, and solar zenith angle. It was found that the distance is the feature that has the most influence [28]. This is consistent with the definition of visibility, and a previous study has also shown that visibility can be estimated using high-frequency information from an image [22]. The region of interest (RoI) has also been manually selected to estimate $PM_{2.5}$ concentrations [28]. However, the estimation performance might be degraded because of such a manually selected RoI. Besides, the computational cost might not be cheap, since a support vector regression model with several features was involved in the estimation. To solve these problems, this paper presents an approach to $PM_{2.5}$ concentration estimation, where only a single feature is used and simple linear regression is employed as an estimator. The main contribution of the proposed approach is to use a series of image processing schemes in $PM_{2.5}$ concentration estimation where the images are taken by a consumer camera. It provides a valuable alternative to estimating $PM_{2.5}$ concentration. The main aims of this image-based approach are as follows: (i) to automatically locate the RoI to replace the manual selection of Liu's work [28]; (ii) to use a single feature for linear regression instead of multiple features in $PM_{2.5}$ concentration estimation, with an acceptable performance; and (iii) to provide a cheaper alternative method with a camera for estimating the $PM_{2.5}$ concentration. This paper is organized as follows. The proposed approach is described in Section 2. In Section 3, real-world data is given to verify the proposed approach. Finally, a conclusion is made in Section 4.

2. The Proposed Approach

There are two stages involved in the proposed approach. In the first stage, a series of image processing schemes are employed to automatically locate the region of interest (RoI) to extract a single feature, which is required in the following stage for $PM_{2.5}$ concentration estimation. In the second stage, a simple linear regression model is used with the training data, which contains pairs of the single feature obtained through the selected RoI and the actual $PM_{2.5}$ concentration measurement. The simple linear regression model is then used in $PM_{2.5}$ concentration estimation with the testing data. The estimated $PM_{2.5}$ concentration is compared with the actual value and evaluated by performance indices. An overall block diagram for the proposed approach is depicted in Figure 1. The details of the proposed approach are described in the following sections. The proposed automatic RoI selection approach is described in Section 2.1, the simple linear regression model is given in Section 2.2, and three performance indices employed to assess the proposed approach are given in Section 2.3.

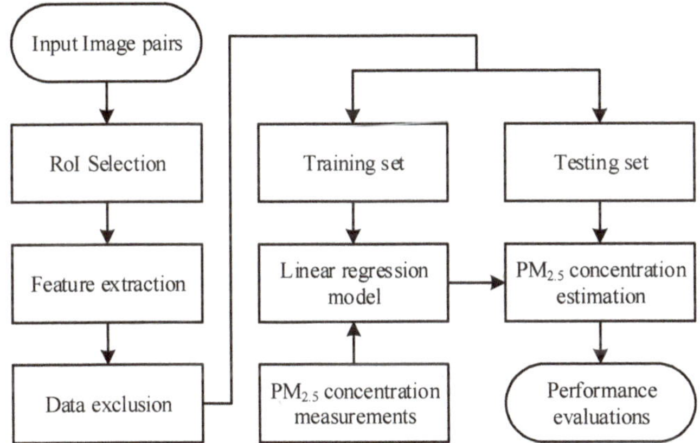

Figure 1. A block diagram of the proposed approach.

2.1. Automatic RoI Selection

It should be noted that not all parts of an image are strongly related to the $PM_{2.5}$ concentration. Therefore, selecting an appropriate RoI to estimate the $PM_{2.5}$ concentration is an important step for the successful application of the proposed approach. It is known that some details in the image will be blurred when the $PM_{2.5}$ concentration is high, compared to when there is a low $PM_{2.5}$ concentration. In other words, the pixel value of the images with a high and low $PM_{2.5}$ concentration is different. This also illustrates that not every feature has a good correlation with the $PM_{2.5}$ concentration. It motivates us to use the differences in image pairs of low and high $PM_{2.5}$ concentrations in automatic RoI selection. A flowchart of the proposed automatic RoI selection is depicted in Figure 2. A pair of images, shown in Figure 3a,b, are given to demonstrate how the proposed automatic RoI selection works. Given a pair of images of low and high $PM_{2.5}$ concentrations, both images are converted into gray-level images. The image of a low $PM_{2.5}$ concentration is denoted as I_1 and the one with a high

$PM_{2.5}$ concentration is denoted as I_2. A series of image processing steps to determine the final RoI is described in the following.

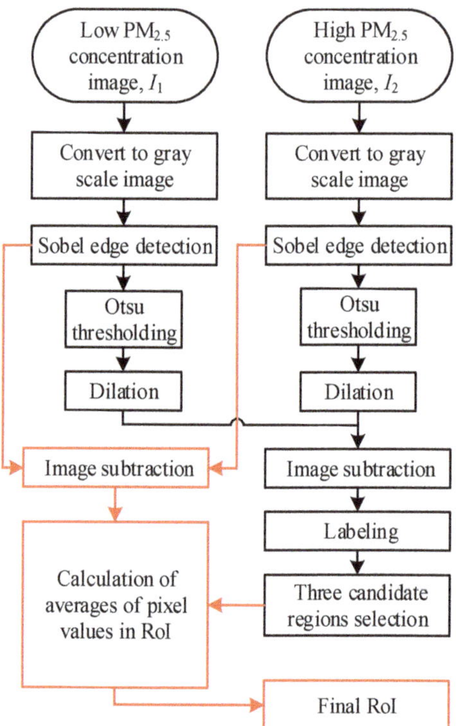

Figure 2. A flowchart of the proposed automatic region of interest (RoI) selection.

(**a**)   (**b**)

Figure 3. A sample image pair. (**a**) I_1 (low $PM_{2.5}$ concentration, 1 µg/m^3); (**b**) I_2 (high $PM_{2.5}$ concentration, 75 µg/m^3).

2.1.1. Sobel Edge Detection

As the first step, Sobel edge detection is applied to the image pair, I_1 and I_2, to extract the high-frequency components [29]. In Sobel edge detection, the gradients used in this approach for the x-axis and y-axis, respectively, are denoted as G_x and G_y, and given as

$$G_x = \begin{bmatrix} -1 & 0 & 1 \\ -2 & 0 & 2 \\ -1 & 0 & 1 \end{bmatrix} \text{ and } G_y = \begin{bmatrix} -1 & -2 & -1 \\ 0 & 0 & 0 \\ 1 & 2 & 1 \end{bmatrix}, \tag{1}$$

where a 3×3 mask is employed. The final magnitude G_{xy} is calculated as

$$G_{xy} = |G_x| + |G_y|. \tag{2}$$

The images produced by Sobel edge detection are denoted as $I_{1,s}$ and $I_{2,s}$, and are shown in Figure 4a,b, respectively. In Figure 4, one can see that the low concentration image I_1, after Sobel edge detection, has more details than I_2. This shows that more high-frequency components are contained in $I_{1,s}$ than $I_{2,s}$. The edge detection results of Figure 3a,b are shown in Figure 4a,b, respectively. We can see that two buildings on the right of Figure 3a do not appear in Figure 3b. This is because the PM$_{2.5}$ concentration of Figure 3b is higher than that of Figure 3a. This means that the edges of the two buildings are invisible in Figure 4b. The difference of Figure 4a,b is shown in Figure 4c. The results show that the PM$_{2.5}$ concentration has a significant effect on the high frequency components of images.

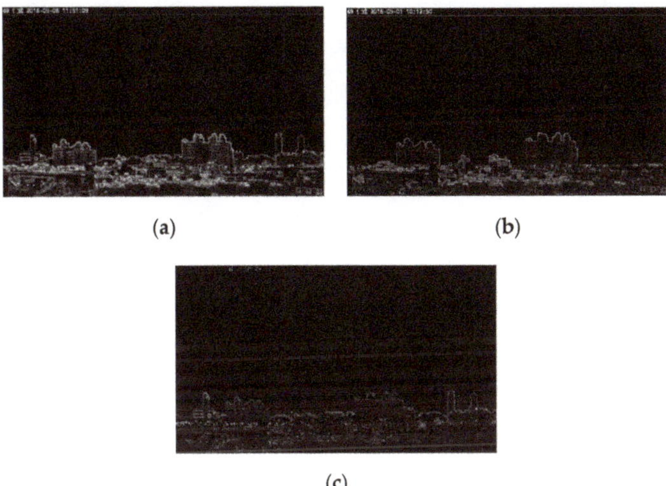

(a)   (b)

(c)

Figure 4. Images after Sobel edge detection. (**a**) $I_{1,s}$ (low PM$_{2.5}$ concentration); (**b**) $I_{2,s}$ (high PM$_{2.5}$ concentration); (**c**) the difference of (**a**) and (**b**).

2.1.2. Otsu Thresholding

After Sobel edge detection, Otsu thresholding [30] is applied to the two images in Figure 4 to obtain binary images. In Otsu thresholding, the pixels in an image are separated into two groups based on the histogram. By employing statistical properties, the optimal threshold, where the variance of each group is minimized and the variance between two groups is maximized, is determined. In Otsu thresholding, the weighted sum of the variance between two groups is found as

$$\sigma_w^2 = w_0(t)\sigma_0^2(t) + w_1(t)\sigma_1^2(t), \tag{3}$$

where $\sigma_0^2(t)$ and $\sigma_1^2(t)$ represent the variance of each group, and $w_0(t)$ and $w_1(t)$ are the weights of two groups separated by the threshold t, respectively. The weights $w_0(t)$ and $w_1(t)$ are obtained, respectively, as

$$w_0(t) = \sum_{i=0}^{t-1} p(i) \tag{4}$$

and

$$w_1(t) = \sum_{i=t}^{L-1} p(i), \tag{5}$$

where $p(i)$ is the probability of the pixel value i and L is the number of gray levels. The variance between two groups is given as

$$\sigma_o^2(t) = \sigma^2 - \sigma_w^2(t),\tag{6}$$

where σ^2 is the variance of the whole image. Equation (6) can be transformed into

$$\sigma_o^2(t) = w_0(t)w_1(t)[\mu_0(t) - \mu_1(t)]^2,\tag{7}$$

where $\mu_0(t)$ and $\mu_1(t)$ are the means of two groups separated by threshold t. The optimal threshold is then found with t, which maximizes $\sigma_o^2(t)$ in Equation (7). The images $I_{1,s}$ and $I_{2,s}$ after Otsu thresholding, are denoted as $I_{1,so}$ and $I_{2,so}$ and shown in Figure 5a and Figure 5b, respectively.

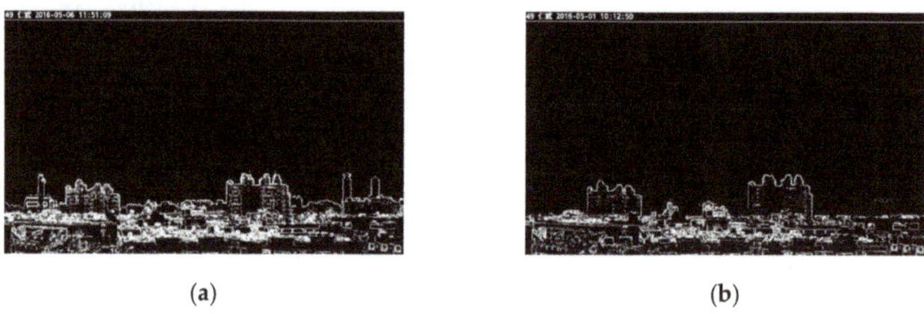

(a)   (b)

Figure 5. Images after Otsu thresholding. (a) $I_{1,so}$ (low $PM_{2.5}$ concentration); (b) $I_{2,so}$ (high $PM_{2.5}$ concentration).

2.1.3. Morphological Dilation

Using the obtained binary images, $I_{1,so}$ and $I_{2,so}$, shown in Figure 5, morphological dilation is applied to expand boundaries and to connect neighborhood pixels. The degree of expansion depends on the size of structuring elements. The equation employed for morphological dilation is given below:

$$A \oplus B = \{\text{white}|B_x \cap A \neq \varnothing\},\tag{8}$$

where A is the image to be processed and B represents the structuring elements.

In the proposed RoI scheme, the 3×3 mask for structuring elements with all white pixels is used. After morphological dilation, the resulting images are denoted as $I_{1,som}$ and $I_{2,som}$ and shown in Figure 6a and Figure 6b, respectively.

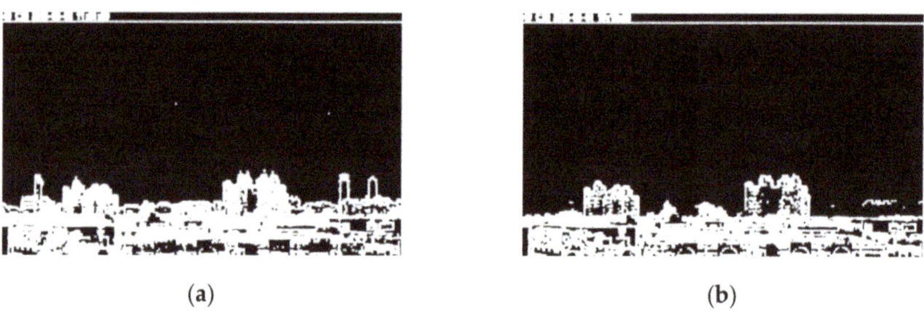

(a)   (b)

Figure 6. Images after morphological dilation. (a) $I_{1,som}$ (low $PM_{2.5}$ concentration); (b) $I_{2,som}$ (high $PM_{2.5}$ concentration).

2.1.4. Image Subtraction and Labeling

In this step, image subtraction is used to obtain the difference image for $I_{1,som}$ and $I_{2,som}$ in Figure 6. Then, a labeling scheme is employed to identify connected pixels. The difference image for $I_{1,som}$ and $I_{2,som}$ is shown in Figure 7, denoted as I_d, where pixels with the same value in the image pair are eliminated and those with different pixel values remain in a white color. In order to distinguish whether pixels are connected, a labeling scheme [31] is applied to mark the connected pixels by colors. The connected neighborhood pixels are marked with the same color. After labeling, the resulting image, denoted as I_{dl}, is as shown in Figure 8. Finally, the labeled regions with the top three largest numbers of pixels are considered as candidate regions of interest.

Figure 7. The difference image I_d after image subtraction.

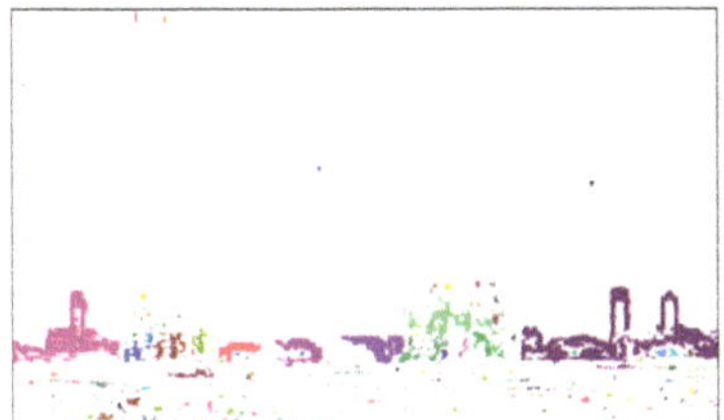

Figure 8. The image I_{dl} after labeling.

2.1.5. Selected RoI in the Given Pair of Images

Now, the red flow path shown in Figure 2 will be described. The difference image, denoted as I_{sd}, for $I_{1,s}$ and $I_{2,s}$ is obtained by image subtraction. Then, the three candidate regions of interest and the difference image I_{sd} are overlapped to select the pixels in the candidate regions of interest. Next, the averages of pixel values in each candidate region of interest are calculated. Then, the RoI with the highest average is determined as the final RoI in the given pair of images, I_1 and I_2. This completes the process of automatic RoI selection given in Figure 2 for the given pair of images.

2.1.6. Final RoI Determination

It needs to be pointed out that the image pair given above is just an example provided to show the process of the proposed automatic RoI selection. In practice, in automatic RoI selection, 30 images with a low PM$_{2.5}$ concentration (≤ 5 µg/m^3) and 120 images with a high PM$_{2.5}$ concentration (≥ 70 µg/m^3) are randomly selected from the training set. In this study, the images with low and high PM$_{2.5}$ concentrations are paired by combinations. In other words, the 30×120 paired images are included in the automatic RoI selection process, as described in Figure 2. By using the averages of 3600 results, the three candidate regions of interest are determined, as shown in Figure 9. The box plot given in Figure 10 shows the range of average pixel values in each candidate RoI. Since Region 1 has the highest average value, it is selected as the final RoI to estimate the PM$_{2.5}$ concentration. The average pixel value within the final RoI will be used as the only single feature for the following simple linear regression model in the proposed approach.

Figure 9. The three candidate regions of interest indicated by red boxes.

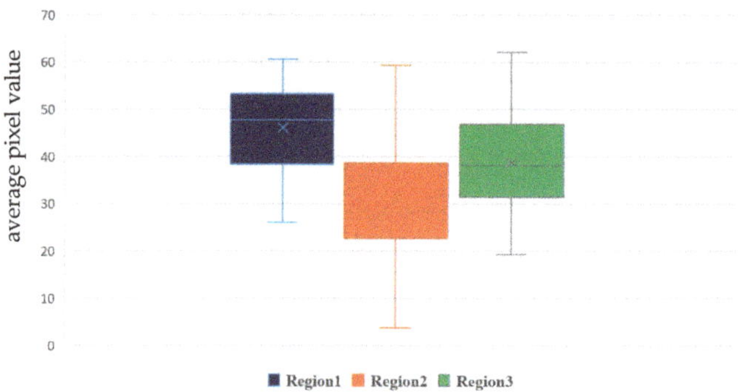

Figure 10. A box plot for three candidate regions of interest.

2.2. Simple Linear Regression Model

A simple linear regression model, which is a statistical analysis scheme [25], will be used to estimate the PM$_{2.5}$ concentration in the proposed approach. x_i is the average pixel value within the final data and y_i is the corresponding PM$_{2.5}$ concentration measurement in the training data (where subscript i denotes the ith sample). It is assumed that these two sequences of data have a linear relation, shown as

$$y_i = \alpha + \beta x_i, \tag{9}$$

where α and β are coefficients to be determined. y_i denotes an estimate of y_i (corresponding PM$_{2.5}$ concentration). The estimation error between y_i and y_i is given as

$$\varepsilon_i = y_i - y_i. \tag{10}$$

Employing the least squares algorithm to minimize the estimation error, coefficients α and β can be found as

$$\alpha = \sum_{i=1}^{N} y_i - \beta \sum_{i=1}^{N} x_i \tag{11}$$

and

$$\beta = \frac{\sum_{i=1}^{N} x_i y_i - \sum_{i=1}^{N} x_i \sum_{i=1}^{N} y_i}{\sum_{i=1}^{N} x_i^2 - \sum_{i=1}^{N} x_i \sum_{i=1}^{N} x_i}, \tag{12}$$

where N is the number of samples. Once the simple linear regression model is obtained, it is employed to estimate the PM$_{2.5}$ concentration with the testing data.

2.3. Performance Indices

Inherently, image-based method cannot analyze the ingredients in the air, as in previous works, thus it is hard to define a parameter to show the performance by error. Instead, three overall performance indices are used to evaluate the proposed approach. The first one is the root mean square error (RMSE). It is used to show the error between the recorded value and the estimated value of the proposed method. RMSE is calculated as

$$\text{RMSE} = \sqrt{\frac{1}{N}\sum_{i=1}^{N}(y_i - y_i)^2}, \tag{13}$$

where y_i and y_i are the true and estimated $PM_{2.5}$ concentrations, respectively. The second performance index is R squared (R^2), which has also been used in previous work [28], and is employed to show the correlation between estimated results and measured values. It is defined as

$$R^2 = 1 - \frac{\sum_{i=1}^{N}(y_i - y_i)^2}{\sum_{i=1}^{N}(y_i - \overline{y})^2}, \tag{14}$$

where $\overline{y}$ is the mean of y_i. R^2 indicates the linearity between y_i and y_i. When it is linear, $R^2 = 1$. The third index is F-test, which is the test statistic for an F-distribution under the null hypothesis [32], where the p-value indicates the statistical significance; that is, it determines whether the result is beyond chance or not. The p-value will be used as an indicator of statistical significance in the following experiments.

3. Experimental Results

In this section, the proposed approach is verified by a real-world data set, which is described later in Section 3.1. Then, the results without and with unreliable data exclusion are shown in Sections 3.2 and 3.3, respectively.

3.1. Experimental Data Sets

In the experiments, the images were taken from Renwu Environmental Monitoring Station, Kaohsiung City, Taiwan. A consumer camera was set up at the station and took one image every ten minutes during the period of 7:00 AM to 5:00 PM. In total, 10,084 images were collected from May to October 2016. We did not exclude sampled images of sunny or rainy days. The image data were divided into training and testing data, of which the proportions were 60% and 40%, respectively. The images shown in Figure 3a,b are examples taken from the data set. Furthermore, the hourly $PM_{2.5}$ concentration and relative humidity (RH) in the corresponding area were obtained from the open data released by the Environmental Protection Administration, Executive Yuan, Taiwan [33]. Using the data, a simple linear regression model was obtained and used to estimate the $PM_{2.5}$ concentration by employing the proposed approach.

3.2. Results with All Data

In this experiment, all of the data set, including 10,084 images, was used. As described in Section 2, three candidate regions of interest were automatically selected and the final RoI was determined by the highest average pixel value among the three candidate regions of interest. Besides, the average pixel value in the final RoI was used as the only single feature. To compare the estimation performances for the whole image, Region 1, Region 2, and Region 3 are presented in Figure 11a–d, which show scattering plots for each case, where the region under consideration is shown in the upper right corner. The three performance indices with all data are displayed in Table 1. Table 1 indicated that Region 1 had a better performance than the other cases. Besides, all results were statistically significant in the F-test. In the case with all data, the highest $R^2 = 0.41$, which was achieved by Region 1. One may see that the performance of the whole image case is inferior to those for candidate regions of interest. When Regions 1 to 3 are considered, the performance index R^2, from high to low, is Region 1, Region 3, and Region 2. The result is consistent with the priority for the proposed automatic RoI selection. In other words, the proposed automatic RoI selection is appropriate for the given data.

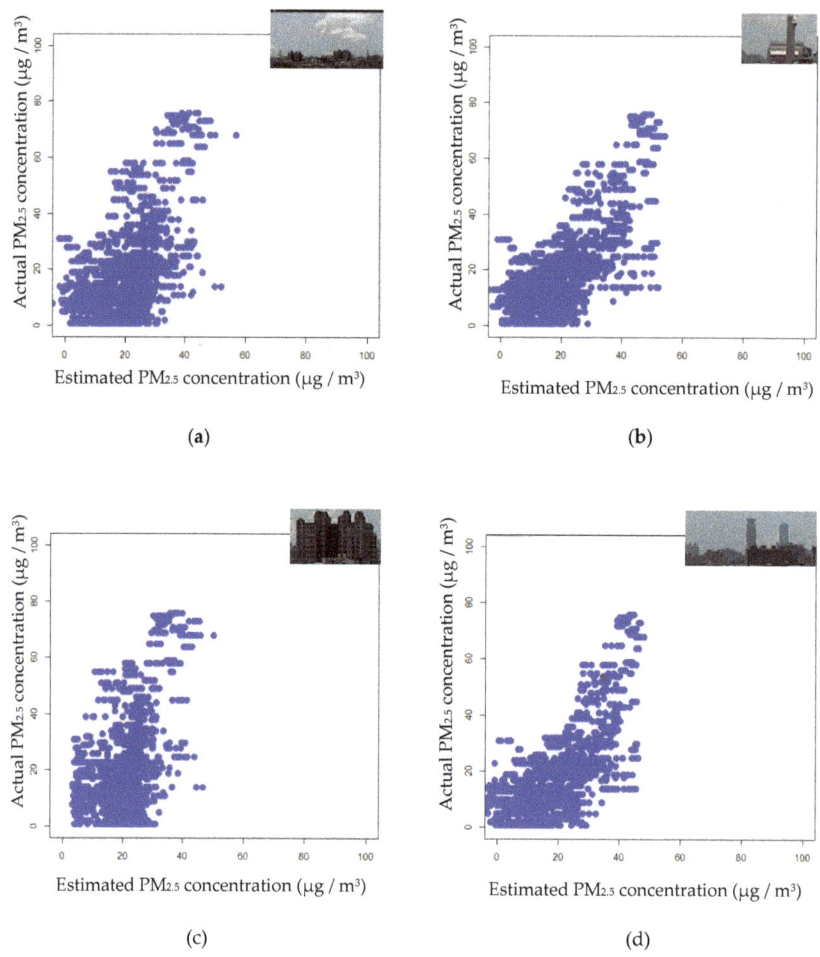

Figure 11. The scatter plots for (**a**) the whole image; (**b**) Region 1 (selected); (**c**) Region 2; (**d**) Region 3.

Table 1. The performance indices (with all data).

	RMSE ($\mu g/m^3$)	R^2	*F*-test
Whole image	14.54	0.11	$p < 0.0001$
Region 1	11.88	0.41	$p < 0.0001$
Region 2	13.53	0.23	$p < 0.0001$
Region 3	12.55	0.34	$p < 0.0001$

3.3. Results with Unreliable Data Exclusion

By conducting experiments, it was observed that two factors may affect the performance of the proposed approach. One is the time difference between the time to take images and the time to measure the $PM_{2.5}$ concentration. For the data set described in Section 3.1, the images were taken every ten minutes, but the $PM_{2.5}$ concentration was collected hourly. In other words, six images were related to only one $PM_{2.5}$ concentration for each hour. When the $PM_{2.5}$ concentration changes within an hour, it might degrade the estimation performance. To solve this problem, the variance of six images taken in the same hour was calculated. When the variance was greater than 1, the images were considered as unreliable data and discarded.

The other factor seen to affect the performance of the proposed approach was the RH. There are many substances, in addition to $PM_{2.5}$, in the atmosphere that affect visibility, such as sulfur oxides, nitrogen oxides, carbon monoxide, and water droplets. It has been observed that $PM_{2.5}$ aerosols are expanded by absorbing water molecules in the air and this affects visibility [34]. It has also been reported that the RH affects $PM_{2.5}$ concentration estimation [28]. Consequently, the effect of RH on $PM_{2.5}$ concentration estimation was considered in the proposed approach.

By conducting experiments, we observed that the estimation performance of the proposed approach was significantly degraded when RH $\geq$ 65%. Consequently, the data was excluded if its corresponding RH $\geq$ 65%. Moreover, it should be noted that human health is mostly endangered by a higher $PM_{2.5}$ concentration, instead of a lower one. Consequently, the data with $PM_{2.5}$ concentrations less than 5 $\mu g/m^3$ were excluded. By employing the criteria RH $\geq$ 65% or $PM_{2.5}$ concentration less than 5 $\mu g/m^3$, 2361 images were excluded from the given data set. With the consideration of data exclusion, the three performance indices were recorded and are presented in Table 2 for all cases, as in Table 1. As seen in Table 2, Region 1 had a better performance than the other cases, as in Table 1. Moreover, all results were statistically significant in the *F*-test. When comparing the results presented in Tables 1 and 2, one can see that the RMSE and R^2 were obviously improved in all cases with data exclusion. Additionally, Region 1 exhibited the most improvement. The RMSE was reduced from 11.88 to 8.67, while the R^2 increased from 0.41 to 0.73. Again, the results implied that the automatically selected RoI was appropriate in the given example. To sum up, the proposed approach with automatic RoI selection and data exclusion is feasible and has an acceptable performance for $PM_{2.5}$ concentration estimation. By Table 2, one may observe that the performance of the whole image case is inferior to those for candidate regions of interest, as in Table 1. According to the results, the performances from high to low are Region 1, Region 3, and Region 2, which is consistent with the priority for the proposed automatic RoI selection, as shown in Figure 10. Again, the results have verified the feasibility of the proposed automatic RoI selection scheme in the given experiments.

Table 2. The performance indices (with unreliable data exclusion).

	RMSE ($\mu g/m^3$)	R^2	*F*-test
Whole image	13.17	0.22	$p < 0.0001$
Region 1	8.67	0.73	$p < 0.0001$
Region 2	11.51	0.34	$p < 0.0001$
Region 3	10.76	0.65	$p < 0.0001$

4. Conclusions

This paper has presented a simple alternative for estimating the $PM_{2.5}$ concentration in which a series of image processing schemes and simple linear regression are employed. The proposed method uses images with a high and low PM2.5 concentration to obtain the difference between these images. The difference is used to find the RoI. Two main stages are involved in this approach. The first stage includes a series of image processing schemes, which are used to automatically select the final RoI, from which only a single feature is extracted and used in a simple linear regression model. The second stage is employed to find a simple linear regression model with the single feature, by applying the final RoI identified in the first stage. Then, $PM_{2.5}$ concentration estimation is performed. Using an image data set and an open $PM_{2.5}$ concentration data set, experiments were conducted to verify the proposed approach. The results indicated that the proposed approach with the automatically selected RoI achieved the best performance, with $R^2 = 0.73$. Although the proposed method is not as direct as chemical schemes used to analyze the composition of air, the aim of this paper has been fulfilled, i.e., to provide a simple alternative approach for $PM_{2.5}$ concentration estimation with an acceptable performance. The proposed approach is not expected to replace component analysis using physical or chemical techniques. However, we hope that the proposed method can provide a cheaper and easier way to conduct $PM_{2.5}$ estimation with an acceptable performance more efficiently. To achieve this, further work will be conducted and can be summarized as follows:

1. Since the proposed method uses a fixed camera to capture images at the same location, the influence of images taken in different locations on the results of this study need to be investigated further;

2. Though we have shown that the performance for each candidate RoI is better than the whole image case, it is still worthy to seek a better way to find the final RoI for the performance improvement;

3. In this study, sunny or rainy days are not considered and they will be researched in the future. Besides, other weather factors, such as solar conditions, will be considered in the $PM_{2.5}$ concentration estimation from a higher dimension aspect.

Author Contributions: Conceptualization, J.-J.L. and C.-H.H.; Formal analysis, J.-J.L. and C.-H.H.; Investigation, J.-J.L.; Methodology, J.-J.L.; Resources, Y.-F.H. and C.-H.L.; Software, D.-C.L.; Supervision, J.-J.L.; Visualization, D.-C.L.; Writing—original draft, J.-J.L. and D.-C.L.; Writing—review and editing, Y.-F.H. and C.-H.H. All authors have read and agreed to the published version of the manuscript.

Funding: This research is partially sponsored by Chaoyang University of Technology (CYUT) and Higher Education Sprout Project, Ministry of Education (MOE), Taiwan, under the project titled "The R&D and the cultivation of talent for health-enhancement products."

Conflicts of Interest: The authors declare no conflict of interest.

References

1. Narasimhan, S.G.; Nayar, S.K. Vision and the atmosphere. *Int. J. Comput. Vis.* **2002**, *48*, 233–254. [CrossRef]

2. Anderson, J.O.; Thundiyil, J.G.; Stolbach, A. Clearing the air: A review of the effects of particulate matter air pollution on human health. *J. Med Toxicol.* **2011**, *8*, 166–175. [CrossRef] [PubMed]

3. Raaschou-Nielsen, O.; Andersen, Z.J.; Beelen, R.; Samoli, E.; Stafoggia, M.; Weinmayr, G.; Hoffmann, B.; Fischer, P.; Nieuwenhuijsen, M.J.; Brunekreef, B.; et al. Air pollution and lung cancer incidence in 17 European cohorts: Prospective analyses from the European Study of Cohorts for Air Pollution Effects (ESCAPE). *Lancet Oncol.* **2013**, *14*, 813–822. [CrossRef]

4. More than 90% of World's Children Breathe Toxic Air, Report Says, as India Prepares for Most Polluted Season. Cable News Network. Available online: https://edition.cnn.com/2018/10/29/health/air-pollution-children-health-who-india-intl/index.html (accessed on 16 August 2019).

5. Melstrom, P.; Koszowski, B.; Thanner, M.H.; Hoh, E.; King, B.; Bunnell, R.; McAfee, T. Measuring $PM_{2.5}$, ultrafine particles, nicotine air and wipe samples following the use of electronic cigarettes. *Nicotine Tob. Res.* **2017**, *19*, 1055–1061. [CrossRef] [PubMed]

6. Zíková, N.; Hopke, P.K.; Ferro, A.R. Evaluation of new low-cost particle monitors for $PM_{2.5}$ concentrations measurements. *J. Aerosol Sci.* **2016**, *105*, 24–34. [CrossRef]

7. Hauck, H.; Berner, A.; Gomiscek, B.; Stopper, S.; Puxbaum, H.; Kundi, M.; Preining, O. On the equivalence of gravimetric PM data with TEOM and beta-attenuation measurements. *Aerosol Sci.* **2004**, *35*, 1135–1149. [CrossRef]

8. Ruppecht, E.; Meyer, M.; Patashnick, H. The tapered element oscillating microbalance as a tool for measuring ambient particulate concentrations in real time. *J. Aerosol Sci.* **1992**, *23*, 635–638. [CrossRef]

9. Macias, E.S.; Husar, R.B. Atmospheric particulate mass measurement with beta attenuation mass monitor. *Environ. Sci. Technol.* **1976**, *10*, 904–907. [CrossRef]

10. Smith, J.D.; Atkinson, D.B. A portable pulsed cavity ring-down transmissometer for measurement of the optical extinction of the atmospheric aerosol. *Analyst* **2001**, *126*, 1216–1220. [CrossRef]

11. Li, C.; He, Q.; Schade, J.; Passig, J.; Zimmermann, R.; Meidan, D.; Laskin, A.; Rudich, Y. Dynamic changes in optical and chemical properties of tar ball aerosols by atmospheric photochemical aging. *Atmos. Chem. Phys.* **2019**, *19*, 139–163. [CrossRef]

12. Sclar, S.; Saikawa, E. Household air pollution in a changing Tibet: A mixed methods ethnography and indoor air quality measurements. *Environ. Manag.* **2019**. [CrossRef] [PubMed]

13. Song, Y.; Huang, B.; He, Q.; Chen, B.; Wei, J.; Mahmood, R. Dynamic assessment of $PM_{2.5}$ exposure and health risk using remote sensing and geo-spatial big data. *Environ. Pollut.* **2019**, *253*, 288–296. [CrossRef] [PubMed]

14. Han, W.; Tong, L. Satellite-Based Estimation of Daily Ground-Level PM2.5 Concentrations over Urban Agglomeration of Chengdu Plain. *Atmosphere* **2019**, *10*, 245. [CrossRef]

15. Zikova, N.; Masiol, M.; Chalupa, D.C.; Rich, D.Q.; Ferro, A.R.; Hopke, P.K. Estimating Hourly Concentrations of PM2.5 across a Metropolitan Area Using Low-Cost Particle Monitors. *Sensors* **2017**, *17*, 1992. [CrossRef] [PubMed]

16. Gao, M.; Cao, J.; Seto, E. A distributed network of low-cost continuous reading sensors to measure spatiotemporal variations of PM2.5 in Xi'an, China. *Environ. Pollut.* **2015**, *199*, 56–65. [CrossRef] [PubMed]

17. Badura, M.; Batog, P.; Drzeniecka-Osiadacz, A.; Modzel, P. Evaluation of Low-Cost Sensors for Ambient PM2.5 Monitoring. *J. Sens.* **2018**, 16. [CrossRef]

18. McCartney, E.J. *Optics of the Atmosphere: Scattering by Molecules and Particles*, 1st ed.; John Wiley & Sons Inc: New York, NY, USA, 1976.

19. Liu, X.; Hui, Y.; Yin, Z.Y.; Wang, Z.; Xie, X.; Fang, J. Deteriorating haze situation and the severe haze episode during December 18–25 of 2013 in Xi'an, China, the worst event on record. *Theor. Appl. Climatol.* **2016**, *125*, 321–335. [CrossRef]

20. Larson, S.M.; Cass, G.R.; Hussey, K.J.; Luce, F. Verification of image processing based visibility models. *Environ. Sci. Technol.* **1988**, *22*, 629–637. [CrossRef]

21. Malm, W.C.; Molenar, J.V. Visibility measurements in National Parks in the western United States. *J. Air Pollut. Control Assoc.* **1984**, *34*, 899–904. [CrossRef]

22. Xie, L.; Chiu, A.; Newsam, S. *Estimating Atmospheric Visibility Using General-Purpose Cameras*; International Symposium on Visual Computing: Las Vegas, NV, USA, 2008.

23. Shih, W.Y. Variations of Urban Fine Suspended Particulate Matter ($PM_{2.5}$) from Various Environmental Factors and Sources and Its Role on Atmospheric Visibility in Taiwan. Master's Thesis, National Central University, Taiwan, 2013.

24. Ren, S.; He, K.; Girshick, R.; Sun, J. Faster R-CNN: Towards real-time object detection with region proposal networks. *IEEE Trans. Pattern Anal. Mach. Intell.* **2017**, *39*, 1137–1149. [CrossRef]

25. Gelman, A.; Hill, J. *Data Analysis Using Regression and Multilevel/Hierarchical Models*, 1st ed.; Cambridge University Press: New York, NY, USA, 2006.

26. Le, T.N.; Sun, X.; Chowdhury, M.; Liu, Z. AlloX: Allocation across Computing Resources for Hybrid CPU/GPU clusters. *ACM Sigmetrics Perform. Eval. Rev.* **2019**, *46*, 87–88. [CrossRef]

27. Weichenthal, S.; Ryswyk, K.V.; Goldstein, A.; Bagg, S.; Shekkarizfard, M.; Hatzopoulou, M. A land use regression model for ambient ultrafine particles in Montreal, Canada: A comparison of linear regression and a machine learning approach. *Environ. Res.* **2016**, *146*, 65–72. [CrossRef] [PubMed]

28. Liu, C.; Tsow, F.; Zou, Y.; Tao, N. Particle pollution estimation based on image analysis. *PLoS ONE* **2016**, *11*, e0145955. [CrossRef] [PubMed]

29. Jin, S.; Kim, W.; Jeong, J. Fine directional de-interlacing algorithm using modified Sobel operation. *IEEE Trans. Consum. Electron.* **2008**, *54*, 587–862. [CrossRef]

30. Goh, T.Y.; Basah, S.N.; Yazid, H.; Safar, M.J.A.; Saad, F.S.A. Performance analysis of image thresholding: Otsu technique. *Measurement* **2018**, *114*, 298–307. [CrossRef]

31. Dougherty, E.R.; Lotufo, R.A. *Hands-on Morphological Image Processing*, 3rd ed.; SPIE Press: Washington, DC, USA, 2003.

32. Markowski, C.A.; Markowski, E.P. Conditions for the effectiveness of a preliminary test of variance. *Am. Stat.* **1990**, *44*, 322–326.

33. Environmental Protection Administration Executive Yuan, R.O.C., Taiwan. Available online: https://taqm.epa.gov.tw/taqm/tw/YearlyDataDownload.aspx (accessed on 16 August 2019).

34. Swietlicki, E.; Zhou, J.; Berg, O.H.; Martinsson, B.; Frank, G.; Cederfelt, S.I.; Ulrike, D.; Berner, A.; Birmili, W.; Wiedensohler, A.; et al. A closure study of sub-micrometer aerosol particle hygroscopic behavior. *Atmos. Res.* **1999**, *50*, 205–240. [CrossRef]

© 2020 by the authors. Licensee MDPI, Basel, Switzerland. This article is an open access article distributed under the terms and conditions of the Creative Commons Attribution (CC BY) license (http://creativecommons.org/licenses/by/4.0/).

Article

A Durable Hybrid RAM Disk with a Rapid Resilience for Sustainable IoT Devices

Sung Hoon Baek [1] and Ki-Woong Park [2],*

[1] Department of Computer System Engineering, Jungwon University, Chungcheongbuk-do 28024, Korea; shbaek@jwu.ac.kr

[2] Department of Computer and Information Security, Sejong University, Seoul 05006, Korea

* Correspondence: woongbak@sejong.ac.kr; Tel.: +82-2-6935-2453

Received: 21 February 2020; Accepted: 9 April 2020 ; Published: 11 April 2020

Abstract: Flash-based storage is considered to be a de facto storage module for sustainable Internet of things (IoT) platforms under a harsh environment due to its relatively fast speed and operational stability compared to disk storage. Although their performance is considerably faster than disk-based mechanical storage devices, the read and write latency still could not catch up with that of Random-access memory (RAM). Therefore, RAM could be used as storage devices or systems for time-critical IoT applications. Despite such advantages of RAM, a RAM-based storage system has limitations in its use for sustainable IoT devices due to its nature of volatile storage. As a remedy to this problem, this paper presents a durable hybrid RAM disk enhanced with a new read interface. The proposed durable hybrid RAM disk is designed for sustainable IoT devices that require not only high read/write performance but also data durability. It includes two performance improvement schemes: rapid resilience with a fast initialization and direct byte read (DBR). The rapid resilience with a fast initialization shortens the long booting time required to initialize the durable hybrid RAM disk. The new read interface, DBR, enables the durable hybrid RAM disk to bypass the disk cache, which is an overhead in RAM-based storages. DBR performs byte–range I/O, whereas direct I/O requires block-range I/O; therefore, it provides a more efficient interface than direct I/O. The presented schemes and device were implemented in the Linux kernel. Experimental evaluations were performed using various benchmarks at the block level till the file level. In workloads where reads and writes were mixed, the durable hybrid RAM disk showed 15 times better performance than that of Solid-state drive (SSD) itself.

Keywords: IoT; sustainability; hybrid RAM disk; direct byte read; secondary storage; operating system

---

1. Introduction

A reliable but responsive storage device is an inevitable concern for realizing sustainable Internet of things (IoT) devices for mission-critical systems [1]. Unlike general consumer devices, sustainable IoT devices must perform their own tasks correctly in a stable manner without failures. As shown in Figure 1a, mission-critical systems need to ensure rapid system recovery and resilience, even in the face of a sudden power failure because one failure inside a mission-critical system may result in a mission failure or a tragedy. Therefore, realizing sustainable IoT devices ensuring low read latency for accessing critical data, as well as data durability for storing critical data, is a critical factor for their sustainability [2]. Short boot times are also an important factor in mission-critical IoT systems because it should be ready for certain mission-critical tasks even after a sudden power failure. Regarding these points, flash-based storage is considered to be a de facto storage module for sustainable IoT platforms under a harsh environment due to its relatively fast speed and operational stability compared to disk storage. To achieve these design goals, usually flash-based Solid-state drive (SSD) is installed with the configuration of single attached or Redundant array of inexpensive disks (RAID)manner, and mostly, IoT operating systems such as Android Things [3], and an I/O subsystem in the operating system, manages the storage device [4]. Although their performance is considerably faster than disk-based mechanical storage devices, the read and write latency still could not catch up with that of Random-access memory (RAM). Therefore, RAM could be used as a storage device or system for time-critical IoT applications.

A traditional storage I/O subsystem, even if it uses flash-based SSD, will have larger latency and lower bandwidth compared to memory-oriented read/write operations. Even though the performance gap diminishes, there still exists several orders of magnitude for latency. Additionally, RAM can access data on a byte-addressable level. There are a lot of studies to improve I/O performance with RAM memory, and most of them use RAM memory as a buffer for storage devices [5–9]. The read and write buffer can reduce write latency by buffering incoming data to a RAM buffer, or reduce read latency to get data directly from a RAM buffer, with appropriate buffer-management algorithms. However, the buffer-based approach still must go through existing file system operations and may have memory management overhead such as page cache.

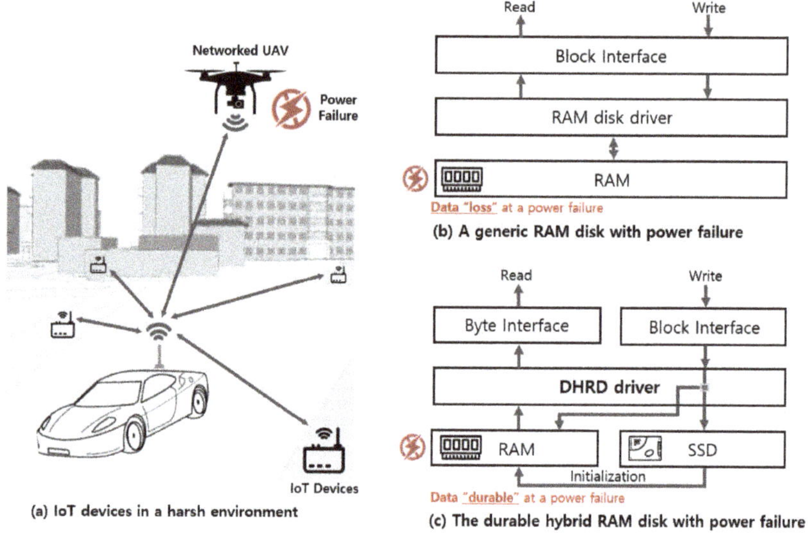

Figure 1. Software stack for the durable hybrid RAM disk (DHRD) with direct byte read (DBR).

On the other hand, the RAM disk is a software program that turns a portion of the main memory into a block device [10–12]. The RAM disk is the most expensive and fastest storage device, in which the RAM memory block device works like a disk drive. It is also referred to as a software RAM drive to differentiate it from a hardware RAM drive. RAM disks can provide fast I/O response and low latency when accessing data. However, it has the disadvantage of data loss in the event of a power failure; therefore, it lacks durability and persistency. To address this problem, many studies have been conducted on various types of systems. The simplest approach to prevent data loss is to asynchronously dump the entire contents of the RAM disk into a dedicated hard disk drive [13]. For better durability, dumping into a hard disk drive can be performed synchronously, but this method would use the inefficient traditional I/O for RAM-based storage [14]. In addition, there is little scope in terms of byte addressability advantage of RAM devices when designing RAM-based storage systems. The byte addressing data access operation provides very low latency.

Figure 1b,c illustrate a basic difference between a RAM disk and the proposed scheme, durable hybrid RAM disk (DHRD). The generic RAM disk has a block interface and loses data at a power failure. However, DHRD with direct byte read (DBR) can improve both durability and read throughput. The DHRD consists of a RAM disk and a non-volatile storage, such as SSD, as a hybrid storage system. With this hybrid approach, the DHRD provides durability, which means it does not suffer data loss during sudden power failure. The DBR DHRD performs I/O operations with the new read scheme that does not use a disk cache, i.e., page cache, for reads and is byte-addressable unlike direct I/O, wherein direct I/O uses a strict block-addressing mode. The byte-addressable feature is more convenient for applications than the use of direct I/O. The new byte-addressable read scheme can be mixed with buffered writes to apply it to the hybrid RAM disk; moreover, it can be applied to existing applications without any modification. In addition to that, the initialization procedure of DBR DHRD can reduce the boot time of the storage device, since it allows general I/O requests during the initialization process itself, while other RAM disk-based storage cannot support general I/Os during the initialization. Various experiments that could be applied to sustainable IoT devices were performed to DBR DHRD and other storage configurations such as SSD and hybrid storage device. The experimental results show that the DBR DHRD gives better I/O performance than others.

The rest of this paper is organized as follows: Section 2 provides an overview of related work, Section 3 presents the design and implementation of DHRD in detail. The performance evaluation of DHRD is presented in Section 4. Finally, Section 5 concludes this paper and presents the relevant future work.

2. Related Work

RAMDisk is a software-based storage device that takes a portion of the system memory and uses it as a disk drive with legacy file system operations. The more RAM your computer has, the larger the RAM disk you can create, but the cost would also be more.

Recently, RAM has been used as a storage system for several high-end computing systems and IoT devices to provide low latency and low I/O overhead. RAM is used for intensive random I/O in various fields, such as in-memory databases [15], large-scale caching systems [16], cloud computing [17–20], virtual desktop infrastructure [21–23], web search engine [24], and mission-critical systems like space applications [25].

Several RAM disk devices were previously developed such as [10–12]. The most traditional application of RAM disk modules is their use as virtual file systems for Linux kernel from the system's boot time. During the system boot time, Linux kernel uses more than one RAM disk file system to mount the kernel image in its root file system. Also, at run time, Linux uses space to store system information or hardware device information in proc file system or sysfs of the RAM disk. The traditional RAM disk file system acts as a regular file system that is mounted in the memory device in a single computing system. The RAM-based file systems used in Linux have no durability, which means that if the system's power turns off, the data of the RAM-based file systems would disappear. Hence, to ensure durability, dumping from RAM disk drive to the hard disk drive should be performed synchronously [14].

The development for RAM-based storage drive has been more revitalized as computing systems require lower latency for single storage I/O operations, especially, applications that use distributed storage systems such as big data databases and cloud computing systems. Distributed RAM storages in cluster environments have been studied [14,17,18]. To overcome the volatility of RAM, RAMcloud [17] provides durability and persistency in a cluster environment, where each node uses RAM as the main storage. Every node replicates each object in the RAM storage and responds to write requests after updating all the replicas. Hence, reliability is ensured even if a node fails. Additionally, modified data are logged to two or more nodes and the logs are asynchronously transferred to a non-volatile storage to achieve durability.

A Solid-State Hybrid Disk (SSHD) is being used on a personal computer, in which SSHD is composed of SSD and Hard disk drive (HDD) inside the storage device. In the SSHD, Several GB of SSDs significantly improve overall performance. Modern state-of-art storage systems employ tiered storage devices [26–28]. For the tiered storage device, the DBR DHRD scheme, proposed in this paper, can replace SSD in SSHD device. It can significantly improve I/O performance with DBR method.

On the other hand, recently, some new memory-oriented devices are released to give better I/O throughput for memory-intensive applications. The memory devices are connected to CPU via PCIe NVMe (non-volatile memory express) interface, and the controller chip manages the hybrid storage of the memory and non-volatile memory SSD. The main operation of the controller is caching other NVMe SSDs connected to the systems for accelerating storage I/Os. It provides fast response time and high throughput for both random as well as sequential reads and writes at block I/O levels. However, it does not provide byte-level addressable I/O in the internal operations as it is connected to the PCIe bridge interface. Moreover, the dedicated hardware device has limitations in terms of using and enhancing internal I/O mechanism at the software level.

The proposed hybrid storage system is different from memory-oriented device and provides advantages such as the use of a legacy storage device without any additional hardware devices to provide fast response time for read requests using the new read scheme, and durability of RAM disk drive for write requests. The proposed read scheme in RAM-based disk driver uses byte addressability of RAM device for fast response time. The concept of byte-addressable I/O was proposed in our prior work [29], but this paper presents a new byte-address read scheme that can be mixed with buffered writes to apply it to the hybrid RAM disk. The read performance can be improved with the help of byte addressability of RAM while providing durability similar to that of non-RAM disk systems.

3. Durable Hybrid RAM Disk

The DHRD is a hybrid storage that consists of volatile memory and non-volatile storage, while providing the same durability as that of non-volatile storage. In addition, it improves the read performance of RAM disks by using a new read interface that is different from buffered I/O and direct I/O. Read requests are served by the volatile memory, while the write operations are performed on both the volatile memory and the non-volatile storage simultaneously. Therefore, it provides the same durability as that of a non-volatile storage. Read performance is determined by the volatile memory but write performance depends on the non-volatile storage. The DHRD can be used in areas where read performance is more important than write performance and data durability is mandatory. For example, it can be applied to read intensive in-memory databases for sustainable IoT devices. The detailed operations of the proposed system are explained in the next subsections.

3.1. Architecture

Figure 2 shows the software architecture of the DHRD to provide data durability. The DHRD consists of high-performance volatile memory and non-volatile storage. The high-performance volatile storage device can be implemented as a RAM disk (RAM disk software). The non-volatile storage device can be implemented with flash-based SSDs. Non-volatile storage devices are generally slower than volatile memory storage devices but do not lose data during power failure. In the proposed system, the volatile memory storage is used as the main storage area, and the updated data in the volatile memory also gets updated to the non-volatile storage device synchronously.

The DHRD is like a mirrored RAID that consists of a RAM disk and a flash-based SSD, where read requests are served only from the RAM disk and write requests are duplicated to both the RAM disk and the SSD. The RAM disk is mirrored with the SSD. Hence, the RAM disk can be recovered from SSD even if the RAM disk loses its data due to a sudden power failure

When the system restarts, the RAM disk is automatically initialized with the data in the SSD. Depending on the capacity of the RAM disk, it might take a long time to load all the data onto the

RAM disk. This paper presents a technique that allows immediate response that takes care of the initialization and hence, continues to serve I/O requests during the long initialization period.

The page cache that is used by block devices exhibits unnecessary memory copy overhead for RAM disks. The proposed solution provides a cache bypassing read like direct I/O. This uses the stringent block-level interface, where the buffer size, buffer address, request size, and request position must be multiples of the logical block size. However, the proposed read scheme used in DHRD provides a byte interface that has no constraints. This new read interface is described in detail in Section 3.4.

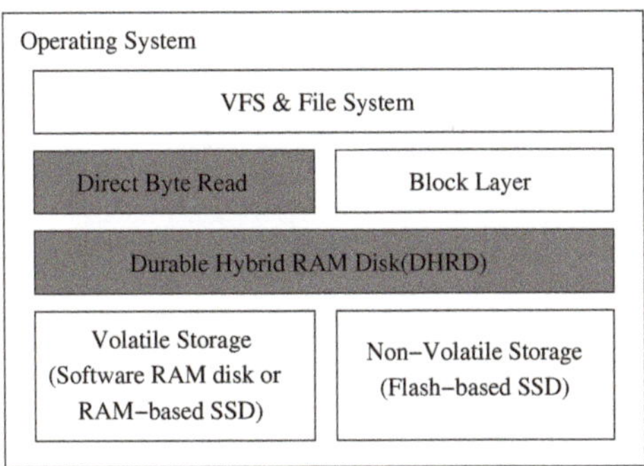

Figure 2. Software stack for the durable hybrid RAM disk (DHRD) with direct byte read(DBR).

3.2. Basic Primitives

There are three primitive operations in DHRD, read, write, and initialization. Figure 3 shows these primitive operations of DHRD. Each primitive operation acts as follows:

- Reads: Figure 3a shows the read operation of DHRD. The data contained in the RAM disk are always the same as those in the SSD; therefore, all read requests delivered to DHRD are forwarded only to the volatile memory. In other words, read operations are performed only in the RAM disk.
- Writes: All write requests delivered to DHRD are sent to both the RAM disk and the SSD. After two writes are completed in these two lower devices, the response for the request is delivered to the upper level of the DHRD. The *endio* in the figure represents the response for the request meaning the I/O is completed. As a summary, the write operation works like a mirrored RAID.
- Initialization: The initialization is performed after the system boots. When the system boots, DHRD copies the contents of the SSD to the RAM disk so that the RAM disk can now become the SSD. Generally, the initialization time is quite long; however, the read and write operations can be performed during this initialization time itself in the DHRD. The detailed read and write operations during the initialization is described in the next subsection.

Data can be completely restored from the SSD despite a power failure because the SSD always keeps up-to-date data. The data recovery efficiency of the DHRD depends on the mounted file system. The read performance is determined by the RAM, but the write performance has a bottleneck in the SSD. Consequently, the DHRD is applicable to a server that requires high durability and higher read performance than write performance.

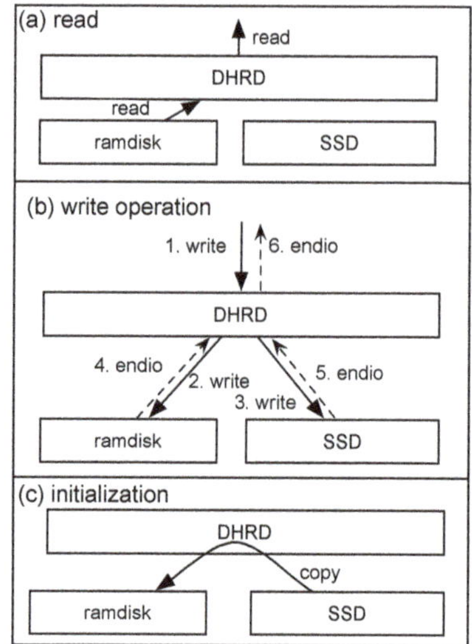

Figure 3. The three primitive operations of the durable hybrid RAM disk; read, write and initialization.

3.3. Rapid Resilience with a Fast Initialization

As soon as a system starts, the RAM disk has no data; however the SSD has valid data in the DHRD, and hence, the RAM disk needs to be filled with the contents of the SSD. The DHRD initializes the RAM disk with the data that are in the SSD so that the RAM disk has the same data as the SSD. It takes a long time for this initialization as it is performed through read sequences from SSD device. The DHRD provides data consistency even if I/O requests are delivered to the DHRD during the copy operations from SSD to RAM disk at initialization. Consequently, it allows rapid resilience with fast boot response. There are two operations during initialization: write and read, and there are several cases for each request. DHRD performs proper policy according to the requests.

3.3.1. Writes During Initialization

Figure 4 shows how write requests are processed during the initialization stage. Data blocks are divided into chunk units. Each chunk consists of multiple sectors. The chunks are sequentially copied from the SSD to the RAM disk. As shown in Figure 4, Chunks 0 to 2 were copied from the SSD to the RAM disk and Chunk 3 is being copied. Chunks 4 to 6 have not been copied yet. Write requests are classified into three cases as follows:

- Case 1: A write request sent to a chunk before being initialized is blocked until the initialization for that chunk completes. When the DHRD has finished copying the chunk to the RAM, all blocked write requests to the chunk are resumed and processed as the initialized chunk.
- Case 2: Write requests to initialized chunks are processed as normal writes. This means that the write requests are delivered to both the SSD and the RAM disk.
- Case 3: A write request to an uninitialized chunk is sent only to the SSD. The data written to the SSD will later be copied to the RAM disk by the initialization process. A write request locks the corresponding chunk and unlocks it after finishing the write operation. When the locked chunk is chosen for initialization, the initialization process is suspended and resumed only when the chunk is unlocked by the completion of the write operation. As shown in Figure 4, while a write request to the uninitialized Chunk 4 is being processed, Chunk 4 is locked, Chunk 3 has finished initialization, and the next initialization for Chunk 4 is blocked. The blocked initialization resumes after all writes for Chunk 4 are completed.

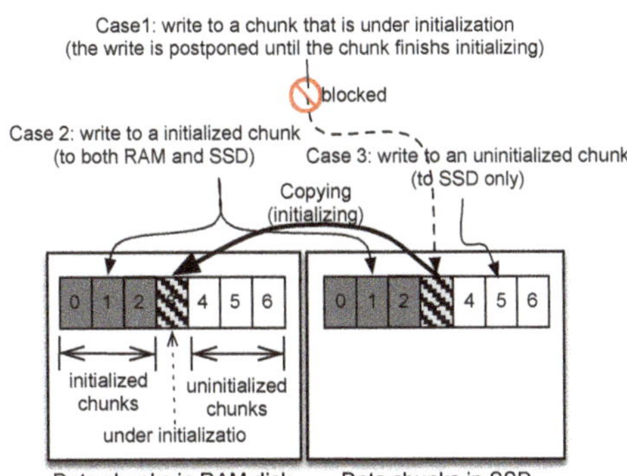

Figure 4. Three write cases during initialization. DHRD ensures data integrity with proper policy for each case.

3.3.2. Reads During Initialization

Read processing is classified into two cases as follows:

- Case 1: Read requests to initialized chunks are processed only on the RAM disk.
- Case 2: Read requests to chunks that are being initialized or were uninitialized are delivered only to the SSD.

This scheme can improve the boot response of the DHRD system. However, requests may not be processed with the best performance during initialization.

3.4. Direct Byte Read

The traditional RAM disk is implemented as a block device that is better suited in the form of disks rather than as RAM disks. The block device causes an additional memory copy from the disk cache, but, on the other hand, the RAM disk does not need this disk cache. Here, the disk cache is integrated with the page cache in the Linux kernel.

The traditional buffered I/O uses the page cache, which degrades the performance of the RAM disk. The traditional direct I/O requires that the request parameters be aligned in the logical block size. We need a new I/O interface that can process byte–range requests without the page cache.

This paper presents a new I/O that is optimized for the DHRD. It can process byte–range read requests that bypasses the page cache and uses the buffered write policy for the SSD. The new I/O requires a modified Virtual File System(VFS) in the Linux operating system and an extended block device interface.

Figure 5 compares redundant memory copy with a direct byte read (DBR). The DHRD without the DBR is presented only as a block device, and performs I/O with the page cache. If the DBR is applied to the DHRD, data can be copied directly from the memory of the RAM disk to the user memory without having to go through the block layer.

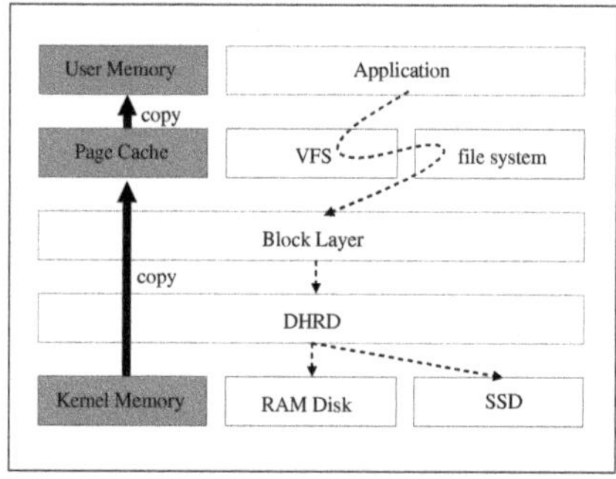 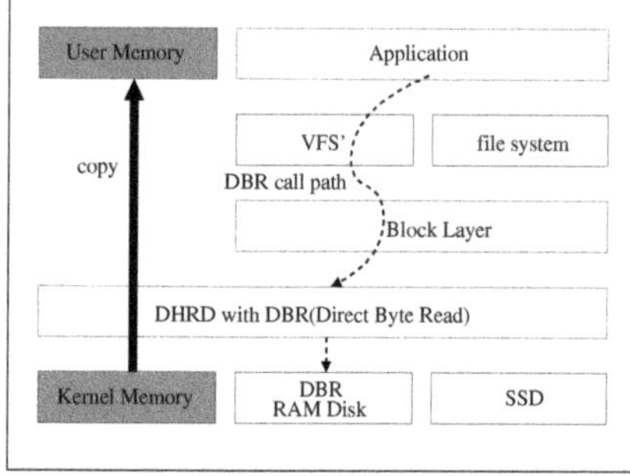

<table>
<tr><td>(a) DHRD without DBR</td><td>(b) DHRD with DBR</td></tr>
</table>

Figure 5. Software stack of DHRD for the cases of redundant memory copy and direct byte read.

3.4.1. Compatible Interface

Applications using buffered I/O can use a DBR without modification. Applications use the conventional buffered I/O interface to use the DBR. For direct I/O, the address of the application buffer memory, size of the application buffer memory, request size, and request position must be aligned in the logical block size. The DBR has no alignment restrictions on request parameters. The DBR processes I/O requests in bytes. There is a requirement for the block devices to provide an additional interface for the DBR, but DBR-enabled block devices are compatible with conventional block devices. Thus, the DBR can use the existing file systems.

The applications use the file position in bytes, the buffer memory in bytes, and the size in bytes for I/O. However, the block device has a block-range interface in which all the parameters are multiples of the logical block size. In the traditional I/O interface, the file system in conjunction with the page cache converts a byte–range request into one or more block-range requests. Thereafter, the converted block-range I/O requests are forwarded to the block device.

The DBR requires a DBR-enabled block device, a DBR-enabled file system, and a DBR module in the Linux kernel. A DBR-enabled block device has the traditional block device interface and an additional function that processes byte–range requests. The DBR-enabled file system also has one additional function for DBR. The DBR-support function in the file system can be simply implemented with the aid of the DBR module.

When the kernel receives an I/O request for a file that is in the DBR-enabled block device, the request is transferred to the DBR function of the DBR-enabled block device through the DBR interface of the file system. Therefore, the byte–range request of the application is passed to the block device without transformation.

3.4.2. Direct Byte Read and Buffered Write

The SSD processes only block-range requests, so the SSD cannot use the new I/O. The SSD is used for write requests in the DHRD, but not for read requests. Therefore, the DHRD processes write requests using the traditional block device interface that involves the page cache, while read requests are processed by the direct byte read (DBR). Figure 5 shows the read path and the write path of the DHRD with the DBR. The DHRD uses a buffered write policy that uses the page cache and DBR, which does not use the page cache. To maintain data integrity when read requests and write requests are delivered to the DHRD simultaneously, the DHRD operates as follows:

- Page not found: When a read request is transferred to the VFS, the VFS checks whether there is buffered data in the page cache. If it is not there, the read request is processed by the DBR.
- Page found: If there is a buffered page that corresponds to the read request, the data in the buffered page is transferred to the application buffer.

This scheme provides data integrity even though byte-level direct reads are mixed with traditional buffered writes.

4. Evaluation

4.1. Experimental Setup

This section describes a system that we build to measure the performance of the DBR, DHRD, and evaluation results of the proposed DHRD in comparison with a legacy system. For the performance evaluation, the proposed DHRD is compared with SSD RAID-0 and a traditional RAM disk. Throughout the section, we will denote the DHRD having DBR capability as 'DBR DHRD' to differentiate it from the basic DHRD. Also, we denote the software RAM disk as RAMDisk.

The system in the experiments uses two SSDs and 128 GB of DDR3 SDRAM 133 MHz and dual 3.4 GHz processors that have a total of 16 cores. Although the performance evaluation has been performed on high-end IoT platform equipped with multicore processor, we note that the performance of DHRD and DBR in terms of IO throughput and bandwidth is not affected by the number of CPU cores because most of the internal operations of DHRD and DBR consists of IO bound operations, not CPU bound operations. The SSD RAID is a RAID level 0 array that consists of two SSDs and provides 1.2 GB/s of bandwidth. A Linux kernel (version 3.10.2) ran on this machine hosting benchmark programs, the XFS filesystem, and the proposed DBR DHRD driver. We developed DHRD modules in the Linux kernel and modified the kernel to support DBR. The DHRD consisted of a RAMDisk and a RAID-0 array consisting of two SSDs. The RAMDisk used 122 GB of the main memory.

We did performance evaluation with various types of benchmark programs to show its feasibility with the aspect of various viewpoints regarding sustainability in IoT-based systems. Those benchmark programs can cover several IoT devices such as Direct Attached Storage(DAS), Personal Cloud Storage Device (PCSD), Solid-State Hybrid Device (SSHD), and Digital Video Recorder and Player(DVR), which requires advanced I/O operations.

4.2. Block-Level Experiments

The first benchmark evaluations are testing for block-level I/O operations. This test is for storage-oriented devices such as DAS, since DAS uses dense block-level I/O operations. In the block-level benchmark, block-level read and write operations without file system operations are done with the benchmark running, then the throughput of the read and write block I/O operations are measured. The results of block-level benchmark evaluation are plotted in Figure 6, where it plots throughputs of random read and random write workloads at block level.

At first, Figure 6a shows the performance of random reads in the block devices without a file system. In the block-level I/O operations, the block devices could be driven by buffered I/O or direct I/O, so these were applied to the SSD RAID-0, RAMDisk, and DHRD, respectively. The DBR DHRD does not distinguish between buffered I/O and direct I/O for reads, instead always treats them as DBR. As shown in the results, the proposed DBR DHRD showed 64 times better read throughput than 'SSD RAID-0', which uses direct I/O. On an average, the write throughput of the DHRD with direct I/O was twice that of the DHRD that used buffered I/O. The DBR DHRD showed 2.8 times better read performance than the DHRD that used direct I/O. DBR is implemented as light weight codes, while direct I/O has more complex computing overhead than DBR that has less locks and has no page cache flush and waiting calls. The DBR DHRD, which has low computing overhead and no redundant memory copy, showed the highest read performance.

The write performance of the DHRD depends on the SSD. As shown in Figure 6b, the write performance of the DHRD and that of the SSD RAID-0 are almost the same, but the write performance of DHRD is 3% lower than that of 'SSD RAID-0' because the DHRD includes additional operation in the RAMDisk. The write performance of the RAMDisk is superior to others. However, unlike the RAMDisk, the DHRD and the SSD provide persistency. For DHRD with direct I/O, the performance was about 5 times higher when the number of processes were 32 than when the number was 1. The reason being that the SSD consists of dozens of NAND chips and several channels so that the maximum performance of the SSD can be achieved by several simultaneous I/O requests. The DHRD with buffered I/O has less impact on the degree of concurrent I/O requests. When an application writes data using buffered I/O, the data is copied to the page cache and an immediate response is sent to the application. Therefore, the accumulated pages are concurrently transferred to the final storage device later, so that this I/O parallelism is better for the SSD of the DHRD.

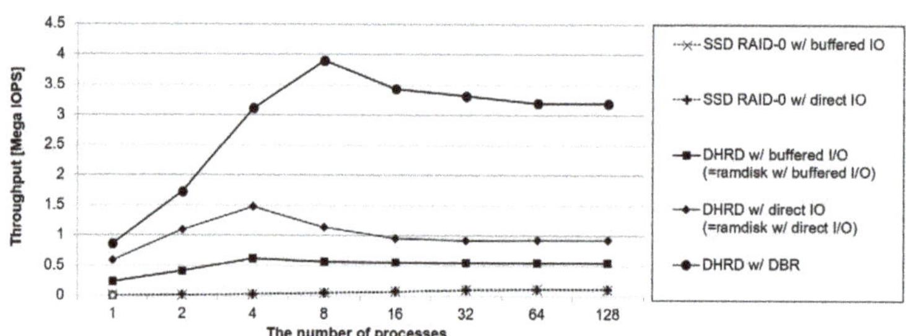

(a) The throughput of random reads at the block level

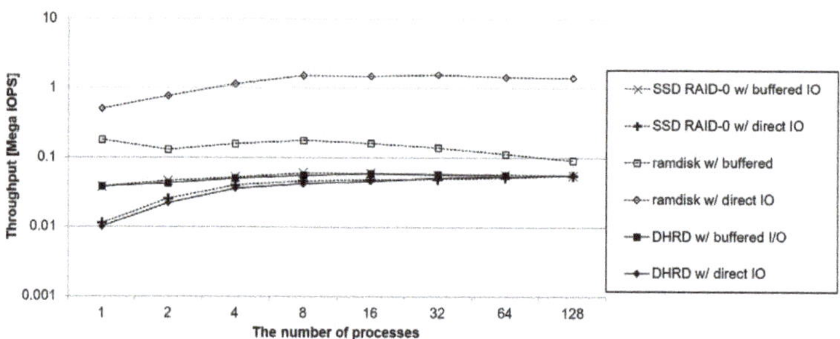

(b) The throughput of random writes at the block level

Figure 6. The results of block-level benchmark evaluation. It plots throughputs of random read and random write workloads at block level.

Figure 7 shows evaluation conducted Storage Performance Council (SPC) traces that consist of two I/O traces from online transaction processing (OLTP) applications running at two large financial institutions and three I/O traces from a popular web search engine [30]. We replayed the SPC traces on the DBR DHRD, DHRD, RAMDisk, and SSD RAID-0 at the block level. The DHRD showed 8% slower performance than the RAMDisk. However, DBR DHRD showed 20% better performance than the RAMDisk and 270% better performance than the SSD RAID-0. The DBR DHRD performed best on SPC workloads that had mixed reads and writes.

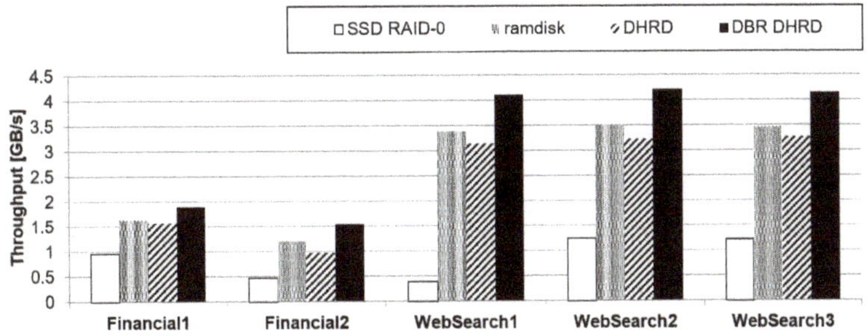

Figure 7. SPC traces: It plots two I/O traces from online transaction processing (OLTP) applications running at two large financial institutions and three I/O traces from a popular search engine.

4.3. File-Level Experiments

Data storage of IoT devices is a kind of remote storage device that lets systems store data and other files for sustainable IoT-based services. In this device, file-level I/O throughput is critical to the system to give best responsiveness. This section presents an evaluation that uses file-level benchmark programs. It exhibited more computing overhead than the block-level benchmarks. In the file-level benchmark running, we do sequential read, sequential write, random read, random write, and mixed pattern of random read/write operations at a file system level with XFS file system [31]. For the sequential benchmark running, a single process does file-level read and write operations, while throughput of random read and write are measured as the number of processes increases to make more complex situations. For each pattern running, DBR DHRD, DHRD, RAMDisk, and SSD RAID-0 are compared. The results of these file-level benchmark evaluation are shown in Figure 8, where throughputs of sequential read and write, random read, random write, and mixed random read/write workloads are plotted.

Figure 8a,b evaluate the sequential and random read/write performance with a 16 GB file on an XFS filesystem. Figure 8a shows sequential read and write performance. As shown in the results, the DBR DHRD gives 3.3 times better sequential read performance than the DHRD in terms of the throughput aspect. It is because the DBR DHRD has half of the memory copy overhead and simpler computing complexity than the DHRD. The write performances of the SSD RAID-0, DHRD, and DBR DHRD were almost the same due to the bottleneck of the SSD as shown in Figure 8d. The performance of the RAMDisk was the best. Figure 8b shows the mixed random reads and random writes, where the ratio of reads and writes was 66:34. Most applications showed similar behavior with this I/O ratios. The DBR DHRD outperformed the DHRD by 16% on average. The DBR DHRD showed 15 times better performance than the SSD RAID-0 on average with the same durability.

Filebench is a file system and storage benchmark that can generate a wide variety of workloads [32]. Unlike typical benchmarks, it is flexible and allows an application's I/O behavior to be specified using its extensive Workload Model Language (WML). In this section, we evaluate them with the predefined file server workloads among various Filebench workloads. The file server workload runs 50 threads simultaneously, and each thread creates an average of 128 KB of files, adds data to the file, or reads a file. We measured throughputs for four system configurations as the number of files varies from 32 k to 512 k.

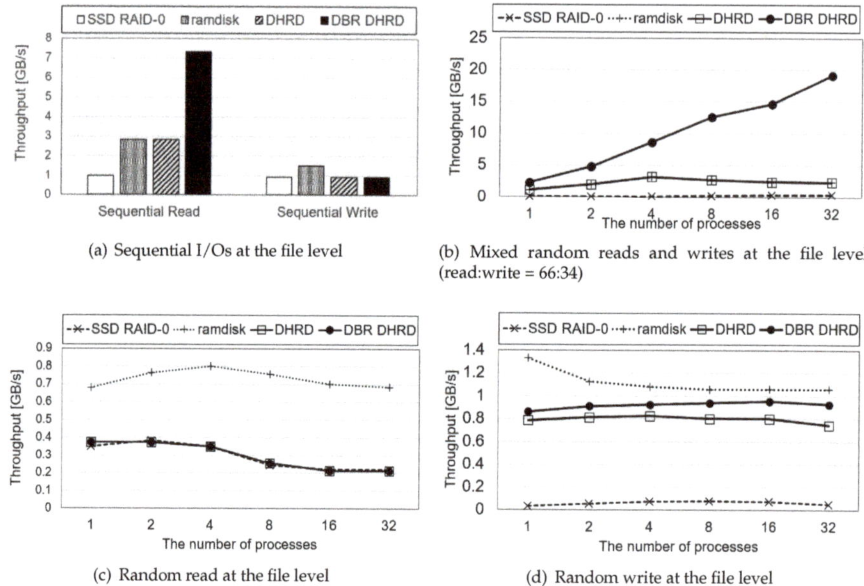

(a) Sequential I/Os at the file level

(b) Mixed random reads and writes at the file level (read:write = 66:34)

(c) Random read at the file level

(d) Random write at the file level

Figure 8. The results of file-level benchmark evaluation. It plots throughputs of sequential I/O, random read, random write, and mixed random read and write workloads at file level.

Figure 9 shows performance results obtained using file server workloads using Filebench. In the figure, the *x*-axis represents the number of files and the *y*-axis represents throughputs of each system running. The file server workload has a 50:50 ratio of reads and writes. As shown in Figure 9, the DBR DHRD showed 28% and 54% better performances than the DHRD and the SSD RAID-0, respectively. As this workload has many writes, the RAMDisk achieved the best performance. Although RAMDisk shows higher throughput than DBR DHRD, the RAMDisk suffers from low durability. Thus, DBR DHRD can be said to show better performance while keeping reasonable durability when RAM and SSD are used together in the computing system.

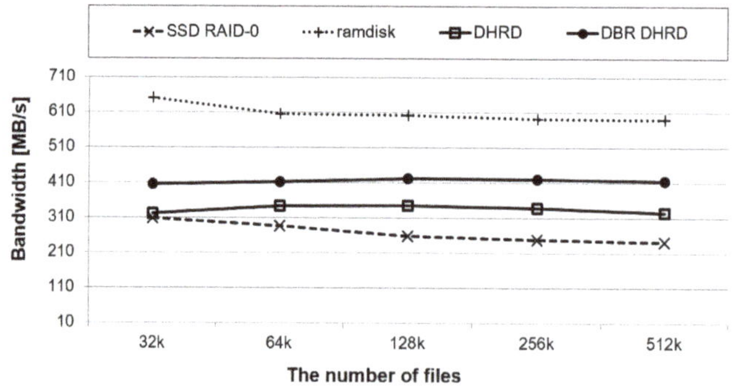

Figure 9. A benchmark using Filebench with fileserver workloads.

4.4. Hybrid Storage Devices and DVR Applications

The tiered storage is a data storage method or system consisting of two or more storage media types. Generally, the frequently used data are served from the fasted storage media such as SSD, and other cold data are accessed from a low-cost media such as HDD, where the first-tier storage as the fasted media is usually performed as a cache for the lower-tier storage. Therefore, the first-tier storage is also called a cache tier.

One of the emerging storage devices is a tiered storage such as SSHD, which is a traditional spinning hard disk with a small amount of fast solid-state storage. BDR DHRD can be applied to the solid-state storage in a SSHD as shown in Figure 10, BDR DHRD can replace the solid-state storage of SSHD, thereby improving the performance of the solid-state storage of a SSHD. To see if the performance of DBR DHRD is improved in a tired storage, we compared the I/O performance of tiered storage devices with DBR DHRD. In this experiment, SSD, HDD, DHRD, and DBR DHRD were configured in tiered storage devices. Three-tiered storage models, SSD+HDD, DHRD + HDD, and DBR DHRD + HDD are considered.

SSHD can be implemented by the flashcache [33] module in Linux. The flashcache can make a tiered storage with SSD and HDD. DHRD is implemented as a general block device, so a DHRD device can replace the SSD of a flashcache device. By this way, we can make a SSHD that consists of DHRD and HDD.

PC Matic Research said that the average memory size of desktop computers is 1 GB in 2008, and 8 GB in 2018 [34]. We can forecast that the average size of PC memory will be 64 GB in 2028. PC motherboards can support up to 128 GB of memory in 2019. In this experiment, the tiered storage used 8 GB of memory, which can be used in the mid-sized to high-end desktop computers.

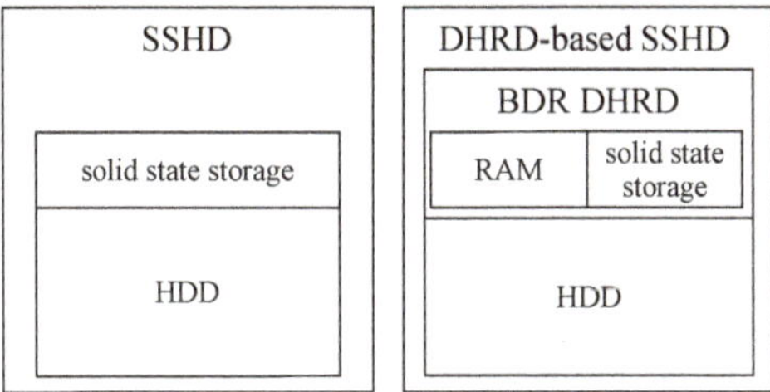

Figure 10. A generic SSHD and a DHRD-based SSHD.

The I/O traces used in the experiment were collected from three general users using a personal computer. One is a system administrator user, two are developers, and their daily I/O traces are collected and used as experimental I/O traces. In those tiered systems, I/O traces collected from users were performed and throughput is estimated. During the experiment, it is assumed that 70% and 80% of all I/O traces are allocated to SSD or DHRD, which is considered to be the cache tier in tiered storage system.

The results are plotted in Figure 11, in that Figure 11a compares three types of tiered storage devices, SSHD(SSD+HDD), DHRD+HDD, and DBR DHRD+HDD, when the hit rate is 70%. Figure 11b compares them when the hit rate of the cache tier is 80%. Both the RAM size and the SSD size are 8 GB, which is a typical size of a commercial SSHD. As shown in the figures, throughput of DBR DHRD+HDD and DHRD+HDD-based tiered storage outperforms SSD+HDD-based tiered storage about several times for each I/O traces. DBR DHRD scheme also outperforms DHRD only, which is the advantage of direct byte-level read operations supported by DBR. If we compare hit ratio of the cache tier in the tiered storage, the higher the cache tier hit ratio, the higher throughput we have when DBR DHRD is used. From the figure, we identify that the throughput DBR DHRD for 80% cache tier hit ratio is increased about 14%.

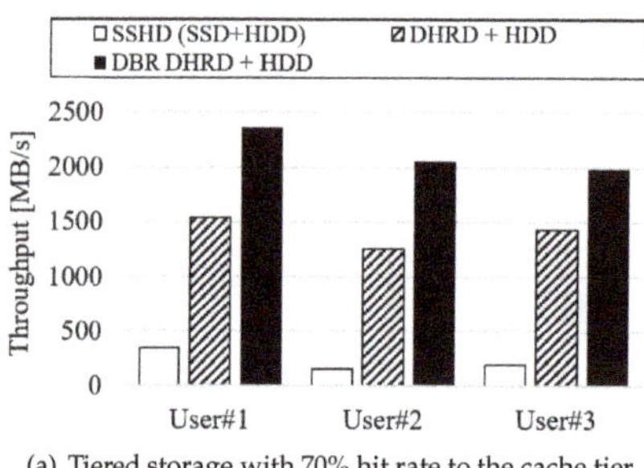
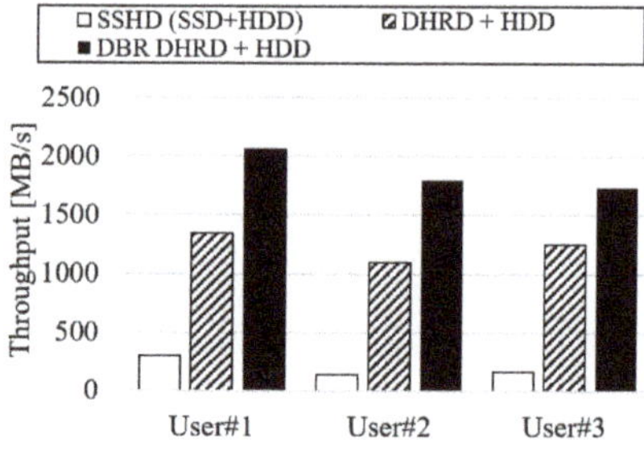

(a) Tiered storage with 70% hit rate to the cache tier
(b) Tiered storage with 80% hit rate to the cache tier

Figure 11. The results of tiered storage in hybrid storage device. It plots I/O throughputs of tiered storage assuming that SSD, DHRD, and DBR DHRD are used as a cache tiered in a tiered storage.

Lastly, we conducted experiments on reading and rewriting video files, which is a kind of experiment applicable to multimedia-oriented IoT applications. In this experiment, 1.8 GB sized video file is read, modified partially, and save it as another file. For each system configuration, i.e., SSD, DHRD, and DBR DHRD, we did those operations three times and measured overall execution time. The results are plotted in Figure 12. As shown in the figure, DBR DHRD and DHRD were 2.26 times faster and 1.74 times faster than SSD, respectively. From the results, we identify that DBR DHRD can be applied to IoT devices that deal with multimedia data.

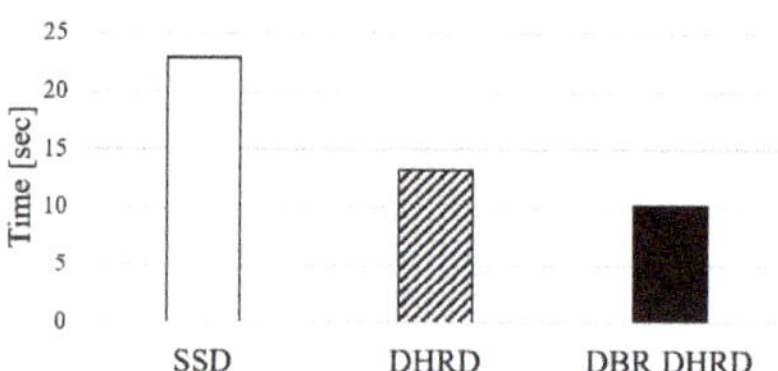

Figure 12. A result of reading and writing for video files.

5. Conclusions

RAM disk is a software-based storage device to provide low latency, which is compatible with legacy file system operations. The traditional RAM disk includes the disk cache; however, the fact is that it does not require disk cache. Another way for a block device to bypass disk cache, Direct I/O is used; however, the parameters must be a multiple of the logical block size for Direct I/O, so a byte-level addressable path from application to storage device does not exist.

This paper introduced the DRB DHRD scheme for hybrid storage systems that is composed of RAM disk and SSD. The proposed DBR-enabled DHRD provides a byte–range interface. It is compatible with existing interfaces and can be used with buffered writes. The initialization procedure of DBR-enabled DHRD can reduce the boot time of the storage device, since it allows general I/O requests during the initialization process itself, while other RAMDisk-based storage cannot support general I/Os during the initialization. Experimental evaluation was performed using various benchmarks that are applicable to various IoT-based systems performing dense I/O operations. In workloads where reads and writes were mixed, the DHRD performed 15 times better than the SSD. The DBR also improved the performance of the DHRD by 2.8 times. For the hybrid storage device, DBR DHRD performed 3 to 5 times faster throughputs than SSHD. Also, DBR DHRD can reduce execution times of multimedia file's read and write processing.

As the next step of this study, we are exploring a more advanced version of DRB DHRD for further features and for performance improvement. A more rigorous comparison of the performance of this DRB DHRD scheme versus others could be an important task to improve the completeness of the proposed system. We set the more rigorous performance evaluations as our further work.

Author Contributions: Conceptualization, K.-W.P.; Data curation, S.H.B.; Funding acquisition, K.-W.P.; Methodology, K.-W.P.; Project administration, K.-W.P.; Resources, S.H.B.; Software, S.H.B. and K.-W.P.; Supervision, K.-W.P. All authors have read and agreed to the published version of the manuscript.

Funding: This research received no external funding.

Acknowledgments: This work was supported by a Jungwon University Research Grant (South Korea) (Management Number: 2017-031).

Conflicts of Interest: The authors declare no conflict of interest.

References

1. Bagatin, M.; Gerardin, S.; Paccagnella, A.; Beltrami, S.; Camerlenghi, E.; Bertuccio, M.; Costantino, A.; Zadeh, A.; Ferlet-Cavrois, V.; Santin, G.; et al. Effects of heavy-ion irradiation on vertical 3-D NAND flash memories. *IEEE Trans. Nucl. Sci.* **2017**, *65*, 318–325. [CrossRef]
2. Fedorchenko, A.; Kotenko, I.V.; Chechulin, A. Integrated Repository of Security Information for Network Security Evaluation. *JoWUA* **2015**, *6*, 41–57.
3. Android Things. Available online: https://developer.android.com/things (accessed on 10 April 2020).
4. Ninglekhu, J.; Krishnan, R.; John, E.; Panday, M. Securing Implantable Cardioverter Defibrillators Using Smartphones. *J. Internet Serv. Inf. Secur. (JISIS)* **2015**, *5*, 47–64.
5. Jo, H.; Kang, J.U.; Park, S.Y.; Kim, J.S.; Lee, J. FAB: Flash-aware buffer management policy for portable media players. *IEEE Trans. Consum. Electron.* **2006**, *52*, 485–493.
6. Kim, H.; Ahn, S. BPLRU: A buffer management scheme for improving random writes in flash storage. *Usenix FAST* **2008**, *8*, 239–252.
7. Li, Z.; Jin, P.; Su, X.; Cui, K.; Yue, L. CCF-LRU: A new buffer replacement algorithm for flash memory. *IEEE Trans. Consum. Electron.* **2009**, *55*, 1351–1359. [CrossRef]
8. Zuolo, L.; Zambelli, C.; Micheloni, R.; Olivo, P. Memory driven design methodologies for optimal SSD performance. In *Inside Solid-State Drives*; Springer: Singapore, 2018; pp. 181–204.
9. Du C.; Yao Y.; Zhou J.; Xu, X. VBBMS: A Novel Buffer Management Strategy for NAND Flash Storage Devices. *IEEE Trans. Consum. Electron.* **2019**, *56*, 1351–1359. [CrossRef]
10. RAM drive, 2020. Wikipedia.Available online: https://en.wikipedia.org/wiki/RAM_drive (accessed on 10 April 2020).
11. Koutoupis, P. The linux ram disk. *LiNUX+ Magzine* **2009**, 36–39.
12. Diehl S.T. System and Method for Persistent Ram Disk. US Patent 7,594,068, 22 September 2009.
13. Baek, S.H. A durable and persistent in-memory storage for virtual desktop infrastures. *J. Korean Insti. Next Gen. Comput.* **2016**, *12*, 23–31.
14. Ousterhout, J.; Gopalan, A.; Gupta, A.; Kejriwal, A.; Lee, C.; Montazeri, B.; Ongaro, D.; Park, S.J.; Qin, H.; Rosenblum, M.; et al. The ramcloud storage system. *ACM Trans. on Comp. Sys. (TOCS)* **2015**, *33*, 1–55. [CrossRef]
15. Garcia-Molina, H.; Salem, K. Main memory database systems: An overview. *IEEE Trans. Knowl. Data Eng.* **1992**, *4*, 509–516. [CrossRef]
16. Fitzpatrick, B. Distributed caching with memcached. *Linux J.* **2004**, *124*, 5.
17. Ousterhout, J.; Agrawal, P.; Erickson, D.; Kozyrakis, C.; Leverich, J.; Mazières, D.; Mitra, S.; Narayanan, A.; Parulkar, G.; Rosenblum, M.; et al. The case for ramclouds: scalable high-performance storage entirely in dram. *Acm Sigops Oper. Syst. Rev.* **2010**, *43*, 92–105. [CrossRef]
18. Zaharia, M.; Chowdhury, M.; Das, T.; Dave, A.; Ma, J.; McCauley, M.; Franklin, M.J.; Shenker, S.; Stoica, I. Resilient distributed datasets: A fault-tolerant abstraction for in-memory cluster computing. In Proceedings of the 9th USENIX Symposium on Networked Systems Design and Implementation, San Jose, CA, USA, 25–27 April 2012.
19. Ma, Z.; Hong, K.; Gu, L. Volume: Enable large-scale in-memory computation on commodity clusters. In Proceedings of the IEEE 5th International Conference on Cloud Computing Technology and Science, Bristol, UK, 2–5 December 2013; pp. 56–63.

20. Kim, J.; Jo, S.; Oh, D.; Noh, J. An analysis on the technology stack of cloud computing. *J. Korean Insti. Next Gen. Comput.* **2015**, *11*, 79–89.

21. Rathod, S.B.; Reddy, V.K. NDynamic framework for secure vm migration over cloud computing. *J. Inf. Proc. Syst.* **2017**, *13*, 476–490.

22. Cisco Systems Press. Diskless Vdi with Cisco Ucs & Atlantis Ilio Eliminating Storage from Vdi Architectures. White Paper Aug. 2011. Available online: http://bit.ly/ciscoatlantisdisklessvdi (accessed on 10 April 2020).

23. Technology, K. Memory and Storage Best Practices for Desktop Virtualization: Balancing User Experience, Cost and Flexibility. White Paper, 2013. Available online: https://media.kingston.com/pdfs/MemoryandStorageBestPracticesforDesktopVirtualization_lr.pdf (accessed on 10 April 2020).

24. Nishtala, R.; Fugal, H.; Grimm, S.;Kwiatkowski, M.; Lee, H.;Li, H.C.;McElroy, R.; Paleczny, M.; Peek, D.; Saab, P.; et al. scaling memcache at facebook. In Proceedings of the 10th USENIX Symposium on Networked Systems Design and Implementation, Seattle, DC, USA, 2–5 April 2013; pp. 385–398.

25. Caramia, M.; Di Carlo, S.; Fabiano, M.; Prinetto, P. Flash-memories in space applications: Trends and challenges. In Proceedings of the East-West Design & Test Symposium (EWDTS), Moscow, Russia, 18–21 September 2009; pp. 18–21.

26. Hoseinzadeh, M. A Survey on Tiering and Caching in High-Performance Storage Systems. *arXiv* **2019**, arXiv:1904.11560.

27. Niu, J.; Xu, J.; Xie, L. Hybrid storage systems: a survey of architectures and algorithms. *IEEE Access* **2018**, *6*, 13385–13406. [CrossRef]

28. Baek, S.H.; Park, K.W. A fully persistent and consistent read/write cache using flash-based general SSDs for desktop workloads. *Inf. Syst.* **2016**, *58*, 24–42. [CrossRef]

29. Baek, S.H. A byte direct i/o for ram-based storages. *Adv. Sci. Lett.* **2017**, *23*, 9506–9510. [CrossRef]

30. Search Engine I/O, uMass Trace Repository. Available online: http://traces.cs.umass.edu/index.php/Storage/Storage (accessed on 10 April 2020).

31. Robbins, D. Common threads: Advanced filesystem implementer's guide, Part 9, Introducing XFS, In *Developer Works*; IBM: Armonk, NY, USA, 2002.

32. Tarasov, V.; Zadok, E.; Shepler, S. Filebench: A flexible framework for file system benchmarking, *Usenix Mag.* **2016**, *41*, 6–12.

33. Kgil, T.; Mudge, T. FlashCache: a NAND flash memory file cache for low power web servers. In Proceedings of the 2006 International Conference on Compilers, Architecture and Synthesis for Embedded Systems, Seoul, Korea, 22–25 October 2006.

34. PC Matic Research, Average PC Memory (RAM) Continues to Clim. Available online: https://techtalk.pcmatic.com/2016/10/05/average-pc-memory-ram-continues-climb/ (accessed on 10 April 2020).

© 2020 by the authors. Licensee MDPI, Basel, Switzerland. This article is an open access article distributed under the terms and conditions of the Creative Commons Attribution (CC BY) license (http://creativecommons.org/licenses/by/4.0/).

Article

The Algorithm and Structure for Digital Normalized Cross-Correlation by Using First-Order Moment [†]

Chao Pan [1], Zhicheng Lv [1], Xia Hua [2,*] and Hongyan Li [1]

[1] School of Information and Communication Engineering, Hubei University of Economics, Wuhan 430205, China; PanChao@hbue.edu.cn (C.P.); lion112@139.com (Z.L.); hongyanli2000@126.com (H.L.)

[2] School of Electrical and Information Engineering, Wuhan Institute of Technology, Wuhan 430205, China

[*] Correspondence: hedahuaxia05021046@163.com

[†] This paper is an extended version of our paper published in the First International Symposium on Future ICT (Future-ICT 2019) in Conjunction with 4th International Symposium on Mobile Internet Security (MobiSec 2019).

Received: 1 February 2020; Accepted: 25 February 2020; Published: 1 March 2020

Abstract: Normalized cross-correlation is an important mathematical tool in digital signal processing. This paper presents a new algorithm and its systolic structure for digital normalized cross-correlation, based on the statistical characteristic of inner-product. We first introduce a relationship between the inner-product in cross-correlation and a first-order moment. Then digital normalized cross-correlation is transformed into a new calculation formula that mainly includes a first-order moment. Finally, by using a fast algorithm for first-order moment, we can compute the first-order moment in this new formula rapidly, and thus develop a fast algorithm for normalized cross-correlation, which contributes to that arbitrary-length digital normalized cross-correlation being performed by a simple procedure and less multiplications. Furthermore, as the algorithm for the first-order moment can be implemented by systolic structure, we design a systolic array for normalized cross-correlation with a seldom multiplier, in order for its fast hardware implementation. The proposed algorithm and systolic array are also improved for reducing their addition complexity. The comparisons with some algorithms and structures have shown the performance of the proposed method.

Keywords: normalized cross-correlation; fast algorithm; first-order moment; systolic array; multiplication complexity

---

1. Introduction

Normalized cross-correlation (NCC) is an important mathematical tool in signal and image processing for feature matching, similarity analysis, motion tracking, object recognition, and so on [1–3]. In order to improve its real-time and efficient performance, digital NCC has been suggested to be implemented by some fast algorithms and hardware structures, due to its high computational complexity [4,5].

Nowadays, since correlation and convolution have similar computation structures, there are mainly three kinds of fast convolution algorithms can be applied for fast NCC [6,7]: (1) the Fast Fourier Transform (FFT)-based algorithm, (2) the polynomial-based algorithm, (3) the decomposition algorithm. However, to our knowledge, each of these algorithms has its applicable limitations. The FFT-based algorithm is not well-suited to the discrete domain. Plus, it involves with complex multiplications [8,9]. Both the polynomial-based algorithm and the decomposition algorithm require complex computational structures, and they often lack commonality for arbitrary-length correlations [10,11].

Furthermore, some special algorithms for fast NCC have been presented [12,13]. The fast cross-correlation of binary sequences can be extended to other types of NCC sequences [14]. The estimation algorithm derives the scaling factor between the signal and the kernel, so it computes NCC using only additions at the cost of small noise [15]. Several methods have been used to assist NCC for reducing its searching and computing times in image matching, such as the pyramid method [3,7]. In addition, many parallel algorithms of the inner-product have been published that can perform fast cross-correlation for NCC [16,17], where the Distributed Arithmetic (DA) with look-up table has not multiplication, but needs much Read-Only Memory (ROM) [18].

To hardware implementation of fast NCC, Very-Large-Scale Integration (VLSI) circuits have been applied, where systolic structures are popular due to their regularity and modularity [19–21]. The integration of the systolic array and the DA technique lead to more efficient VLSI implementation of cross-correlation, although they use many ROMs and address decoders [22,23]. The Residue Number System-based DA can reduce ROMs and enhance throughput, while extra encoding processes in the residue domain are necessary [24].

In this paper, we present a new algorithm and structure to implement digital NCC with a simple and fast procedure. It is a breakthrough that an NCC formula expressed in terms of a first-order moment is designed according to the relationship between the inner-product and the first-order moment, so the computational complexity of NCC is transformed into that of a first-order moment. For performing an arbitrary-length digital NCC, our algorithm would first establish the NCC formula based on a first-order moment for correlation sequences, and then introduce a fast algorithm without multiplication from [25,26] to compute this first-order moment in the new NCC formula rapidly. For the hardware implementation of NCC, we develop a simple and scalable systolic array derived from the proposed algorithm, due to the fact that the fast algorithm for the first-order moment is easily performed by systolic structure [27]. The proposed algorithm and systolic array are also improved to reduce their addition complexity, according to an even-odd relationship in the computation of the first-order moment.

The rest of the paper is organized as follows. Section 2 establishes the NCC formula based on a first-order moment. Section 3 introduces a fast algorithm and its systolic implementation for first-order moment. Sections 4 and 5 discuss the fast algorithm and the systolic array inspired by Section 3 to perform the NCC formula in Section 2 rapidly. Comparison and analysis are presented in Section 6 to demonstrate the feasibility of the proposed algorithm and structure. Finally, Section 7 gives the conclusion.

2. Normalized Cross-Correlation Based on First-Order Moment

Being the most complex operation in NCC, the inner-product of two correlation sequences would be transformed into a first-order moment for decreasing computational complexity in fast NCCs. To do this, let us assume two N-point digital sequences $\{f(i)\}$ and $\{g(i)\}$, where $\{f(i)\}$ is an arbitrary input sequence, and $\{g(i)\}$ is the fixed correlation kernel with the value range $g(i)\in\{0, 1, 2, \ldots, L\}$. This section establishes an NCC formula for these two sequences that mainly includes a first-order and a zero-order moment. The aim is to replace the complex computation of cross-correlation in NCC with an easy computation of a first-order moment.

2.1. Cross-Correlation

Cross-correlation is an inner-product between two digital sequences. It is defined as

$$c(n) = f(n) \circ g(n) = \sum_{i=0}^{N-1} f(n+i)g(i) \tag{1}$$

Using mathematical transformation, this Equation (1) could be transformed into a first-order moment by means of the statistical characteristics of the inner-product operation. To do this, we define some subsets S_k ($k = 0, 1, 2, \dots, L$) that divide the index set $i \in \{0, 1, \dots, N-1\}$ into L subsets, depending on the max value in the correlation kernel $\{ g(i) \}$. Specifically,

$$S_k = \left\{ i \,\middle|\, g(i) = k, \quad i \in \{0, 1, 2, \cdots, N-1\} \right\} \tag{2}$$

where $k = 0, 1, 2, \dots, L$. In other words, S_k is a set of indices i that corresponds to $g(i) = k$ in actual. Then a new $(L+1)$-point sequence $\{ a_k(n) \}$ is defined by subsets S_k [28], which is

$$a_k(n) = \begin{cases} \sum_{i \in S_k} f(n+i) & \text{where } S_k \neq \Phi \\ 0 & \text{otherwise} \end{cases}, \tag{3}$$

where $k = 0, 1, 2, \dots, L$, and "Φ" denotes an empty set.

The $a_k(n)$ could be acted as the sum of elements in the sequence $\{ f(n+i) \}$ while the parameter i corresponds to $g(i) = k$. The computation of the $\{ a_k(n) \}$ is actually a statistics procedure for counting how much k would be accumulated in the computation of the $c(n)$. Therefore, the relationship between $\{ f(n+i) \}$ and $\{ a_k(n) \}$ can be described as:

$$\sum_{i=0}^{N-1} f(n+i) = \sum_{k=0}^{L} a_k(n), \tag{4a}$$

$$\sum_{i=0}^{N-1} f(n+i)g(i) = \sum_{k=0}^{L} a_k(n)k = \sum_{k=1}^{L} a_k(n)k \tag{4b}$$

It is obvious that $\sum_{k-1}^{L} a_k(n)$ in Equation (4a) is a zero-order moment of $\{ a_k(n) \}$, and $\sum_{k=1}^{L} a_k(n)k$ in Equation (4b) is a first-order moment of $\{ a_k(n) \}$. As a result, the Equation (1) can be transformed into:

$$c(n) = \sum_{k=1}^{L} a_k(n)k \tag{5}$$

From Equation (5), we obtain a new calculation formula for cross-correlation based on a first-order moment.

2.2. Normalized Cross-Correlation

Normalized cross-correlation is more complex than cross-correlation, because it includes an inner-product between two difference sequences from $\{ f(i) \}$, $\{ g(i) \}$ and their mean value. It is defined as

$$\rho(n) = \frac{\sum_{i=0}^{N-1} [f(n+i) - \overline{f}(n)][g(i) - \overline{g}]}{\left\{ \sum_{i=0}^{N-1} [f(n+i) - \overline{f}(n)]^2 \sum_{i=0}^{N-1} [g(i) - \overline{g}]^2 \right\}^{\frac{1}{2}}}, \tag{6}$$

where $\overline{f}(n) = \frac{1}{N} \sum_{i=0}^{N-1} f(n+i)$ and $\overline{g} = \frac{1}{N} \sum_{i=0}^{N-1} g(i)$.

This Equation (6) can be rewritten as

$$\rho(n) = \frac{\sum\limits_{i=0}^{N-1} f(n+i)g(i) - \overline{g}\sum\limits_{i=0}^{N-1} f(n+i) - \overline{f}(n)\sum\limits_{i=0}^{N-1} g(i) + N[\overline{f}(n)\overline{g}]}{\left\{\sum\limits_{i=0}^{N-1}[f(n+i)^2 - 2f(n+i)\overline{f}(n) + \overline{f}(n)^2]\sum\limits_{i=0}^{N-1}[g(i)-\overline{g}]^2\right\}^{\frac{1}{2}}}$$

$$= \frac{\sum\limits_{i=0}^{N-1} f(n+i)g(i) - \frac{1}{N}\sum\limits_{i=0}^{N-1} f(n+i)\sum\limits_{i=0}^{N-1} g(i)}{\left\{\{\sum\limits_{i=0}^{N-1} f(n+i)^2 - \frac{1}{N}[\sum\limits_{i=0}^{N-1} f(n+i)]^2\}\sum\limits_{i=0}^{N-1}[g(i)-\overline{g}]^2\right\}^{\frac{1}{2}}} \tag{7}$$

If we set

$$b(n) = \sum_{i=0}^{N-1} [f(n+i)]^2 \tag{8}$$

and substitute Equations (4a), (4b) and (8) into Equation (7), the NCC expressed by Equation (6) can be converted to

$$\rho(n) = \frac{\sum\limits_{k=1}^{L} a_k(n)k - \overline{g}\sum\limits_{k=0}^{L} a_k(n)}{\left\{\{b(n) - \frac{1}{N}[\sum\limits_{k=0}^{L} a_k(n)]^2\}\sum\limits_{i=0}^{N-1}[g(i)-\overline{g}]^2\right\}^{\frac{1}{2}}}. \tag{9}$$

From Equation (9), we develop a new calculation formula for NCC based on a first-order moment $\sum a_k(n)k$ and a zero-order moment $\sum a_k(n)$. It is obvious that the computation complexity of this NCC formula depends heavily upon the complexity of $\sum a_k(n)k$ and $b(n)$. Therefore, for a fast implementation of Equation (9), we introduce a fast algorithm and structure for $\sum a_k(n)k$ in Section 3, and an optimization method for $b(n)$ in Section 4.1.

3. The Fast Algorithm and Systolic Array for First-Order Moment

Liu et al. presented an algorithm and its systolic array for first-order moment in [25–27]. Their method is suitable to compute the first-order and the zero-order moment in Equation (4) rapidly. In this section, we introduce this algorithm and systolic array that aims to implement fast NCC by using Equation (9). In addition, because the introduced algorithm and array request many additions as the result of removing all multiplications, we also improve them in order for lower addition complexity.

3.1. The Fast Algorithm for First-Order Moment

According to [25], we illustrate a simple 1-network shown in Figure 1 that represents a map of transforming the two-dimensional vector $(1, x)$ into the vector $(1, (1 + x))$. This map is denoted by F that is

$$F(1, x) = (1, (1+x)).$$

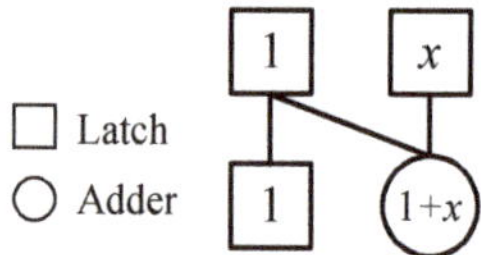

Figure 1. The 1-network.

Some characteristic equations obtained from F are

$$F(a, ax) = (a, a(1+x)), \quad F(a+b, a+b) = F(a, a) + F(b, b) \tag{10}$$

Also,

$$F^2(1,\ x) = F(F(1,\ x)) = F(1,\ (1+x)) = (1,\ 2+x)$$

and by induction

$$F^{L-1}(1,\ x) = F(\ldots F \ldots F(1,\ x)) = (1,\ (L-1+x)).$$

Hence, we have

$$F^{L-1}(1,\ 1) = F(\ldots F \ldots F(1,\ 1)) = (1,\ L),$$

$$F^{L-1}(a,\ a) = (a,\ La). \tag{11}$$

To compute first-order moment by this 1-network, let

$$\mathbf{a}_k = (\ a_k(n),\ a_k(n)\ ) \quad (k = 1,\ 2,\ \ldots, L),$$

so, Equations (10) and (11) are yielded by

$$F(F(\mathbf{a}_k) + \mathbf{a}_{k-1}) = F(F(\mathbf{a}_k)) + F(\mathbf{a}_{k-1}) = F^2(\mathbf{a}_k) + F(\mathbf{a}_{k-1})$$
$$= (\ a_k(n) + a_{k-1}(n),\ 3a_k(n) + 2a_{k-1}(n)\ )$$

Generally, the above equation is expanded into

$$F(F \ldots F(F(F(\mathbf{a}_L) + \mathbf{a}_{L-1}) + \ldots) + \mathbf{a}_2) + \mathbf{a}_1 = F^{L-1}(\mathbf{a}_L) + \ldots + F^2(\mathbf{a}_3) + F(\mathbf{a}_2) + \mathbf{a}_1$$
$$= (\ \sum_{k=1}^{L} a_k(n),\ \sum_{k=1}^{L} a_k(n)k) \tag{12}$$

From Equation (12), $\sum a_k(n)$ in Equation (4a) and $\sum a_k(n)k$ in Equation (4b) can both be obtained from an iterative implementation of the map F. This computational flow uses the $(L-1)$ recursive process of map F that includes $3L$ additions and 0 multiplications [26]. Therefore, the fast algorithm for first-order moment by Equation (12) can be described in Algorithm 1 as a subroutine Moment [29]. Its computational structure is also shown in Figure 2, which is an iterative structure of a 1-network with six adders and three latches. Its total addition number to compute N-point first-order moments $\sum a_k(n)k$ $(n = 0, 1, \ldots, N-1)$ is $3NL$.

Algorithm 1 Moment $(a_L(n), a_{L-1}(n), \ldots, a_0(n))$

Define the array **a** with two elements
Initial $\mathbf{a} \leftarrow (\ a_L(n), a_L(n)\ )$
for each $k \in [2, L]$ **do**   // Equation (12)
   $\mathbf{a}[1] \leftarrow \mathbf{a}[1] + \mathbf{a}[0]$ // 1-network $F(\mathbf{a})$
   $\mathbf{a}[1] \leftarrow \mathbf{a}[1] + a_{L-k+1}(n)$
   $\mathbf{a}[0] \leftarrow \mathbf{a}[0] + a_{L-k+1}(n)$
end for
$\mathbf{a}[0] \leftarrow \mathbf{a}[0] + a_0(n)$
return a

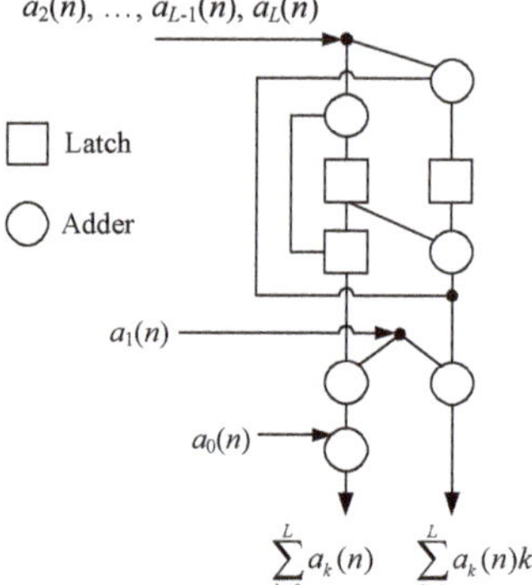

Figure 2. The computational structure for first-order moment.

3.2. The Systolic Array for First-Order Moment

The Equation (12) can be implemented by a systolic array for continuously generating a set of $\sum a_k(n)$ and $\sum a_k(n)k$ in parallel [27]. This systolic array is shown in Figure 3, which is actually a serial arrangement of $(L-1)$ 1-networks extended from Figure 2. It uses $3L-2$ adders, $L+2$ latch, and 0 multiplier. In each clock cycle, we should input a sequence $\{ a_k(n) \}$ into this systolic array and get a $(\sum a_k(n), \sum a_k(n)k)$.

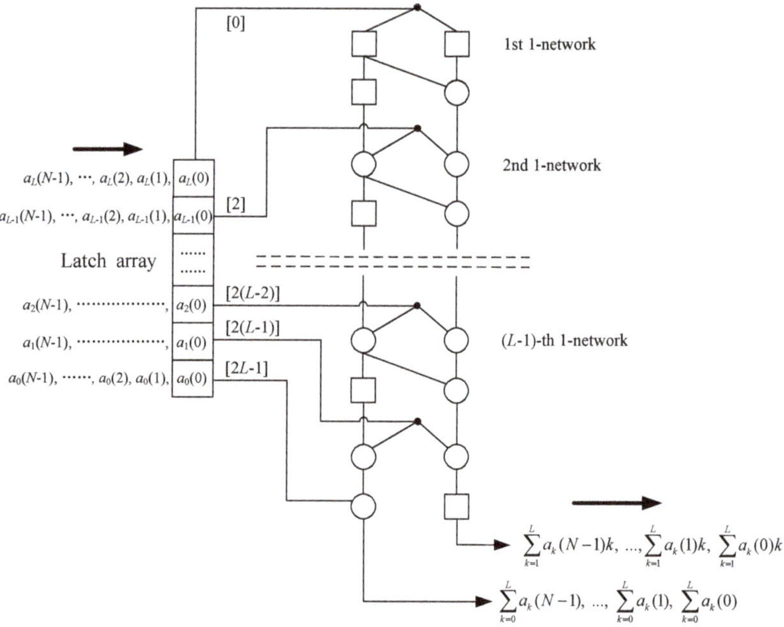

Figure 3. The systolic array for first-order moment.

Especially, to keep an operation synchronization for this parallel structure, the $(L-1)$-point $a_k(n)$ $(k = 2, \ldots, L)$ should be input into the $(L-1)$ 1-networks respectively rather than simultaneously. Generally, a single $a_k(n)$ $(k > 0)$ is input into the $(L-k)$-th 1-network with a latency $n + 2(L-1-k)$ clock cycle. Hence, in Figure 3, we use the extra latch array to generate latency for $a_k(n)$ before it is input into the corresponding 1-network. The number of latch array and latency time is shown in the note "[ ]", which leads to the occurrence that different $a_k(n)$ are input into the different 1-networks at regular intervals. As a result, the total execution time of this systolic array to compute N-point $\sum a_k(n)k$ ($n = 0, 1, \ldots, N - 1$) is that

$$2L - 1 + 1 + N - 1 = 2L + N - 1$$

clock cycles.

3.3. The Improvement of the Fast Algorithm and Systolic Array for First-Order Moment

The algorithm in Section 3.1 requires many additions that are computationally expensive when N is larger. In order to reduce its addition number, this algorithm is improved by means of an even-odd relationship that divides the first-moment of sequence $\{ a_k(n) \}$ into two smaller moments. This even–odd relationship is illustrated as:

$$\sum_{k=0}^{L} a_k(n) = \sum_{k=1}^{L/2} [a_{2k-1}(n) + a_{2k}(n)] + a_0(n), \tag{13a}$$

$$\sum_{k=1}^{L} a_k(n)k = \sum_{k=1}^{L/2} a_{2k-1}(n) \cdot (2k - 1) + \sum_{k=1}^{L/2} a_{2k}(n) \cdot 2k = 2\sum_{k=1}^{L/2} [a_{2k-1}(n) + a_{2k}(n)]k - \sum_{k=1}^{L/2} a_{2k-1}(n). \tag{13b}$$

According to Equation (13), the fast algorithm described by Figure 2 can be improved to the new structure shown in Figure 4. This improved algorithm firstly adds $L/2$ additions to obtain the sequence $\{ a_{2k-1}(n) + a_{2k}(n) \}$ as well as $L/2 - 1$ addition to accumulate $\sum a_{2k-1}(n)$. Then each $a_{2k-1}(n) + a_{2k}(n)$ is input into map F successively for performing $L/2 - 1$ iterations. Finally, a left-shift operation and 1 subtraction are applied to generate $\sum a_k(n)k$. The improved algorithm requires $5L/2 - 1$ additions that are superior to Figure 2, even though its structure is more complex at the cost of decreasing $L/2$ additions. Although the sequence $\{ a_{2k-1}(n) + a_{2k}(n) \}$ could be continually divided by the even-odd relationship for further reducing additions, the fast algorithm's structure would become very complex and unworthy.

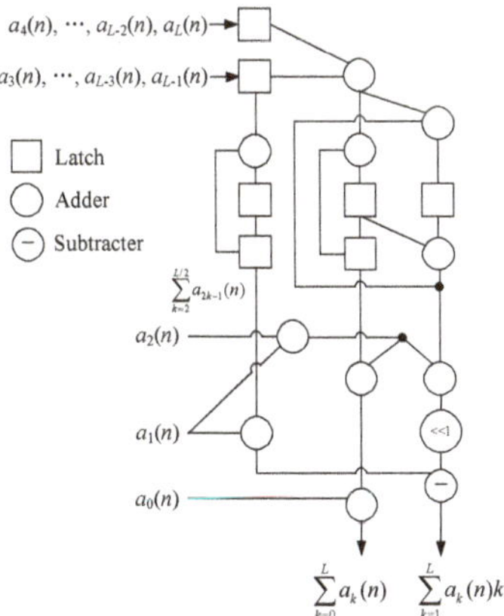

Figure 4. The improved computational structure for first-order moment.

Similarly, the systolic array in Figure 3 can be improved to the structure shown in Figure 5. This improved systolic array is a serial arrangement of the $L/2 - 1$ 1-networks extended from Figure 4. It requires $5L/2 - 3$ adders and $L/2 + 3$ latches that are superior to Figure 3, even though its structure is more complex. As a result, the total execution time of this systolic array to compute N-point $\sum a_k(n)k$ $(n = 0, 1, \ldots , N - 1)$ is decreased to

$$L + 1 + 1 + N - 1 = N + L + 1$$

clock cycles.

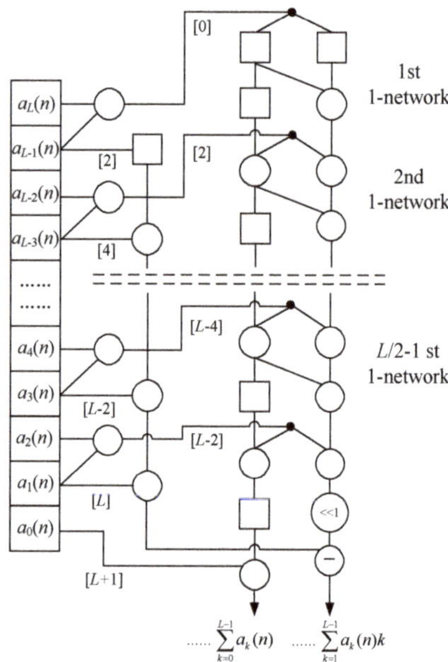

Figure 5. The improved systolic array for first-order moment.

4. The Fast Algorithm for Normalized Cross-Correlation

We apply the improved fast algorithm in Section 3.3 to compute the first-order and the zero-order moments in Equation (9). Thus, the fast algorithm for NCC is presented that can remove most of its multiplications. At first, some optimization methods are introduced in Section 4.1 to further reduce its additions.

4.1. The Optimization Methods

As the sequence { $g(i)$ } is a fixed correlation kernel in general, both $\overline{g} = \sum g(i)/N$ and $\sum [g(i) - \overline{g}]^2$ in Equation (9) could be pre-computed and reused for avoiding their repeated computations [30].

Although $b(n)$ in Equation (8) involves many additions and complex squares, it could also be computed by a simple function with the previous $b(n-1)$, where

$$\begin{aligned} b(n) &= \sum_{i=0}^{N-1} f(n-1+i)^2 + [f(n+N-1)^2 - f(n-1)^2] \\ &= b(n-1) + [f(n+N-1) + f(n-1)][f(n+N-1) - f(n-1)] \end{aligned} \tag{14}$$

We only need to directly compute the first $b(0)$ by N multiplication and $N - 1$ additions, where the square is performed by multiplication. Then, the following $b(n)$ $(n = 1, 2, \ldots , N - 1)$ would be obtained from Equation (14) by only 1 multiplication, 2 additions and 1 subtraction.

4.2. The Step of the Fast Algorithm for NCC

The proposed fast algorithm for NCC would include five steps:

Step 1 Initializing all $a_k(n) = 0$ ($k = 0, 1, \ldots, L$), where $a_0(n)$ is indispensable for $\sum a_k(n)$.

Step 2 Implementing Equation (3) to acquire the sequence $\{ a_k(n) \}$ using N addition.

Step 3 Computing $\sum a_k(n)$, $\sum a_k(n)k$ by Equation (13) and Figure 4 with $5L/2 - 1$ additions.

Step 4 Computing $b(n)$ by Equation (14) with 1 multiplication, 2 additions and 1 subtraction.

Step 5 Inputting $\sum a_k(n)$, $\sum a_k(n)k$ and $b(n)$ into Equation (9) for a NCC $\rho(n)$, which need 2 subtractions, 4 multiplications, 1 division and 1 square root calculation.

The computational flow of this algorithm is illustrated in Algorithm 2. It includes $N + 5L/2 + 1$ additions, 3 subtractions and 5 multiplications per output an NCC $\rho(n)$. Therefore, to compute N-point NCC, it requires $N - 1 + N (N + 5L/2 + 1) - 2 = N (N + 5L/2 + 2) - 3$ additions, and only $N + N - 1 + 4N = 6N - 1$ multiplications.

Algorithm 2 Computing NCC ($n, f, g, b(n\text{-}1)$)

---

  for each a_k in the sequence $\{ a_k \}$: $a_k \leftarrow 0$
  for each $i \in [0, N\text{-}1]$ do   // Equation (3)
    $k \leftarrow g(i)$
    $a_k \leftarrow a_k + f(n + i)$
  end for
  for each $k \in [1, L/2]$ do   // Equation (13a)
    $s \leftarrow s + a_{2k-1}$
    $a_k \leftarrow a_{2k-1} + a_{2k}$
  end for
  $a \leftarrow$ **Moment** ($a_{L/2}, a_{L/2-1}, \ldots, a_2, a_1, a_0$)   // Algorithm 1
  $a[1] \leftarrow a[1] << 1 - s$ // Equation (13b)
  Compute $b(n)$ by $b(n\text{-}1)$, $f(n + N - 1)$ and $f(n - 1)$ // Equation (14)
  Compute $\rho(n)$ by $a[0]$, $a[1]$ and $b(n)$ // Equation (9)
  return $\rho(n)$

---

5. The Systolic Array for Normalized Cross-Correlation

We apply the improved systolic array in Figure 5 to design a hardware structure for fast NCC in parallel. Figure 6 shows this systolic structure that mainly includes three parts: the module **A** to compute $\{ a_{2k-1}(n) + a_{2k}(n) \}$, the module **M** to compute the first-order and zero-order moment of $\{ a_k(n) \}$, and the module **S** to compute $b(n)$. In each cycle, we simultaneously input N-point $f(n + i)$ into this systolic array and get an NCC result $\rho(n)$. At first, since the direct computation for $\{ a_{2k-1}(n) + a_{2k}(n) \}$ needs many adders, a simplified structure for the module **A** is discussed in Section 5.1.

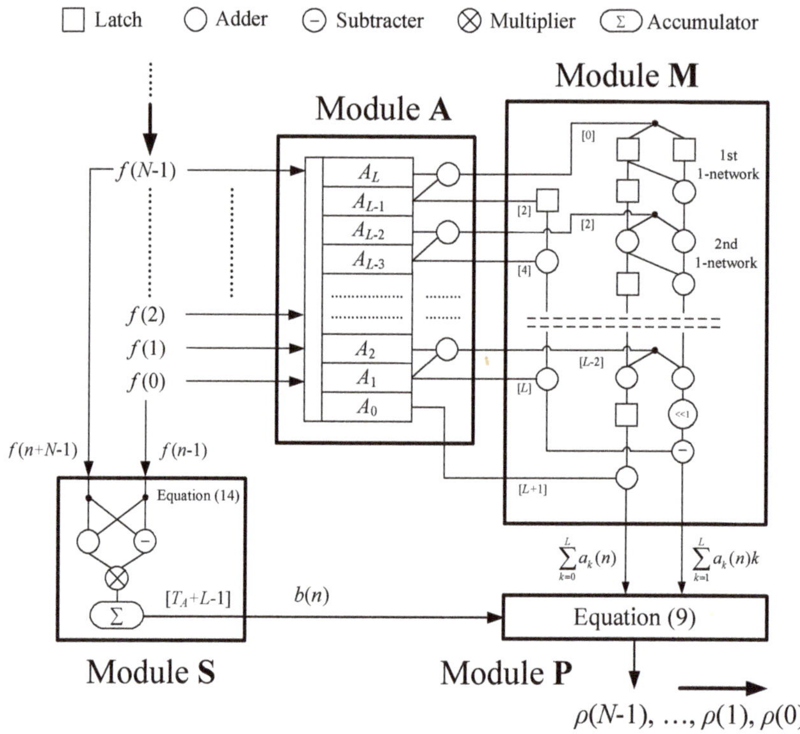

Figure 6. The systolic array for fast normalized cross-correlations (NCCs).

5.1. The Module A

The module **A** is to acquire an $L/2$-point sequence $\{\, a_{2k-1}(n) + a_{2k}(n)\,\}$ according to Equations (3) and (13) in every clock cycle. It includes $L + 1$ sub-modules A_k ($k = 0, 1, 2, \ldots, L$) that firstly count $\{\, f(n + i)\,\}$ to generate corresponding $\{\, a_k(n)\,\}$, and then sum up the two adjacent $a_k(n)$ to obtain $\{\, a_{2k-1}(n) + a_{2k}(n)\,\}$. We assume the execution time of the module **A** is T_A clock cycles. The N-point $f(n + i)$ should be inputted into the sub-modules $\{\, A_k\,\}$ in a gradual way.

Since the correlation kernel $\{\, g(i)\,\}$ is so invariable that the computational strategy for Equations (3) and (13) are known in advance, we could simplify the structure of A_k for less adder and data transfer. For example, for $N = 4, L = 4$ and $\{\, g(i)\,\} = \{\, 1, 2, 3, 4\,\}$, the module **A** could be simplified as shown in Figure 7 with 2 adder and $T_A = 1$. However, for $N = 4, L = 4$ and $\{\, g(i)\,\} = \{\, 2, 1, 4, 2\,\}$, the module A would be re-designed as shown in Figure 8 with 2 adder, 3 latches and $T_A = \log_2 4 = 2$. Therefore, the structure of the module **A** should be not fixed, but changed with different sequences $\{\, g(i)\,\}$ to reduce its hardware complexity. We also show the module **A** using maximum adders when $\{\, g(i)\,\} = \{\, 4, 4, 4, 4\,\}$ in Figure 9a, and the module **A** using 0 adders when $\{\, g(i)\,\} = \{\, 2, 4, 6, 8\,\}$ in Figure 9b. From Figures 7–9, it can be obtained the adder number of the module **A** is from 0 to $N - 1$, and the latency T_A is from 0 to $\log_2 N$.

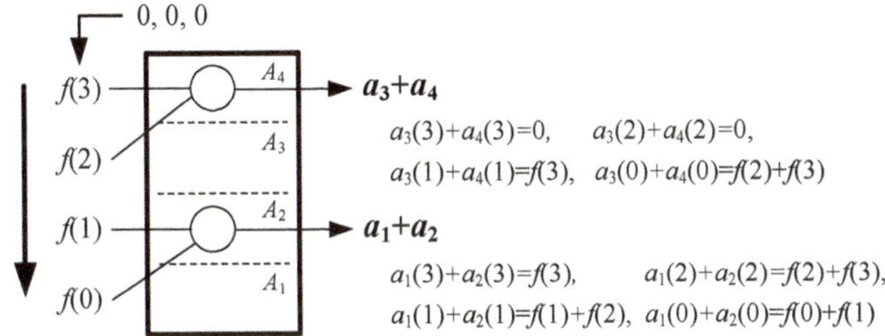

Figure 7. The module **A** for $\{\,g(i)\,\} = \{1, 2, 3, 4\}$.

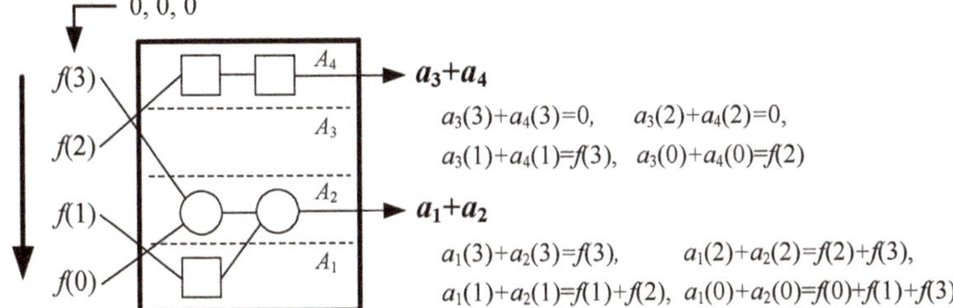

Figure 8. The module **A** for $\{\,g(i)\,\} = \{2, 1, 4, 2\}$.

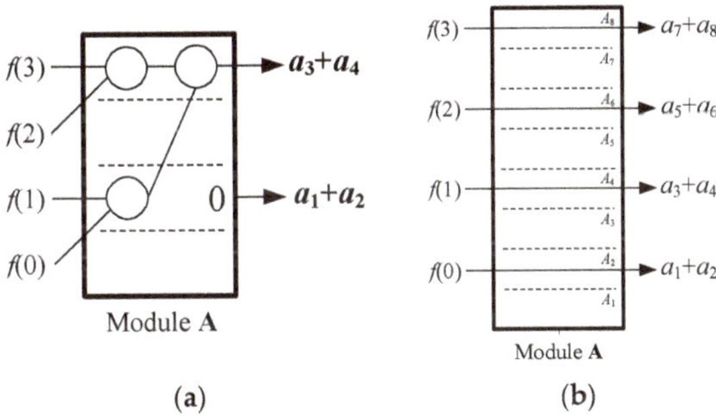

Figure 9. The module **A** using different adders: (**a**) $\{\,g(i)\,\} = \{4, 4, 4, 4\}$; (**b**) $\{\,g(i)\,\} = \{2, 4, 6, 8\}$.

5.2. The Model **P**

The Model **P** is to implement Equation (9) with 4 multipliers, 1 divider and 1 square root extractor. It receives a $\sum a_k(n)k$ and a $b(n)$, and output a corresponding $\rho(n)$ in each cycle. Some fast methods can be applied for the square root operation. In addition, the fixed $\overline{g}$ and $\sum [g(i) - \overline{g}]^2$ are saved in advance against repeated computation.

5.3. The Systolic Array

The systolic array in Figure 6 uses various modules to perform Equations (3), (9), (13) and (14), respectively, for NCC. Some latches are indispensable to connect these modules for assuring their mutual and parallel operation. The latch number has been shown in the note "[ ]". The module M from Figure 5 is to compute first-order moments and zero-order moments based on Equation (13). The module **S** implements Equation (14) and generates $b(n)$ by 1 multiplier, 1 accumulator and 1 subtractor. Finally, the module **P** generates NCC $\rho(n)$. The systolic array's total adder number is ranged from $2L - 2$ to $2L + N - 3$, and its multiplier number is 5.

The initial value of the accumulator in the module **S** is set as $b(0)$. In the n-th clock cycle, $f(n + N - 1)$ and $f(n - 1)$ would be input into the module **S** to get $b(n)$ with three clock cycles. Then

$b(n)$ is output from the module **S** to the module **P** with a latency $T_A + L - 1$. The aim is that $b(n)$, $\sum a_k(n)$ and $\sum a_k(n)k$ can arrive in the **P** at the same time.

6. Comparisons

The proposed algorithm and systolic structure are compared with some existing methods to verify their effectiveness. These compared methods are also focused on reducing their multiplication numbers.

6.1. Algorithm Comparison

Because correlation and convolution can share fast algorithms, we compare the proposed algorithm in Section 4 with some convolution algorithms, as well as a fast NCC algorithm to compute an N-point cyclic NCC. The computational complexity of these algorithms are displayed in Table 1, where we set a complex multiplication, which is equivalent to three real multiplications and three real additions, an "AND" operation is equivalent to an addition [31], and a subtraction is also equivalent to an addition.

From Table 1, the multiplication and addition complexity of the FFT-based algorithm are both $O(N \log_2 N)$, the DA-based algorithm is the least addition complexity, and the fast NCC algorithm has zero multiplication. The proposed algorithm uses $O(N^2)$ additions that are more than the FFT-based and the DA-based algorithm, and $O(N)$ multiplications that are more than the fast NCC algorithm. However, the FFT-based algorithm needs float addition and multiplication operations that are more complex than integer operations, the DA-based algorithm requires tedious decode address and very large memories, as well as that the fast NCC algorithm is the most addition complexity and not suitable for high-precision matching [15]. Figure 10 shows the four algorithms' multiplication and addition number increasing along with N. It is obviously that the proposed algorithm's multiplication number is lower than both the FFT-based algorithm's and the DA-based algorithm's, and its addition number is lower than the fast NCC algorithm's when $N > 320$.

Table 1. The compassions of computational complexity.

Algorithm	Multiplication	Addition
Direct calculation	$2N(N + 1)$	$3N(N + 1)$
FFT-based algorithm [8,9]	$(3/2)\,N\log_2 N - (3/2)\,N + 16$	$(7/2)\,N\log_2 N - N/2 + 15$
DA-based algorithm [22]	$7N - 1$	$(5N - 2)\log_2 L + 8N - 1$
Fast NCC algorithm [15]	0	$3N(N + 1)$
The proposed algorithm	$6N - 1$	$N(N + 5L/2 + 5) - 4$

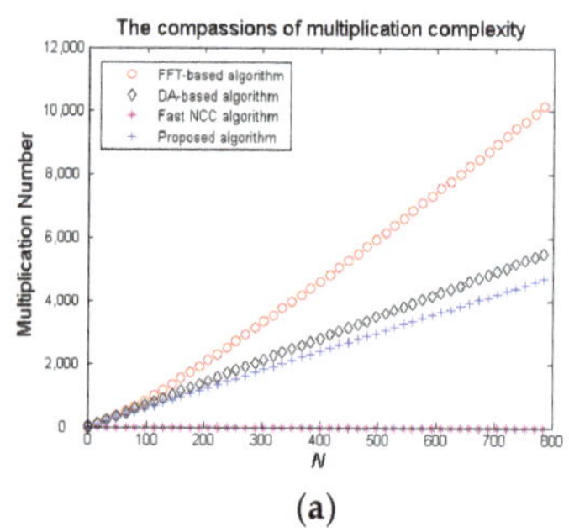
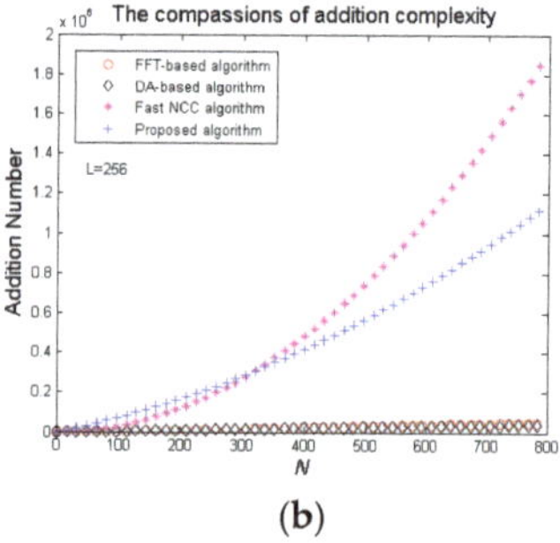

Figure 10. The four algorithm's multiplication and addition number: (**a**) Multiplication (**b**) Addition.

The wireless sensor and communication is an important application field for the proposed algorithm. Therefore, we compare the execution time of the five algorithms from Table 1 by using a mobile phone with the type "HUAWEI nova 2s (HWI-AL00)" and the operation system "Android 9". Figure 11 shows these algorithms' execution time to compute a cyclic NCC by the phone with N

from 100 to 6000. The growth curve of the FFT-based algorithm's time is similar to a step curve, in that the length of FFT needs to be extended from N to $2^{\lceil \log_2 N \rceil}$. Although the DA-based algorithm can use the least time, it needs too much memory to make it worthwhile. From the Figure 11, the proposed algorithm's execution time is less than the FFT-based algorithm's when $N < 5500$, and is very close to the fast NCC algorithm, but not involved with noise.

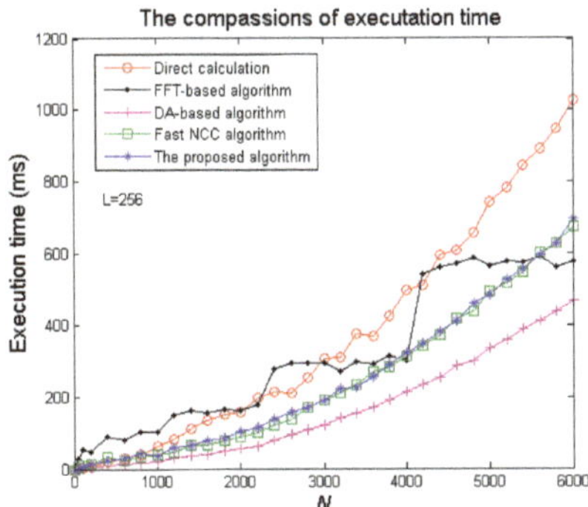

Figure 11. The comparisons of the five algorithm's execution time (ms).

In addition, it is important that the proposed algorithm has five advantages, as follows:

(1)   With less multiplications and memory.
(2)   Simple computational structure due to its simple implementation.
(3)   Precision and Fit to discrete domain as it uses integer operations [32].
(4)   Without limitations on the length of NCC.
(5)   Implementation by simple systolic structure.

6.2. Structure Comparison

We compare the proposed systolic array in Section 5 with some existing hardware structures. Table 2 shows the hardware complexity of these structures to implement an N-point cyclic NCC, where $N = PM$ (P and M are two positive integers derived from [33]). Because the proposed array's adder number and latency are not fixed, but varied with the sequence { $g(i)$ }, we only display their value range according to Section 5.1. The execution time of the model **P** is assumed as three clock cycles.

Table 2. The compassions of hardware complexity.

Complexity	The Proposed Structure	Structure in [22]	Structure in [33]
ROM number	0	$[(N-1)/2] \, [2^{3(N-1)/7}]$	$2^M P \log_2 L$
Adder Number	$2L-2$ To $2L+N-3$	$2N$	$(P+1)\log_2 L$
Latency	$L+5$ To $\log_2 N + L + 5$	$2\log_2 L$	$\log_2 L + P$
Throughput	1	$N/2 \log_2 L$	1

From Table 2, it is an advantage that the proposed systolic structure does not need ROMs, while the other two structures use $O(2^N)$ ROMs that are hardware-expensive when $N > 16$. The structure [22] has minimum latency, but its throughput is more than 1. The structure [33] needs the $O(P)$ adder and latency that would increase rapidly with N.

The proposed structure's hardware complexity is dependent upon L. Furthermore, for long NCCs, or two-dimension NCCs when N and P are larger than L, the adder number of the proposed structure is lower than that of the structure [22], and the latency of the proposed structure is lower than that of the structure [33]. Figure 12 shows the three structures' adder number and latency increasing along with N, where the proposed structure adopts maximum adder and latency to perform comparisons. It is obvious that the proposed structure's adder number is least when $N > 1800$, and its latency is lower than the structure [33] when $N > 1500$. Therefore, although additional $O(L)$ latches are required for data store and transfer, the proposed systolic array could be more efficient in digital signal and image domain where the maximum value of L is less than 256 in general [34].

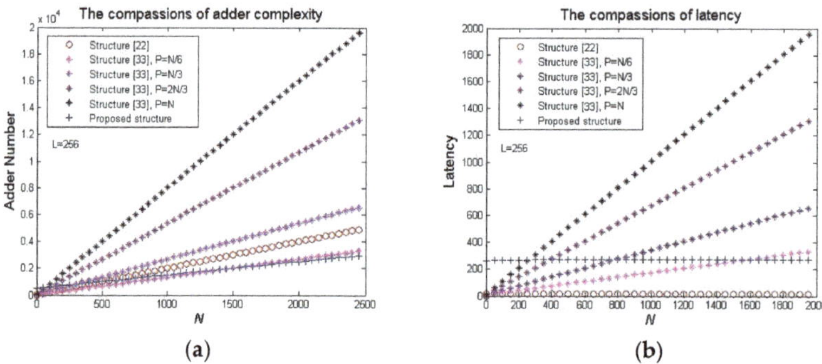

Figure 12. The three structure's adder number and latency: (**a**) Adder (**b**) Latency.

7. Conclusions

It is suggested that digital NCCs be implemented by efficient algorithms and hardware structures for decreasing their high multiplication complexity [35]. With the assist of fast computation for first-order moment, this paper presents an algorithm and a systolic array for fast NCCs that aim to reduce multiplication as much as possible. To do this, the key is to transform the complex inner-product in the NCC into a simple first-order moment according to the statistical properties of the digital inner-product, and then a new NCC formula based on a first-order moment is established in order for eliminating inner-product operations. As a result, by introducing an algorithm without multiplication into the computation of the first-order moment in NCC, we proposed a fast algorithm for NCC with the advantages of simple implementation, less multiplication, no length limitation, and so on. Especially, as the introduced algorithm for first-order moment requests many additions, we also improved it by means of an even-odd relationship to reduce addition complexity and execution time. It is an advantage that the introduced algorithm for the first-order moment can be implemented by systolic structure, so a systolic array composed of latches and adders is designed for implementing fast NCC in parallel. This systolic array is hardware-efficient due to its parallel operation, simple structures and seldom multiplier. This paper analyzes the computational and the hardware complexity for the proposed algorithm and systolic array, and compares them with some existing methods to prove their efficiency. The proposed algorithm and array could also be applied for digital filter and various transforms [36].

There are still many additions in the proposed algorithm and systolic structure. Future studies will focus on further reducing their additions.

Author Contributions: The work presented in this paper was completed with collaboration among all authors. Conceptualization, C.P.; Methodology, C.P., Z.L. and X.H.; Formal analysis, Z.L.; Writing—original draft preparation, C.P.; Writing-review and editing, Z.L., X.H., H.L.; Supervision, H.L.; Funding acquisition, X.H. All authors have read and agreed to the published version of the manuscript.

Funding: This research was funded by the National Natural Science Foundation of China (Grant No. 61801337, 61572012), the Natural Science Foundation of Hubei Province of China (Grant No. 2018CFB661, 2017CFB677).

Conflicts of Interest: The authors declare no conflicts of interest.

References

1. Annaby, M.H.; Fouda, Y.M.; Rushdi, M.A. Improved Normalized Cross-Correlation for Defect Detection in Printed-Circuit Boards. *IEEE Trans. Semicond. Manuf.* **2019**, *32*, 199–211. [CrossRef]
2. Duong, D.H.; Chen, C.S.; Chen, L.C. Absolute Depth Measurement Using Multiphase Normalized Cross-Correlation for Precise Optical Profilometry. *Sensors* **2019**, *19*, 1–20. [CrossRef] [PubMed]
3. Banharnsakun, A. Feature point matching based on ABC-NCC algorithm. *Evol. Syst.* **2018**, *9*, 71–80. [CrossRef]
4. Kotenko, I.; Saenko, I.; Branitskiy, A. Applying Big Data Processing and Machine Learning Methods for Mobile Internet of Things Security Monitoring. *J. Internet Serv. Inf. Secur.* **2018**, *8*, 54–63.
5. Sridevi, M.; Sankaranarayanan, N.; Jyothish, A.; Vats, A.; Lalwani, M. Automatic traffic sign recognition system using fast normalized cross correlation and parallel processing. In Proceedings of the 2017 International Conference on Intelligent Communication and Computational Techniques, Jaipur, India, 22–23 December 2017; pp. 200–204.
6. Bovik, A.C. Basic Tools for Image Fourier Analysis. In *The Essential Guide to Image Processing*; Academic: San Diego, CA, USA, 2009.
7. Wu, P.; Li, W.; Song, W.L. Fast, accurate normalized cross-correlation image matching. *J. Intell. Fuzzy Syst.* **2019**, *37*, 4431–4436. [CrossRef]
8. Liu, G.Q.; Kreinovich, V. Fast convolution and Fast Fourier Transform under interval and fuzzy. *J. Comput. Syst. Sci.* **2010**, *76*, 63–76. [CrossRef]
9. Kaso, A.; Li, Y. Computation of the normalized cross-correlation by fast Fourier transform. *PLoS ONE* **2018**, *13*, e0203434. [CrossRef]
10. Narasimha, M.J. Linear Convolution Using Skew-Cyclic Convolutions. *IEEE Signal. Process. Lett.* **2010**, *14*, 173–176. [CrossRef]
11. Cheng, L.Z.; Jiang, Z.R. An efficient algorithm for cyclic convolution based on fast-polynomial and fast-W transforms. *Circuits Syst. Signal. Process.* **2001**, *20*, 77–88.
12. Li, H.; Lee, W.S.; Wang, K. Immature green citrus fruit detection and counting based on fast normalized cross correlation (FNCC) using natural outdoor colour images. *Precis. Agric.* **2016**, *17*, 678–697. [CrossRef]
13. Tsai, D.M.; Lin, C.T. Fast normalized cross correlation for defect detection. *Pattern Recognit. Lett.* **2003**, *24*, 2625–2631. [CrossRef]
14. Byard, K. Application of fast cross-correlation algorithms. *Electron. Lett.* **2015**, *51*, 242–244. [CrossRef]
15. Yoo, J.C.; Choi, B.D.; Choi, H.K. 1-D fast normalized cross-correlation using additions. *Digit. Signal. Process.* **2010**, *20*, 1482–1493. [CrossRef]
16. Ismail, L.; Guerchi, D. Performance Evaluation of Convolution on the Cell Broadband Engine Processor. *IEEE Trans. Parallel Distrib. Syst.* **2011**, *22*, 337–351. [CrossRef]
17. Chaudhari, R.E.; Dhok, S.B. An Optimized Approach to Pipelined Architecture for Fast 2D Normalized Cross-Correlation. *J. Circuits Syst. Comput.* **2019**, *28*, 1950211. [CrossRef]
18. Mehendale, M.; Sharma, M.; Peher, P.K. DA-Based Circuits for Inner-Product Computation. In *Arithmetic Circuits for DSP Application*; John Wiley & Sons, Inc.: Hoboken, NJ, USA, 2017; pp. 77–112.
19. Cao, L.; Liu, J.G.; Xiong, J.; Zhang, J. Novel structures for cyclic convolution using improved first-order moment algorithm. *IEEE Trans. Circuits Syst. I Regul. Pap.* **2014**, *61*, 2370–2379. [CrossRef]
20. Carranza, C.; Llamocca, D.; Pattichis, M. Fast 2D Convolutions and Cross-Correlations Using Scalable Architectures. *IEEE Trans. Image Process.* **2017**, *26*, 2230–2245. [CrossRef]
21. Meher, P.K. Efficient Systolization of Cyclic Convolutions Using Low-Complexity Rectangular Transform Algorithms. In Proceedings of the 2007 International Symposium on Signals, Circuits and Systems ISSCS '07, Iasi, Romania, 13–14 July 2007; pp. 1–4.
22. Chen, H.C.; Guo, J.I.; Chang, T.S.; Jen, C.W. A Memory-Efficient Realization of Cyclic Convolution and Its Application to Discrete Cosine Transform. *IEEE Trans. Circuits Syst. Video Technol.* **2005**, *15*, 445–453. [CrossRef]
23. Syed, N.A.A.; Meher, P.K.; Vinod, A.P. Efficient Cross-Correlation Algorithm and Architecture for Robust Synchronization in Frame-Based Communication Systems. *Circuits Syst. Signal. Process.* **2018**, *37*, 2548–2573. [CrossRef]

24. Vun, C.H.; Premkumar, A.B.; Zhang, W. A New RNS based DA Approach for Inner Product Computation. *IEEE Trans. Circuits Syst. I Regul. Pap.* **2013**, *60*, 2139–2152. [CrossRef]
25. Liu, J.G.; Pan, C.; Liu, Z.B. Novel Convolutions using First-order Moments. *IEEE Trans. Comput.* **2012**, *61*, 1050–1056. [CrossRef]
26. Hua, X.; Liu, J.G. A Novel Fast Algorithm for the Pseudo Winger-Ville Distribution. *J. Commun. Technol. Electron.* **2015**, *60*, 1238–1247. [CrossRef]
27. Liu, J.G.; Liu, Y.Z.; Wang, G.Y. Fast Discrete W Transforms via Computation of Moments. *IEEE Trans. Signal. Process* **2005**, *53*, 654–659. [CrossRef]
28. Yazdanpanah, H.; Diniz, P.S.R.; Lima, M.V.S. Low-Complexity Feature Stochastic Gradient Algorithm for Block-Lowpass Systems. *IEEE Access* **2019**, *7*, 141587–141593. [CrossRef]
29. Viticchie, A.; Basile, C.; Valenza, F.; Lioy, A. On the impossibility of effectively using likely-invariants for software attestation purposes. *J. Wirel. Mob. Netw. Ubiquitous Comput. Dependable Appl.* **2018**, *9*, 1–25.
30. Adhikari, G.; Sahu, S.; Sahani, S.K.; Das, B.K. Fast normalized cross correlation with early elimination condition. In Proceedings of the 2012 International Conference on Recent Trends in Information Technology, Chennai, India, 19–21 April 2012; pp. 136–140.
31. Blahut, R.E. *Fast Algorithms for Digital Signal Processing*; Addison-Wesley: Reading, MA, USA, 1984.
32. Yuan, Y.; Qin, Z.; Xiong, C.Y. Digital image correlation based on a fast convolution strategy. *Opt. Lasers Eng.* **2017**, *97*, 52–61. [CrossRef]
33. Meher, P.K.; Park, S.Y. A novel DA-based architecture for efficient computation of inner-product of variable vectors. In Proceedings of the 2014 IEEE International Symposium on Circuits and Systems, Melbourne, Australia, 1–5 June 2014; pp. 369–372.
34. Mukherjee, D.; Mukhopadhyay, S. Fast Hardware Architecture for 2-D Separable Convolution Operation. *IEEE Trans. Circuits Syst. II Exp. Briefs* **2018**, *65*, 2042–2046. [CrossRef]
35. Yang, Y.J.; Zhang, Y.H.; Li, D.M.; Wang, Z.J. Parallel Correlation Filters for Real-Time Visual Tracking. *Sensors* **2019**, *19*, 1–22. [CrossRef]
36. Hartl, A.; Annessi, R.; Zseby, T. Subliminal Channels in High-Speed Signature. *J. Wirel. Mob. Netw. Ubiquitous Comput. Dependable Appl.* **2018**, *9*, 30–53.

© 2020 by the authors. Licensee MDPI, Basel, Switzerland. This article is an open access article distributed under the terms and conditions of the Creative Commons Attribution (CC BY) license (http://creativecommons.org/licenses/by/4.0/).

Letter

A Low-Power WSN Protocol with ADR and TP Hybrid Control

Chung-Wen Hung *, Hao-Jun Zhang, Wen-Ting Hsu and Yi-Da Zhuang

Department of Electrical Engineering, National Yunlin University of Science and Technology, 123 University Road, Section 3, Douliou, Yunlin 64002, Taiwan; m10512027@yuntech.edu.tw (H.-J.Z.); m10712051@yuntech.edu.tw (W.-T.H.); b10512118@yuntech.edu.tw (Y.-D.Z.)
* Correspondence: wenhung@yuntech.edu.tw

Received: 31 August 2020; Accepted: 10 October 2020; Published: 12 October 2020

Abstract: Most Internet of Things (IoT) systems are based on the wireless sensor network (WSN) due to the reduction of the cable layout cost. However, the battery life of nodes is a key issue when the node is powered by a battery. A Low-Power WSN Protocol with ADR and TP Hybrid Control is proposed in this paper to improve battery life significantly. Besides, techniques including the Sub-1GHz star topology network with Time Division Multiple Access (TDMA), adaptive data rate (ADR), and transmission power control (TPC) are also used. The long-term testing results show that the nodes with the proposed algorithm can balance the communication quality and low power consumption simultaneously. The experimental results also show that the power consumption of the node with the algorithm was reduced by 38.46-54.44% compared with the control group. If using AAA battery with 1200 mAh, the node could run approximately 4.2 years with the proposed hybrid control algorithm with an acquisition period of under 5 s.

Keywords: adaptive data rate (ADR); transmit power control (TPC); time division multiple access (TDMA); wireless sensor network (WSN); power consumption; Internet of Things (IoT)

1. Introduction

The wireless sensor network (WSN) is the one of bases in Internet of Things (IoT), and most nodes in the WSN are powered by battery. Extending battery life, saving the maintenance fee, and raising system reliability are the motivations of this paper. In addition to the battery technology improvement and power capacity increase, the low-power technology of the device is also significant. In many applications, IoT device's security and power consumption are significant issues [1]. The battery power is usually used in the application of IoT devices, and the battery life is a troublesome problem [2]. Therefore, achieving extremely low power consumption on battery IoT devices has also become a big challenge.

Moreover, wireless transmission is the highest power-consuming process in communication devices. A study pointed out that the power consumption of nodes in the wireless sensing network is mostly concentrated in the process of wireless communication [3]. Therefore, maintaining reliable communication quality and reducing the power consumption of the device through wireless communication optimization is the focus of this paper. A WSN structure-integrated time division multiple access (TDMA), transmission power control (TPC), and adaptive data rate (ADR) are proposed in this paper to reduce the power consumption of wireless communication.

This paper is based on previously published research by this paper's authors, which has discussed the relationship and performance analysis of the transmission power and data rate [4]. This paper then implements an ultra-low-power WSN IoT with transmission power and data rate hybrid control and introduces the details. The predecessor of this paper aimed to evaluate the performance difference between the transmission power and the data rate with the same packet error rate (PER) of 1% through sensitivity measurement, thereby achieving a comparison basis of hybrid control between data rate and transmission power [4]. In this paper, TDMA scheduling detail, data rate parameter settings, low current selection of transmission power, a simplified hybrid control algorithm, and practical application are discussed. Finally, this paper also places multiple sensing nodes and measures the energy-saving effect and PER state of the nodes, and the proposed hybrid control algorithm is expected to achieve reliable wireless communication and extremely low power consumption simultaneously.

In summary, this paper provides the following contributions:

1. We propose a hybrid control algorithm combined with TPC and ADR that could adapt the environmental interferences.
2. Experimental results analysis show that the proposed algorithm achieved energy-saving with stable communication quality.

2. Related Research

The architecture of the WSN in IoT application and the selection of communication frequency bands have been discussed in the following studies. The designs for power consumption reduction in the wireless network, such as media access control (MAC), transmission power, and data rate control, have also been also discussed in the following literature.

There are two data processing methods that have been proposed, centralization and distribution data fusion, which each have different benefits. Centralized data fusion processes all the data on a central node, while the nodes in distributed system process their own data [5]. In order to keep high maintainability and easy data processing, we adapted star topology network in this paper.

Compared with the 2.4 GHz or 5 GHz frequency band, the transmission distance of the Sub-1 GHz wireless communication is farther, so its coverage is wider and its power consumption is lower [6–8]. The method of multinodes communication in the WSN includes competition-based [9–12] and scheduling-based MAC [13–16]. Compared with competition-based MAC, scheduling-based MAC network throughput is not good, but the design of scheduling-based MAC is simpler. The authors of [17] proposed a TDMA structure-based MAC protocol for short and long-range networks, and the sensor node can run for about 3 years. TDMA could reduce dramatic the power consumption of the WSN, as shown in [17]. However, the proposed algorithm in this paper including ADR and TP control would lower the power requirement further. Considering the system complexity and low-power design, TDMA was adopted in this paper.

In wireless communication, transmission power is a major factor for power consumption. A method called TPC minimizes the transmission power as possible when the communication quality can be maintained [16–24]. If the environment is better, TPC can achieve more power-saving effects. In addition to the transmission power, the data rate is also a major factor affecting power consumption. In the case of packet transmission, if the transmission data rate is faster, the wireless transmission time is shorter. However, the cost of a faster data rate is transmission quality reduction. The appropriate data rate is selected based on the relative relationship between the frame delivery ratio (FDR) and the received signal strength indicator (RSSI) [25–27]. Sodhro, A.H. et al., [28] proposed an energy-efficient transmission power control (ETPC) algorithm that was based on a wireless channel estimation. The channel estimation is an important issue for communication quality and power consumption. The results of the wearable electrocardiogram, demonstrated in [28], have been validated successfully and make a great contribution in energy-saving and signal-processing. However, the complex control should be considered. The methods of combining data rate and transmission power control to achieve power saving were proposed in [29,30]. Due to the fact that ADR and TP hybrid control is complex, and there are a lack of lectures discussing it, there are research gaps to be covered in this paper. The authors of [31] also considered DSSS and MFSK for communication quality and power consumption. However, the proposed algorithm was based on the decided communication to select a suitable combination of DSSS and MFSK. In order to adapt the environmental interference, a data rate and transmission power-integrated control algorithm was proposed in this paper. The control algorithm can adapt to the environment in which the sensor node is located and choose the best data rate and transmission power. However, the previous works have discussed the TP and ADR control individually, but there is a lack of discussion on combination control due to its complexity. Moreover, because ADR control under TDMA is more difficult because of the complicated communication handshaking, few researchers have discussed the topic. In this paper, the TP or ADR combination control TDMA are detailed.

3. System Implementation

3.1. System Architeture

Figure 1 is the implementation platform of Texas Instruments CC430F6137. The network architecture is a simple star scheme, as shown in Figure 2, and the frequency-shift keying (FSK)/Gaussian frequency-shift keying (GFSK) was selected as the radio modulation method. The network consisted of multiple sensing nodes and a central bridge. The bridge was supplied by grid power and the sensor nodes were supplied by the battery. The placements of the bridge and nodes are shown in Figure 3. The bridge received packet from the nodes, and it was connected by cable to a gateway. The exact position is shown in Section 6.1. The TDMA protocol was proposed in this system, and the data rate and transmission power control algorithms were used to reduce the power consumption of the node to extend the battery life.

3.2. TDMA Protocol

Figure 4 is a TDMA diagram proposed in this paper. The concept is to predetermine the communication time slot and turn off the wireless communication function when nodes are idle. The nodes enter the sleep mode for the rest of the time to save power. First, Trigger Time (TTrig.) is the synchronous trigger signal sent by the bridge. Next, Sensing Time (Tsensing) is the time interruption required by the node for sensing, and the time length depends on the processing time required by the installed sensors. Waiting for Require Time (TWRn) is the interval required by the nth node to wait for a bridge command. Response Time (TRes.n) is required to reply to the sensing information packet. Moreover, Delay Time (TDelay) is the slot for packet parsing, wireless communication reception, transmission mode switching, and radio wave calibration for the nodes and bridges. Finally, Acquisition Time (TAcq.) indicates the period from the start of the triggering to the end of the node polling.

The node ID is assigned before the network constructed, and the sequence of the TDMA time slot is dependent on this ID. When a connected sensor node is turned off, the bridge skips the slot after a couple reconnections. The reservation slot is reserved in the last portion of acquisition duration, and it is reserved for the connection of a new sensor node with lowest data rate and highest power. If the node is checked by the bridge, the time slot will be arranged into the dedicated ID slot in acquisition duration.

4. Measurement and Analysis of Radio Frequency

4.1. Parameters of Data Rate

The crystal oscillator used in this paper had a crystal oscillator error of 10 ppm. Texas Instruments provided the SmartRF Studio tool, which can set and select the appropriate frequency deviation and the receiving channel bandwidth at a specific data rate. The node update period of 5 s was used in this paper, the design goal was 255 nodes. Therefore, the slowest data rate could only be set to 26 kbps. In view of the above factors, the data rate of this paper was divided into nine segments, from fastest (250 kHz) to slowest (50 kHz), to obtain the parameters in Table 1. The paper adopted this method to set the required data rate and the correlation coefficient.

Figure 1. CC430F6137 development board.

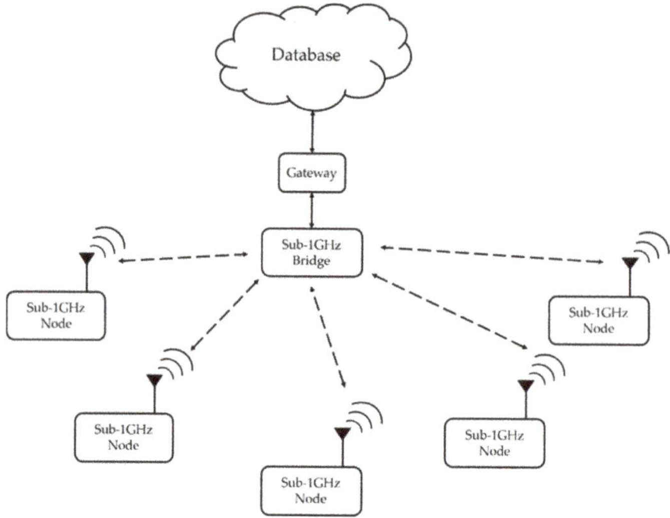

Figure 2. Star network architecture.

(**a**) (**b**)

Figure 3. (**a**) The placement of the bridge; (**b**) the placement of node 2 and node 7.

4.2. Receiver Sensitivity and Transmission Time in Different Data Rate

A sensitivity experiment was proposed in this paper, and the detail is shown in Figure 5. Two CC430 RF devices were used in this experiment: The transmitter and the receiver. On the transmitter, adjusting the transmission power is used to change the RSSI, 1000 packets are sent in a fixed data rate, and then average RSSI and PER are calculated on the receiver. In this paper, the corresponding RSSI when the PER was 1% was called sensitivity, and each data rate had a sensitivity. The relationship between data rate, RSSI, and PER is shown in Figure 6. The RSSI closest to one percent PER at each data rate was taken as the receiver sensitivity of the data rate, as shown in Table 2. The above Figure 6 relationship and Table 2 sensitivity table were used to analyze the wireless performance of different data rates.

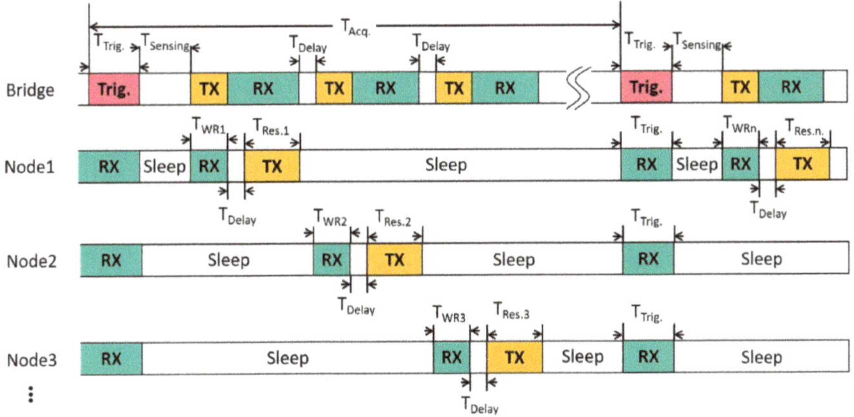

Figure 4. Diagram of TDMA architecture.

Table 1. Parameters of different data rates.

Data Rate	Frequency Deviation	RX BW
250 kbps	126.953125 kHz	541.666667 kHz
225 kbps	114.257812 kHz	464.285714 kHz
200 kbps	101.562500 kHz	406.250000 kHz
175 kbps	88.867188 kHz	406.250000 kHz
150 kbps	76.171875 kHz	325.000000 kHz
125 kbps	63.476562 kHz	270.833333 kHz
100 kbps	50.781250 kHz	232.142857 kHz
75 kbps	38.085938 kHz	162.500000 kHz
50 kbps	25.390625 kHz	116.071429 kHz

4.3. Current Consumption in Different Transmission Power

The transmission powers of 121 segments are provided by the CC430F6137, Sub-1GHz wireless communication chip, while only 41 segments were selected in this paper. Due to the transmission power table provided by the original manufacturer, the actual value did not correspond to the 920 MHz band used in this paper. Therefore, the 121-segment transmission power values were measured at 920 MHz by the Rohde & Schwarz RTO2044 digital oscilloscope with a bandwidth of 4 GHz and a sampling rate of 20 GSa/s, and the more suitable 41 segment settings were chosen in this paper.

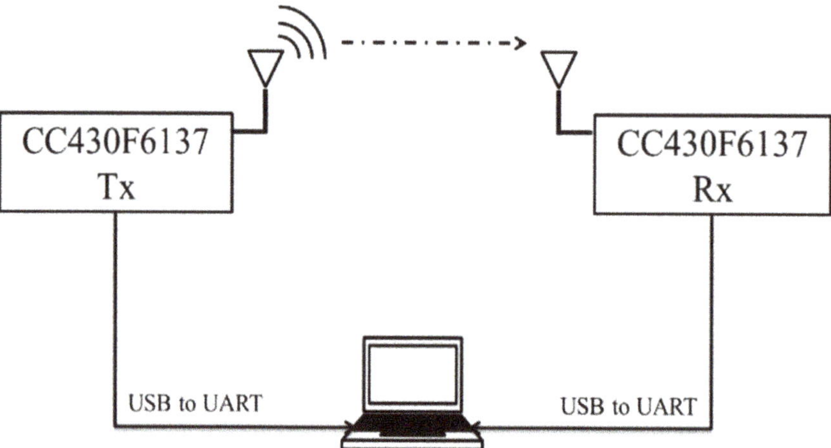

Figure 5. Experimental architecture of receiver sensitivity.

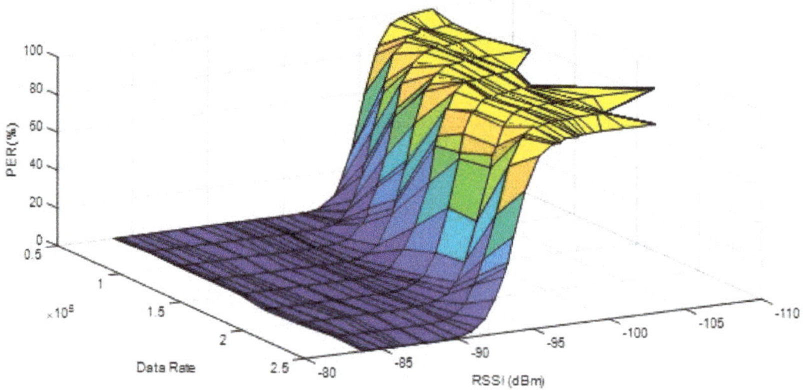

Figure 6. The relation between RSSI, data rate, and PER.

Table 2. RSSI when the PER was 1% at different data rates.

Data Rate	Sensitivity
50 kbps	−96.93 dBm
75 kbps	−95.22 dBm
100 kbps	−94.36 dBm
125 kbps	−93.69 dBm
150 kbps	−93.12 dBm
175 kbps	−91.96 dBm
200 kbps	−91.53 dBm
225 kbps	−90.25 dBm
250 kbps	−90.06 dBm

Figure 7 is a chart comparing the measurement results of the current consumption corresponding to each transmission power with the original manufacturer. When the transmission power was larger, the difference between the values of the datasheet and the measurements was larger. In the interval where the transmit power was −9 dBm to 6 dBm, regardless of the values of datasheet or the measurements, the consumption variation was suddenly increased. Therefore, the appropriate 41 segment settings from the set value of 121 were selected, as shown in Table 3.

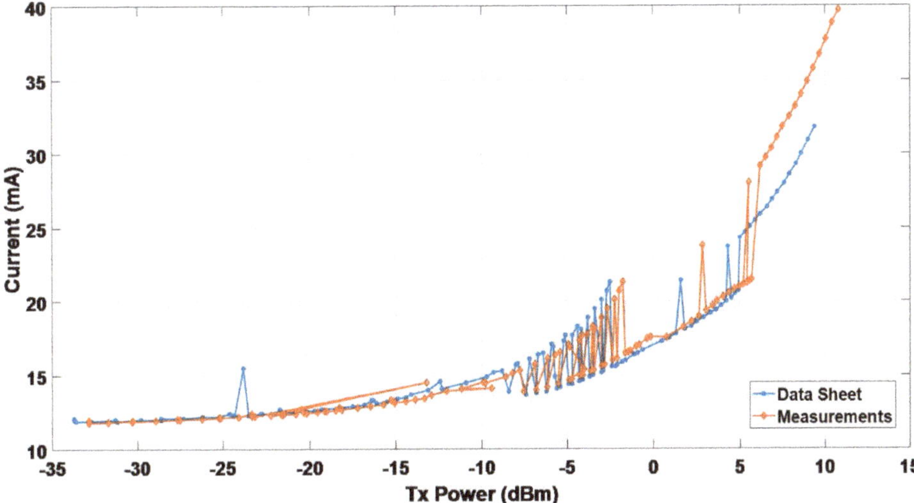

Figure 7. Comparison of the datasheet and measured values in the 121 segment transmission power.

Table 3. Current consumptions in 41-segment transmission power.

Transmit Power (dBm)	Current (mA)	Transmit Power (dBm)	Current (mA)	Transmit Power (dBm)	Current (mA)
10.062	37.739	−1.3157	16.637	−15.688	13.005
9.3152	35.763	−2.0975	16.144	−17.207	12.791
8.2542	33.223	−2.9932	15.633	−18.484	12.612
7.1839	31.071	−4.0892	15.027	−19.539	12.509
6.8546	30.347	−4.7187	14.729	−20.784	12.395
6.2026	29.153	−5.5103	14.42	−21.604	12.356
5.5377	28.042	−6.8004	14.012	−22.212	12.299
4.3618	20.562	−7.5317	13.913	−23.3	12.246
3.6703	20.005	−8.5849	14.893	−24.073	12.173
3.0454	19.349	−9.7606	14.474	−25.125	12.116
2.2062	18.613	−11.155	14.077	−26.202	12.059
0.78139	17.556	−12.904	13.65	−27.653	12.021
−0.14598	17.583	−13.856	13.325	−28.899	11.941
−0.98536	16.942	−14.407	13.219		

4.4. Total Power Consumption of Data Rate and Transmission Power

In Equation (1), I(TxPower) is the current consumption of the transmit power, and the relationship between the transmit power, data rate, and power consumption is shown in Figure 8 on the condition that the packet length is 33 Bytes. Figure 8 results were used for the method of energy efficiency comparisons in Section 5.

$$Power\ Consumption = I(TxPower) \times Bit\ number \times \frac{1}{Data\ Rate} \tag{1}$$

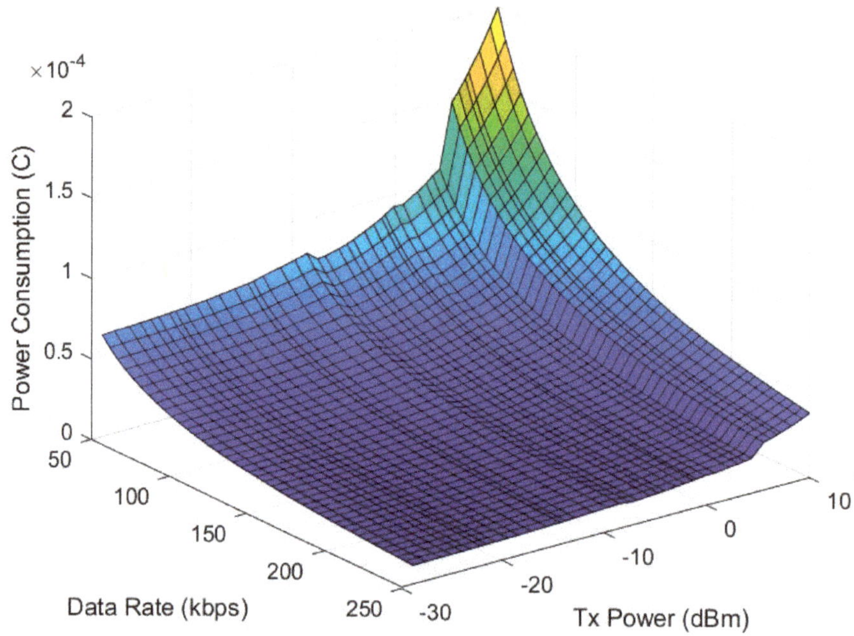

Figure 8. Relation of data rate, transmission power, and power consumption.

5. Control Algorithm

Algorithm of Transmission Power and Data Rate Hybrid Control

The transmission power and data rate hybrid control algorithm was proposed in this paper, and this algorithm was used to balance wireless quality and low power consumption to achieve a PER less than 1% and more power-saving. The control architecture between the bridge and sensor nodes is shown in Figure 9, and the hybrid control algorithm was run in Bridge. In Figure 9, the bridge received the RSSI feedback from sensor nodes' transmission signals and calculated the PER using the packet error interval algorithm. After the hybrid control algorithm was complete, the new transmission power and data rate control command that generated by adaptive algorithm was sent to the sensor nodes from the bridge. The detailed flow of the hybrid control algorithm is shown in Figure 10 and its pseudocode is shown in Figure 11. In this system, input includes the real-time RSSI feedback, PER record, sensitivity table of different data rates, and power consumption table, and output includes the data rate control and transmission power control.

The communication quality target of this paper was set at a PER below 1%. In the algorithm of the packet error interval, the threshold of data rate and transmission power is adjusted by 128 packet durations. If there are no errors in the continuous 128 packets, it means that the PER is less than 1%. The data rate will be increased, or the transmission power will be reduced. However, if there is only one incorrect packet in a 128-packet period, it means that the PER is 1%, and the data rate and transmission power will not be changed.

Then, if two errors have occurred before 128 packets have been completed, the data rate will be reduced or the transmission power will be increased. In this algorithm, when the error interval is short, the larger amplitude of the transmission power is set, because it is necessary to react immediately when an error occurs. Otherwise, when the error interval is long, the lower amplitude of the transmission power is adjusted, because the environment may be stabilized and the error does not occur easily.

Finally, by the error interval method, we determined how much to set the N-grade of data rate and transmission power for the next transmission. After the above N-order adjustment, based on the database of Figures 6 and 8, the lowest power consumption combination of data rate and transmission power was selected as the result of the final adjustment. The method flow is shown in Figure 12.

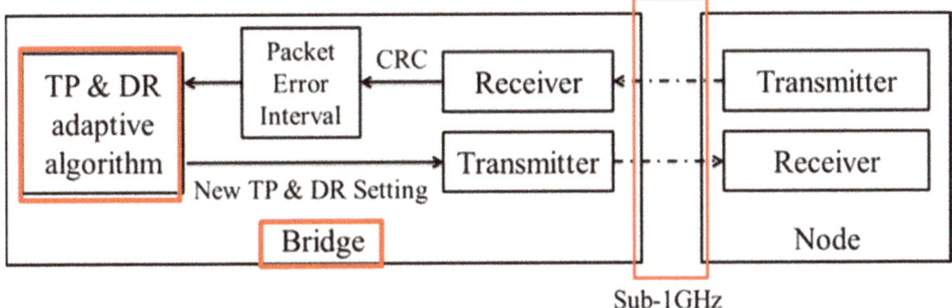

Figure 9. Control architecture between the bridge and sensor nodes.

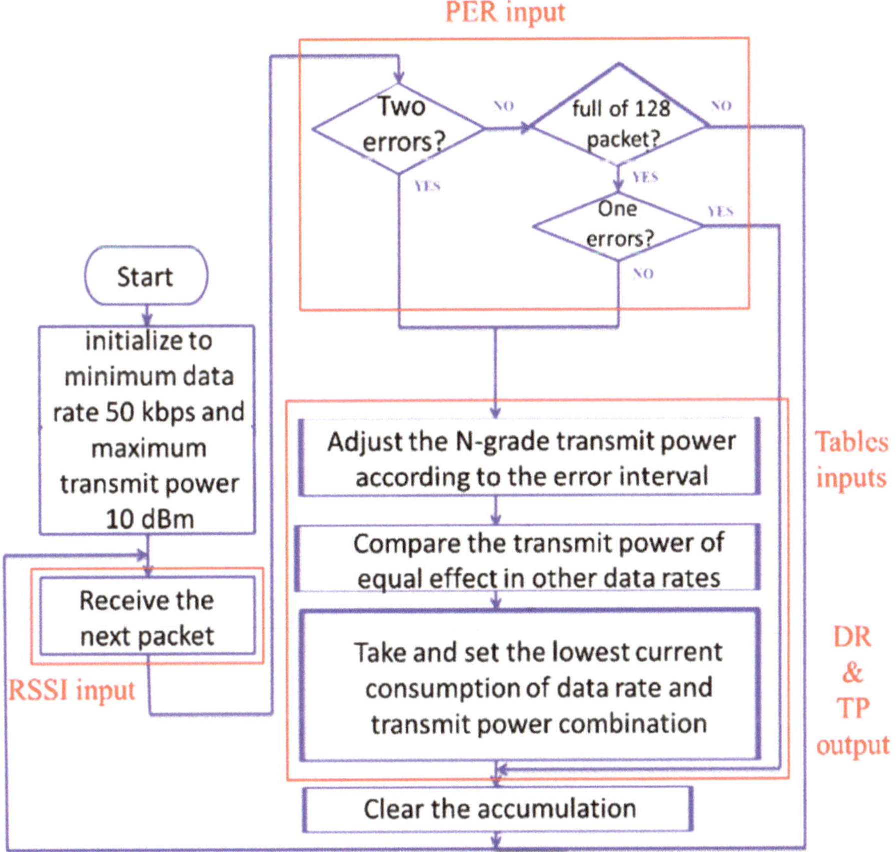

Figure 10. Flowchart of error interval algorithm controlling data rate and transmission power.

1. initial TP & DR params

2. receive next packet with RSSI node info, pak_cnt++

3. if pak_err >= 2, then goto #6

4. if pak_cnt < 128, then goto #2

5. if pak_err == 1, then goto #7

6. adjust TP & DR based on RSSI & PER

7. let pak_cnt and pak_err to be zero

8. goto #2

Figure 11. The pseudocode of control algorithm.

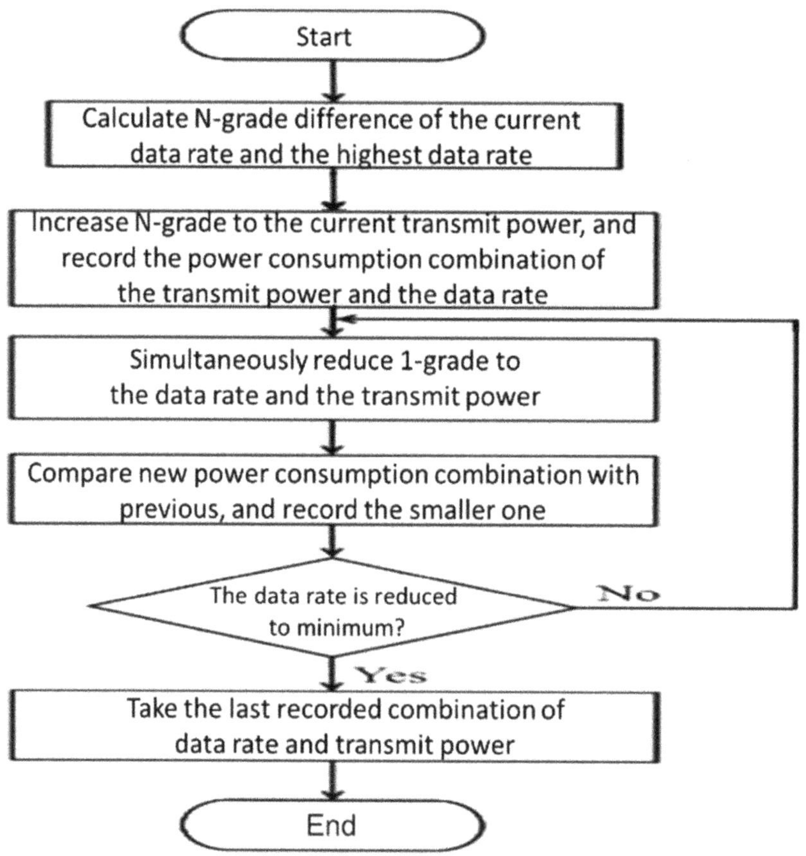

Figure 12. Flowchart for selecting the best energy efficient combination of data rate and transmit power.

6. Result

6.1. Experimental Method

The experimental location of this paper is the Sixth Hall of Engineering, National Yunlin University of Science and Technology, Taiwan. A total of one bridge and ten nodes were set up in the experiment, as Figure 13 shows.

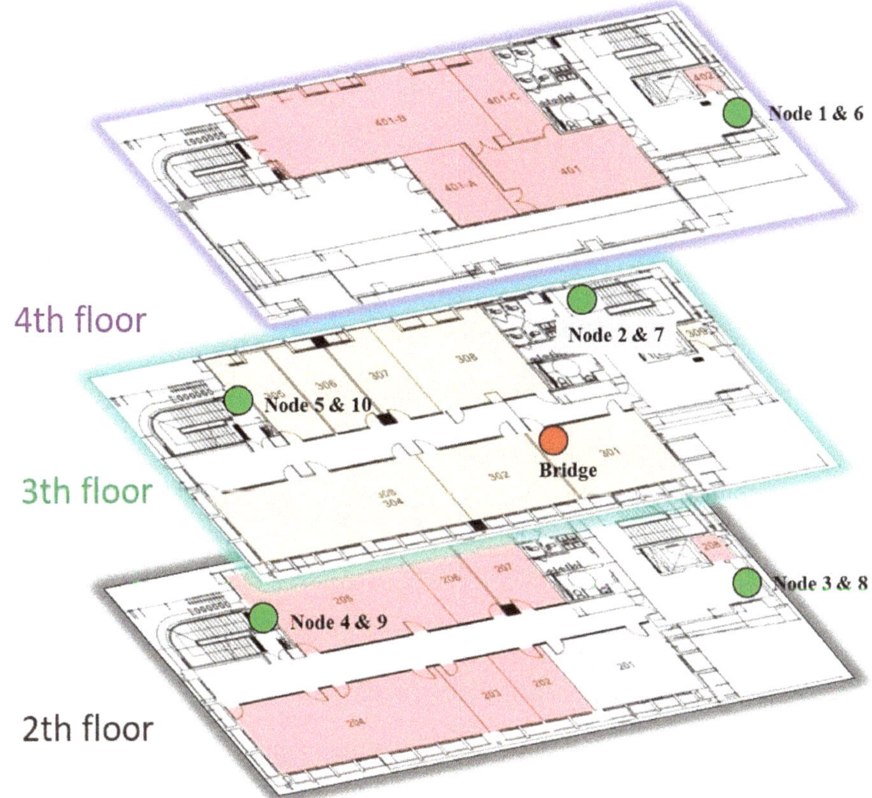

Figure 13. Experimental bridge and nodes placement map.

Every two nodes were placed in the same position. One node had the ADR and TPC control algorithms, and the other node fixed the data rate to the lowest (50 kbps) and the transmission power to the maximum. Through long-term testing, the power consumption and PER of nodes placed at the same position were compared to verify the effect of the algorithm. Nodes 1–5 were the nodes that had the algorithms, and nodes 6–10 comprised the experimental control group. A total of ten nodes were located in five different locations. Note that all nodes ran in the TDMA mode to save a lot of power for the sensing node. However, the power consumption of TDMA is not discussed later.

6.2. Results and Analysis

The experimental data of all nodes is organized as shown in Table 4. Since nodes 6 to 10 were without algorithms, the transmission power was set to a maximum of 10 dBm and the data rate was set to the lowest at 50 kbps. The PERs of nodes 6 to 10 were lower than the PERs of nodes 1 to 5 under the data rate and the transmission power control algorithm. However, the nodes with algorithms maintained a PER of less than 1% except for node 1, and the average current consumption was much lower than that the nodes without algorithms.

Table 4. Node experimental results.

	Packet Error Number	PER (%)	Response (TX) Average Current Consumption	Overall Average Current Consumption
Node1	7989	1.4035	8.4740 uA	35.924 uA
Node6	213	0.0374	39.852 uA	74.281 uA
Node2	3599	0.6322	10.493 uA	38.310 uA
Node7	864	0.1518	39.852 uA	74.281 uA
Node3	5426	0.9532	11.760 uA	39.845 uA
Node8	519	0.0912	39.852 uA	74.281 uA
Node4	4798	0.8429	16.106 uA	44.991 uA
Node9	1896	0.3331	39.852 uA	74.281 uA
Node5	2283	0.4011	5.6440 uA	32.516 uA
Node10	1922	0.3376	39.852 uA	74.281 uA

Comparing node 1 with node 6, the PER of node 1 was higher than node 6. However, node 1 saved 78.74% more power consumption than node 6 in response packet. In the overall average current consumption, node 1 saved 51.64% more of the energy than node 6.

Node 2 and node 7 were the nearest nodes for the bridge node. In Table 4, the PER of node 2 was 0.6322%, and node 7 was 0.1518%. Although the PER of node 2 was larger than that of node 7, its overall PER was still less than 1%. In the response packet, node 2 saved 73.67% more power consumption than node 7, and in the overall average current consumption, it saved 48.43% more power consumption.

The PER of node 3 was 0.9532%, and node 8 was 0.0912%. In the response packet, node 3 saved 70.49% more power consumption than node 8, and in the overall average current consumption, it saved 46.36% more power consumption.

The experimental result is showed in Figure 14. The PER of node 4 was 0.8429%, and node 9 was 0.3331%. In the overall average current consumption, node 4 was 44.991 uA, and node 9 was 74.281 uA. In the response packet, node 4 saved 59.59% more power consumption than node 9, and in the overall average current consumption, it saved 39.43% more power consumption.

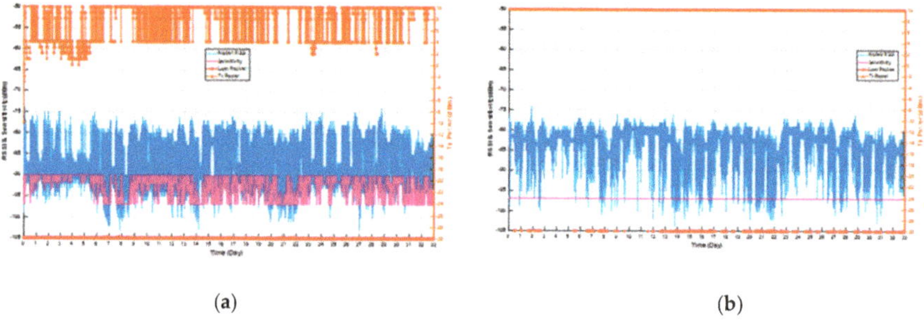

(a)   (b)

Figure 14. Experimental results of (**a**) node 4 and (**b**) node 9.

The PER of node 5 was 0.4011%, and node 10 was 0.3376%. In the overall average current consumption, node 5 was 32.516 uA, and node 10 was 74.281 uA. In the response packet, node 5 saved 85.83% more power consumption than node 10, and in the overall average current consumption, it saved 56.23% more power consumption.

Obviously, the RSSI values of nodes with algorithms are close to the sensitivity values corresponding to the current data rate from the experimental results. If the RSSI value is lower than the sensitivity value, the probability of packet error will increase. Moreover, since the position of the nodes is affected by the people in the office and the class, the RSSI value of each node floats dramatically during the daytime and the probability of packet error is high. However, at night, the RSSI value is so stable that the probability of packet error is low. If the node is supplied by AAA battery with 1200 mAh, the execution duration could approach about 4.21 years with the proposed algorithm. The nodes without the proposed algorithm could run only for 1.8 years.

The data of Table 4 is drawn with a bar graph of the nodes' average current consumption as Figure 15. It is clear the response packet of the nodes with the algorithm saved the most significant energy in the TX Mode, and the ranking of the power saving was sequentially ranked as nodes 5, 1, 2, 3, and 4. The reason for the difference in the amount of power consumed by each node was assumed to be the positional relationship.

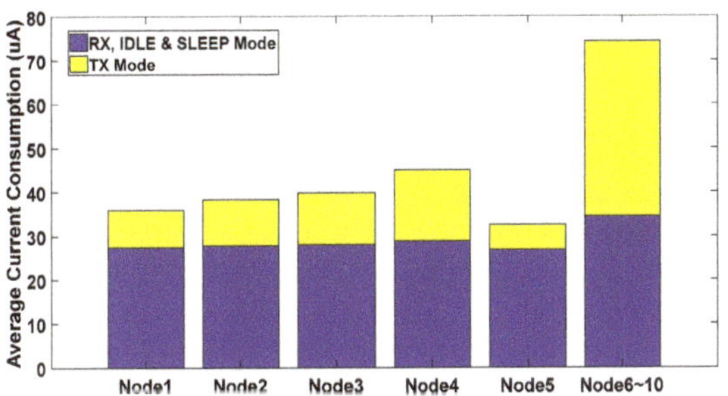

Figure 15. Nodes overall average current consumption.

For justification of the proposed algorithm, the experimental results for long-term testing are presented in Figure 16 and Table 5. Figure 16 shows the battery voltage variation in node 4 and node 9, which ran with and without control algorithm for 69 days of execution. The battery voltage of the node 9 in Table 5 was obviously lower than that of node 4, and the results also verify the proposed algorithm workable.

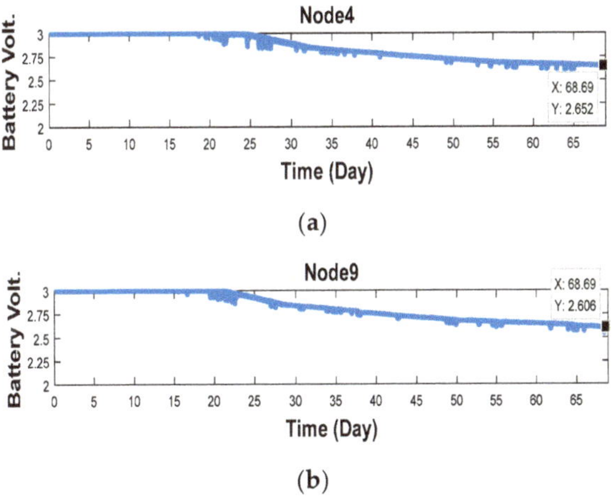

Figure 16. The battery voltage of (**a**) node 4 and (**b**) node 9 within 69 days.

Table 5. The battery voltage of each node at the 69th day.

Experimental Group	Node 1	Node 2	Node 3	Node 4	Node 5
Battery voltage	2.659 V	2.65 V	2.66 V	2.652 V	2.669 V
Control group	Node 6	Node 7	Node 8	Node 9	Node 10
Battery voltage	2.618 V	2.614 V	2.617 V	2.606 V	2.624 V

7. Conclusions

A wireless sensor network based on Sub1G-Hz and star topology was constructed in this paper, and the TDMA wireless communication protocol and the transmission power and data rate control algorithm were proposed to reduce power consumption on sensing nodes usefully. According to the dynamic environment, sensing nodes with hybrid control algorithms automatically adapt the transmission power and data rate to achieve good communication quality and low power consumption simultaneously. The above algorithms will increase the performance and reduce power consumption on wireless communication. Then, communication devices could be operated at very low power consumption when using wireless communication.

The experimental results show that PER states of nodes can effectively be controlled near the target value, 1%, which can prove the good reliability of communication. In addition, because all nodes run in the TDMA architecture's wireless protocol, TDMA can enable a wireless transmission in a low duty cycle. The average current consumption of the node without the hybrid control algorithm was calculated as 74.281 uA, and the power consumptions of algorithm nodes were different and depended on the positions of the nodes. According to the experimental results, when the power consumption of the response packet in the transmission mode was compared, the power consumption saved up to 85.83%. The overall consumption saved up to 56.23% of the power consumption, which indicates that the algorithms proposed in this paper actually have an energy-saving effect for wireless communication. If the node is powered by AAA battery with 1200 mAh, the node could run approximately 4.21 years with proposed algorithm. The other TDMA is discussed in [17] for the LoRa system but not included about ADR and TP, and the authors suggest that the battery life is about 3 years.

In summary, the proposed hybrid control algorithm is complex, and the payload and node number in the WSN are also limited. However, the system architecture and control algorithm proposed in this paper could lower several important things such as the power consumption, system complexity, maintenance fee, etc.

Author Contributions: All authors contributed ideas, discussed the results. C.-W.H. and H.-J.Z. designed the experiment, and performed most of the analysis. H.-J.Z. compiled the data. C.-W.H. and W.-T.H. wrote most of the main text. Y.-D.Z. supported the important revision of the article. All authors have read and agreed to the published version of the manuscript.

Funding: This research was funded by the Ministry of Science and Technology, ROC, grant number under contract No. MOST 109-2221-E-224-023-, 108-2221-E-224-045- and 108-2218-E-150-004-.

Conflicts of Interest: To the best of our knowledge, the authors have no conflict of interest, financial or otherwise.

References

1. Hung, C.W.; Hsu, W.T. Power consumption and calculation requirement analysis of AES for WSN IoT. *Sensors* **2018**, *18*, 1675. [CrossRef] [PubMed]
2. Navarro-Ortiz, J.; Sendra, S.; Ameigeiras, P.; Lopez-Soler, J.M. Integration of LoRaWAN and 4G/5G for the Industrial Internet of Things. *IEEE Commun. Mag.* **2018**, *56*, 60–67. [CrossRef]
3. Bachir, A.; Dohler, M.; Watteyne, T.; Leung, K.K. MAC essentials for wireless sensor networks. *IEEE Commun. Surv. Tutor.* **2010**, *12*, 222–248. [CrossRef]
4. Hung, C.W.; Zhang, H.J.; Hsu, W.T. Analysis of Data Rate and Transmission Power Hybrid Control in WSN IoT. In Proceedings of the 4th International Symposium on Mobile Internet Security MobiSec 2019, Taichung, Taiwan, 17–19 October 2019.

5. Sodhro, A.H.; Muzammal, M.; Talat, R.; Pirbhulal, S. A multi-sensor data fusion enabled ensemble approach for medical data from body sensor networks. *Inf. Fusion* **2020**, *53*, 155–164.

6. Aust, S.; Ito, T. Sub 1GHz wireless LAN propagation path loss models for urban smart grid applications. In Proceedings of the 2012 International Conference on Computing, Networking and Communications (ICNC), IEEE, Maui, HI, USA, 30 January–2 February 2012; pp. 116–120.

7. Aust, S.; Ito, T. Sub 1GHz wireless LAN deployment scenarios and design implications in rural areas. In Proceedings of the 2011 IEEE GLOBECOM Workshops (GC Wkshps), IEEE, Houston, TX, USA, 5–9 December 2011; pp. 1045–1049.

8. Aust, S.; Prasad, R.V.; Niemegeers, I.G. IEEE 802.11 ah: Advantages in standards and further challenges for sub 1 GHz Wi-Fi. In Proceedings of the 2012 IEEE international conference on communications (ICC), IEEE, Ottawa, ON, Canada, 10–15 June 2012; pp. 6885–6889.

9. Sahoo, P.K.; Pattanaik, S.R.; Wu, S.L. Design and analysis of a low latency deterministic network MAC for wireless sensor networks. *Sensors* **2017**, *17*, 2185. [CrossRef] [PubMed]

10. Iala, I.; Quodou, M.; Aboutajdine, D.; Zytoune, O. Energy based collision avoidance at the mac layer for wireless sensor network. In Proceedings of the 2017 International Conference on Advanced Technologies for Signal and Image Processing (ATSIP), IEEE, Fez, Morocco, 22–24 May 2017; pp. 1–5.

11. Pande, H.; Kharat, M.U. Modified WiseMAC protocol for energy efficient wireless sensor networks with better throughput. In Proceedings of the 2016 International Conference on Internet of Things and Applications (IOTA), IEEE, Pune, India, 22–24 January 2016; pp. 364–367.

12. Zhang, D.G.; Zhou, S.; Tang, Y.M. A low duty cycle efficient MAC protocol based on self-adaption and predictive strategy. *Mob. Netw. Appl.* **2018**, *23*, 828–839. [CrossRef]

13. Lu, Y.; Qiu, Z.; Luo, Y.; Wei, L.; Lin, S.; Liu, X. A Modified TDMA Algorithm Based on Adaptive Timeslot Exchange in Ad Hoc Network. In Proceedings of the 2018 IEEE 4th International Conference on Computer and Communications (ICCC), IEEE, Chengdu, China, 7–10 December 2018; pp. 457–461.

14. Lin, C.; Cai, X.; Su, Y.; Ni, P.; Shi, H. A dynamic slot assignment algorithm of TDMA for the distribution class protocol using node neighborhood information. In Proceedings of the 2017 11th IEEE International Conference on Anti-counterfeiting, Security, and Identification (ASID), IEEE, Xiamen, China, 27–29 October 2017; pp. 138–141.

15. Bhatia, A.; Hansdah, R.C. RD-TDMA: A randomized distributed TDMA scheduling for correlated contention in WSNs. In Proceedings of the 2014 28th International Conference on Advanced Information Networking and Applications Workshops, IEEE, Victoria, BC, Canada, 13–16 May 2014; pp. 378–384.

16. Hao, X.; Yao, N.; Wang, L.; Wang, J. Joint resource allocation algorithm based on multi-objective optimization for wireless sensor networks. *Appl. Soft Comput.* **2020**, *94*, 106470. [CrossRef]

17. Piyare, R.; Murpgy, A.L.; Magno, M.; Benini, L. On-demand TDMA for energy efficient data collection with LoRa and wake-up receiver. In Proceedings of the 2018 14th International Conference on Wireless and Mobile Computing, Networking and Communications (WiMob), IEEE, Limassol, Cyprus, 15–17 October 2018; pp. 1–4.

18. Hung, C.W.; Chang, H.T.; Hsia, K.H.; Lai, Y.H. Transmission power control for wireless sensor network. *J. Robot. Netw. Artif. Life* **2017**, *3*, 279–282.

19. Ikram, W.; Petersen, S.; Orten, P.; Thornhill, N.F. Adaptive multi-channel transmission power control for industrial wireless instrumentation. *IEEE Trans. Ind. Inform.* **2014**, *10*, 978–990. [CrossRef]

20. Ramakrishnan, S.; Krishna, B.T. Closed loop fuzzy logic based transmission power control for energy efficiency in wireless sensor networks. In Proceedings of the IEEE International Conference on Computer Communication and Systems ICCCS14, IEEE, Chennai, India, 20–21 February 2014; pp. 195–200.

21. Wu, J.; Luo, J. Research on multi-rate in wireless sensor network based on real platform. In Proceedings of the 2012 2nd International Conference on Consumer Electronics, Communications and Networks (CECNet), IEEE, Yichang, China, 21–23 April 2012; pp. 1240–1243.

22. Hughes, J.B.; Lazaridis, P.; Glover, I.; Ball, A. A survey of link quality properties related to transmission power control protocols in wireless sensor networks. In Proceedings of the 2017 23rd International Conference on Automation and Computing (ICAC), IEEE, Huddersfield, UK, 7–8 September 2017; pp. 1–5.

23. Chincoli, M.; Liotta, A. Self-learning power control in wireless sensor networks. *Sensors* **2018**, *18*, 375. [CrossRef] [PubMed]

24. Gong, X.; Plets, D.; Tanghe, E.; Pessemier, T.D.; Martens, L.; Joseph, W. An efficient genetic algorithm for large-scale transmit power control of dense and robust wireless networks in harsh industrial environments. *Appl. Soft Comput.* **2018**, *65*, 243–259. [CrossRef]
25. Dou, Z.; Zhao, Z.Z.; Jin, Q.; Zhang, L.; Shu, Y.; Yang, O. Energy-efficient rate adaptation for outdoor long distance WiFi links. In Proceedings of the 2011 IEEE Conference on Computer Communications Workshops (INFOCOM WKSHPS), IEEE, Shanghai, China, 10–15 April 2011; pp. 271–276.
26. Yao, M.; Liu, J.; Qian, Y.; Qiu, Z.; Kwak, K.S.; Hanlim, I. Adaptive Rate Control and Frame Length Adjustment for IEEE 802.11 n Wireless Networks. In Proceedings of the 2017 IEEE Wireless Communications and Networking Conference (WCNC), IEEE, San Francisco, CA, USA, 19–22 March 2017; pp. 1–6.
27. Yin, W.; Hu, P.; Indulska, J.; Portmann, M.; Mao, Y. MAC-layer rate control for 802.11 networks: A survey. *Wirel. Netw.* **2020**, *26*, 3793–3830. [CrossRef]
28. Sodhro, A.H.; Sangaiah, A.K.; Sodhro, G.H.; Lohano, S.; Pirbhulal, S. An energy-efficient algorithm for wearable electrocardiogram signal processing in ubiquitous healthcare applications. *Sensors* **2018**, *18*, 923. [CrossRef] [PubMed]
29. Qin, F.; Chen, Y.; Dai, X. Utilize Adaptive Spreading Code Length to Increase Energy Efficiency for WSN. In Proceedings of the 2013 IEEE 77th Vehicular Technology Conference (VTC Spring), IEEE, Dresden, Germany, 2–5 June 2013; pp. 1–5.
30. Zhang, Q. Energy saving efficiency comparison of transmit power control and link adaptation in BANs. In Proceedings of the 2013 IEEE International Conference on Communications (ICC), IEEE, Budapest, Hungary, 9–13 June 2013; pp. 1672–1677.
31. Hung, C.W.; Hsu, W.T.; Hsia, K.H. Using Adaptive Data Rate with DSSS Optimization and Transmission Power Control for Ultra-Low Power WSN. In Proceedings of the 2019 12th International Conference on Developments in eSystems Engineering (DeSE), IEEE, Kazan, Russia, 7–10 October 2019; pp. 611–614.

© 2020 by the authors. Licensee MDPI, Basel, Switzerland. This article is an open access article distributed under the terms and conditions of the Creative Commons Attribution (CC BY) license (http://creativecommons.org/licenses/by/4.0/).

MDPI

St. Alban-Anlage 66

4052 Basel

Switzerland

Tel. +41 61 683 77 34

Fax +41 61 302 89 18

www.mdpi.com

Sensors Editorial Office

E-mail: sensors@mdpi.com

www.mdpi.com/journal/sensors

www.ingramcontent.com/pod-product-compliance
Lightning Source LLC
LaVergne TN
LVHW071307170726
843515LV00006BA/1510